CW00467858

MACHINE LEARNING METHODS IN THE ENVIRONMENTAL SCIENCES

Neural Networks and Kernels

William W. Hsieh

Machine learning methods, having originated from computational intelligence (i.e. artificial intelligence), are now ubiquitous in the environmental sciences. This is the first single-authored textbook to give a unified treatment of machine learning methods and their applications in the environmental sciences.

Machine learning methods began to infiltrate the environmental sciences in the 1990s. Today, thanks to their powerful nonlinear modelling capability, they are no longer an exotic fringe species, as they are heavily used in satellite data processing, in general circulation models (GCM), in weather and climate prediction, air quality forecasting, analysis and modelling of environmental data, oceanographic and hydrological forecasting, ecological modelling, and in the monitoring of snow, ice and forests, etc. End-of-chapter review questions are included, allowing readers to develop their problem-solving skills and monitor their understanding of the material presented. An appendix lists websites available for downloading computer code and data sources. A resources website is available containing datasets for exercises, and additional material to keep the book completely up-to-date.

This book presents machine learning methods and their applications in the environmental sciences (including satellite remote sensing, atmospheric science, climate science, oceanography, hydrology and ecology), written at a level suitable for beginning graduate students and advanced undergraduates. It is also valuable for researchers and practitioners in environmental sciences interested in applying these new methods to their own work.

William W. Hsieh is a Professor in the Department of Earth and Ocean Sciences and in the Department of Physics and Astronomy, as well as Chair of the Atmospheric Science Programme, at the University of British Columbia. He is internationally known for his pioneering work in developing and applying machine learning methods in the environmental sciences. He has published over 80 peer-reviewed journal publications covering areas of climate variability, machine learning, oceanography, atmospheric science and hydrology.

MACHINE LEARNING METHODS IN THE ENVIRONMENTAL SCIENCES

Neural Networks and Kernels

WILLIAM W. HSIEH

University of British Columbia
Vancouver, BC, Canada

CAMBRIDGE UNIVERSITY PRESS
Cambridge, New York, Melbourne, Madrid, Cape Town, Singapore, São Paulo, Delhi

Cambridge University Press
The Edinburgh Building, Cambridge CB2 8RU, UK

Published in the United States of America by Cambridge University Press, New York

www.cambridge.org
Information on this title: www.cambridge.org/9780521791922

First published 2009

Printed in the United Kingdom at the University Press, Cambridge

A catalogue record for this publication is available from the British Library

ISBN 978-0-521-79192-2 hardback

Contents

Preface

Machine learning is a major sub-field in computational intelligence (also called artificial intelligence). Its main objective is to use computational methods to extract information from data. Machine learning has a wide spectrum of applications including handwriting and speech recognition, object recognition in computer vision, robotics and computer games, natural language processing, brain–machine interfaces, medical diagnosis, DNA classification, search engines, spam and fraud detection, and stock market analysis. Neural network methods, generally regarded as forming the first wave of breakthrough in machine learning, became popular in the late 1980s, while kernel methods arrived in a second wave in the second half of the 1990s.

In the 1990s, machine learning methods began to infiltrate the environmental sciences. Today, they are no longer an exotic fringe species, since their presence is ubiquitous in the environmental sciences, as illustrated by the lengthy References section of this book. They are heavily used in satellite data processing, in general circulation models (GCM) for emulating physics, in post-processing of GCM model outputs, in weather and climate prediction, air quality forecasting, analysis and modelling of environmental data, oceanographic and hydrological forecasting, ecological modelling, and in monitoring of snow, ice and forests, etc.

This book presents machine learning methods (mainly neural network and kernel methods) and their applications in the environmental sciences, written at a level suitable for beginning graduate students and advanced undergraduates. It is also aimed at researchers and practitioners in environmental sciences, who having been intrigued by exotic terms like neural networks, support vector machines, self-organizing maps, evolutionary computation, etc., are motivated to learn more about these new methods and to use them in their own work. The reader is assumed to know multivariate calculus, linear algebra and basic probability.

Chapters 1–3, intended mainly as background material for students, cover the standard statistical methods used in environmental sciences. The machine learning methods of later chapters provide powerful nonlinear generalizations for many of these standard linear statistical methods. The reader already familiar with the background material of Chapters 1–3 can start directly with Chapter 4, which introduces neural network methods. While Chapter 5 is a relatively technical chapter on nonlinear optimization algorithms, Chapter 6 on learning and generalization is essential to the proper use of machine learning methods – in particular, Section 6.10 explains why a nonlinear machine learning method often outperforms a linear method in weather applications but fails to do so in climate applications. Kernel methods are introduced in Chapter 7. Chapter 8 covers nonlinear classification, Chapter 9, nonlinear regression, Chapter 10, nonlinear principal component analysis, and Chapter 11, nonlinear canonical correlation analysis. Chapter 12 broadly surveys applications of machine learning methods in the environmental sciences (remote sensing, atmospheric science, oceanography, hydrology, ecology, etc.). For exercises, the student could test the methods on data from their own area or from some of the websites listed in Appendix A. Codes for many machine learning methods are also available from sites listed in Appendix A. The book website `www.cambridge.org/hsieh` also provides datasets for some of the exercises given at the ends of the chapters.

On a personal note, writing this book has been both exhilarating and gruelling. When I first became intrigued by neural networks through discussions with Dr Benyang Tang in 1992, I recognized that the new machine learning methods would have a major impact on the environmental sciences. However, I also realized that I had a steep learning curve ahead of me, as my background training was in physics, mathematics and environmental sciences, but not in statistics nor computer science. By the late 1990s I became convinced that the best way for me to learn more about machine learning was to write a book. What I thought would take a couple of years turned into a marathon of over eight years, as I desperately tried to keep pace with a rapidly expanding research field. I managed to limp past the finish line in pain, as repetitive strain injury from overusage of keyboard and mouse struck in the final months of intensive writing!

I have been fortunate in having supervised numerous talented graduate students, post-doctoral fellows and research associates, many of whom taught me far more than I taught them. I received helpful editorial assistance from the staff at the Cambridge University Press and from Max Ng. I am grateful for the support from my two university departments (Earth and Ocean Sciences, and Physics and Astronomy), the Peter Wall Institute of Advanced Studies, the Natural Sciences and Engineering Research Council of Canada and the Canadian Foundation for Climate and Atmospheric Sciences.

Without the loving support from my family (my wife Jean and my daughters, Teresa and Serena), and the strong educational roots planted decades ago by my parents and my teachers, I could not have written this book.

Notation used in this book

In general, vectors are denoted by lower case bold letters (e.g. $\mathbf{v}$), matrices by upper case bold letters (e.g. $\mathbf{A}$) and scalar variables by italics (e.g. x or J). A column vector is denoted by $\mathbf{v}$, while its transpose $\mathbf{v}^{\mathrm{T}}$ is a row vector, i.e. $\mathbf{v}^{\mathrm{T}} = (v_1, v_2, \ldots, v_n)$ and $\mathbf{v} = (v_1, v_2, \ldots, v_n)^{\mathrm{T}}$, and the inner or dot product of two vectors $\mathbf{a} \cdot \mathbf{b} = \mathbf{a}^{\mathrm{T}}\mathbf{b} = \mathbf{b}^{\mathrm{T}}\mathbf{a}$. The elements of a matrix $\mathbf{A}$ are written as A_{ij} or $(\mathbf{A})_{ij}$. The probability for discrete variables is denoted by upper case P, whereas the probability density for continuous variables is denoted by lower case p. The expectation is denoted by $\mathrm{E}[\ldots]$ or $\langle \ldots \rangle$. The natural logarithm is denoted by ln or log.

Abbreviations

AO = Arctic Oscillation
BNN = Bayesian neural network
CART = classification and regression tree
CCA = canonical correlation analysis
CDN = conditional density network
EC = evolutionary computation
EEOF = extended empirical orthogonal function
ENSO = El Niño-Southern Oscillation
EOF = empirical orthogonal function
GA = genetic algorithm
GCM = general circulation model (or global climate model)
GP = Gaussian process model
IC = information criterion
LP = linear projection
MAE = mean absolute error
MCA = maximum covariance analysis
MJO = Madden–Julian Oscillation
MLP = multi-layer perceptron neural network
MLR = multiple linear regression
MOS = model output statistics
MSE = mean squared error
MSSA = multichannel singular spectrum analysis
NAO = North Atlantic Oscillation
NLCCA = nonlinear canonical correlation analysis
NLCPCA = nonlinear complex PCA
NN = neural network
NLPC = nonlinear principal component
NLPCA = nonlinear principal component analysis

NLSSA = nonlinear singular spectrum analysis
PC = principal component
PCA = principal component analysis
PNA = Pacific-North American teleconnection
POP = principal oscillation pattern
QBO = Quasi-Biennial Oscillation
RBF = radial basis function
RMSE = root mean squared error
SLP = sea level pressure
SOM = self-organizing map
SSA = singular spectrum analysis
SST = sea surface temperature (sum of squares in Chapter 1)
SVD = singular value decomposition
SVM = support vector machine
SVR = support vector regression

1

Basic notions in classical data analysis

The goal of data analysis is to discover relations in a dataset. The basic ideas of probability distributions, and the mean and variance of a random variable are introduced first. The relations between two variables are then explored with correlation and regression analysis. Other basic notions introduced in this chapter include Bayes theorem, discriminant functions, classification and clustering.

1.1 Expectation and mean

Let x be a random variable which takes on discrete values. For example, x can be the outcome of a die cast, where the possible values are $x_i = i$, with $i = 1, \ldots, 6$. The *expectation* or expected value of x from a population is given by

$$\mathrm{E}[x] = \sum_i x_i P_i, \tag{1.1}$$

where P_i is the probability of x_i occurring. If the die is fair, $P_i = 1/6$ for all i, so $\mathrm{E}[x]$ is 3.5. We also write

$$\mathrm{E}[x] = \mu_x, \tag{1.2}$$

with μ_x denoting the *mean* of x for the population.

The expectation of a sum of random variables satisfies

$$\mathrm{E}[ax + by + c] = a\,\mathrm{E}[x] + b\,\mathrm{E}[y] + c, \tag{1.3}$$

where x and y are random variables, and a, b and c are constants.

For a random variable x which takes on continuous values over a domain Ω, the expection is given by an integral,

$$\mathrm{E}[x] = \int_\Omega x p(x)\,\mathrm{d}x, \tag{1.4}$$

where $p(x)$ is the *probability density* function. For any function $f(x)$, the expectation is

$$\mathrm{E}[f(x)] = \int_\Omega f(x)p(x)\,\mathrm{d}x \quad \text{(continuous case)}$$
$$= \sum_i f(x_i)P_i \quad \text{(discrete case).} \tag{1.5}$$

In practice, one can sample only N measurements of x $(x_1, \ldots, x_N)$ from the population. The *sample mean* $\overline{x}$ or $\langle x \rangle$ is calculated as

$$\overline{x} \equiv \langle x \rangle = \frac{1}{N}\sum_{i=1}^{N} x_i, \tag{1.6}$$

which is in general different from the population mean μ_x. As the sample size increases, the sample mean approaches the population mean.

1.2 Variance and covariance

Fluctuation about the mean value is commonly characterized by the variance of the population,

$$\mathrm{var}(x) \equiv \mathrm{E}[(x-\mu_x)^2] = \mathrm{E}[x^2 - 2x\mu_x + \mu_x^2] = \mathrm{E}[x^2] - \mu_x^2, \tag{1.7}$$

where (1.3) and (1.2) have been invoked. The standard deviation s is the positive square root of the population variance, i.e.

$$s^2 = \mathrm{var}(x). \tag{1.8}$$

The sample standard deviation σ is the positive square root of the sample variance, given by

$$\sigma^2 = \frac{1}{N-1}\sum_{i=1}^{N}(x_i - \overline{x})^2. \tag{1.9}$$

As the sample size increases, the sample variance approaches the population variance. For large N, distinction is often not made between having $N-1$ or N in the denominator of (1.9).

Often one would like to compare two very different variables, e.g. sea surface temperature and fish population. To avoid comparing apples with oranges, one usually standardizes the variables before making the comparison. The *standardized variable*

$$x_s = (x - \overline{x})/\sigma, \tag{1.10}$$

is obtained from the original variable by subtracting the sample mean and dividing by the sample standard deviation. The standardized variable is also called the *normalized* variable or the *standardized anomaly* (where *anomaly* means the deviation from the mean value).

For two random variables x and y, with mean μ_x and μ_y respectively, their *covariance* is given by

$$\mathrm{cov}(x, y) = \mathrm{E}[(x - \mu_x)(y - \mu_y)]. \tag{1.11}$$

The variance is simply a special case of the covariance, with

$$\mathrm{var}(x) = \mathrm{cov}(x, x). \tag{1.12}$$

The sample covariance is computed as

$$\mathrm{cov}(x, y) = \frac{1}{N-1}\sum_{i=1}^{N}(x_i - \overline{x})(y_i - \overline{y}). \tag{1.13}$$

1.3 Correlation

The (Pearson) correlation coefficient, widely used to represent the strength of the linear relationship between two variables x and y, is defined as

$$\hat{\rho}_{xy} = \frac{\mathrm{cov}(x, y)}{s_x s_y}, \tag{1.14}$$

where s_x and s_y are the population standard deviations for x and y, respectively.

For a sample containing N pairs of (x, y) measurements or observations, the *sample correlation* is computed by

$$\rho \equiv \rho_{xy} = \frac{\displaystyle\sum_{i=1}^{N}(x_i - \overline{x})(y_i - \overline{y})}{\left[\displaystyle\sum_{i=1}^{N}(x_i - \overline{x})^2\right]^{\frac{1}{2}}\left[\displaystyle\sum_{i=1}^{N}(y_i - \overline{y})^2\right]^{\frac{1}{2}}}, \tag{1.15}$$

which lies between -1 and $+1$. At the value $+1$, x and y will show a perfect straight-line relation with a positive slope; whereas at -1, the perfect straight line will have a negative slope. With increasing noise in the data, the sample correlation moves towards 0.

An important question is whether the obtained sample correlation can be considered significantly different from 0 – this is also called a test of the null (i.e. $\hat{\rho}_{xy} = 0$) hypothesis. A common approach involves transforming to the variable

$$t = \rho \sqrt{\frac{N-2}{1-\rho^2}}, \tag{1.16}$$

which in the null case is distributed as the Student's t distribution, with $\nu = N - 2$ degrees of freedom.

For example, with $N = 32$ data pairs, ρ was found to be 0.36. Is this correlation significant at the 5% level? In other words, if the true correlation is zero ($\hat{\rho}_{xy} = 0$), is there less than 5% chance that we could obtain $\rho \geq 0.36$ for our sample? To answer this, we need to find the value $t_{0.975}$ in the t-distribution, where $t > t_{0.975}$ occur less than 2.5% of the time and $t < -t_{0.975}$ occur less than 2.5% of the time (as the t-distribution is symmetrical), so altogether $|t| > t_{0.975}$ occur less than 5% of the time. From t-distribution tables, we find that with $\nu = 32 - 2 = 30$, $t_{0.975} =$ 2.04.

From (1.16), we have

$$\rho^2 = \frac{t^2}{N - 2 + t^2}, \tag{1.17}$$

so substituting in $t_{0.975} = 2.04$ yields $\rho_{0.05} = 0.349$, i.e. less than 5% of the sample correlation values will indeed exceed $\rho_{0.05}$ in magnitude if $\hat{\rho}_{xy} = 0$. Hence our $\rho = 0.36 > \rho_{0.05}$ is significant at the 5% level based on a '2-tailed' t test. For moderately large N ($N \geq 10$), an alternative test involves using Fisher's z-transformation (Bickel and Doksum, 1977).

Often the observations are measurements at regular time intervals, i.e. time series, and there is *autocorrelation* in the time series – i.e. neighbouring data points in the time series are correlated. Autocorrelation is well illustrated by persistence in weather patterns, e.g. if it rains one day, it increases the probability of rain the following day. With autocorrelation, the effective sample size may be far smaller than the actual number of observations in the sample, and the value of N used in the significance tests will have to be adjusted to represent the effective sample size.

A statistical measure is said to be *robust* if the measure gives reasonable results even when the model assumptions (e.g. data obeying Gaussian distribution) are not satisfied. A statistical measure is said to be *resistant* if the measure gives reasonable results even when the dataset contains one or a few outliers (an *outlier* being an extreme data value arising from a measurement or recording error, or from an abnormal event).

Correlation assumes a linear relation between x and y; however, the sample correlation is not *robust* to deviations from linearity in the relation, as illustrated in Fig. 1.1a where $\rho \approx 0$ even though there is a strong (nonlinear) relationship between the two variables. Thus the correlation can be misleading when the underlying relation is nonlinear. Furthermore, the sample correlation is not *resistant* to

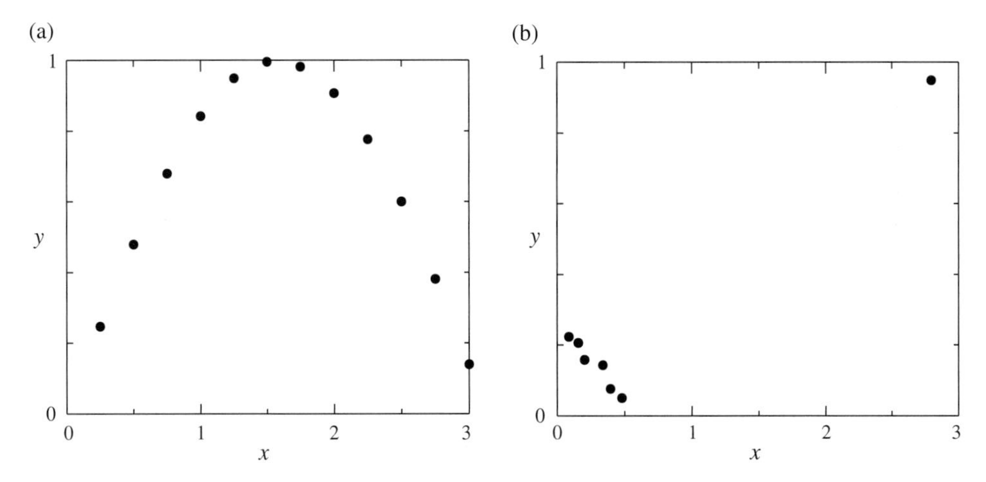

Fig. 1.1 (a) An example showing that correlation is not robust to deviations from linearity. Here the strong nonlinear relation between x and y is completely missed by the near-zero correlation coefficient. (b) An example showing that correlation is not resistant to outliers. Without the single outlier, the correlation coefficient changes from positive to negative.

outliers, where in Fig. 1.1b if the outlier datum is removed, ρ changes from being positive to negative.

1.3.1 Rank correlation

For the correlation to be more robust and resistant to outliers, the Spearman rank correlation is often used instead. If the data $\{x_1, \ldots, x_N\}$ are rearranged in order according to their size (starting with the smallest), and if x is the nth member, then $\text{rank}(x) \equiv r_x = n$. The correlation is then calculated for r_x and r_y instead, which can be shown to simplify to

$$\rho_{\text{rank}} = \rho_{r_x r_y} = 1 - \frac{6 \sum_{i=1}^{N} (r_{x_i} - r_{y_i})^2}{N(N^2 - 1)}. \tag{1.18}$$

For example, if six measurements of x yielded the values $1, 3, 0, 5, 3, 6$ then the corresponding r_x values are $2, 3.5, 1, 5, 3.5, 6$, (where the tied values were all assigned an averaged rank). If measurements of y yielded $2, 3, -1, 5, 4, -99$ (an outlier), then the corresponding r_y values are $3, 4, 2, 6, 5, 1$. The Spearman rank correlation is $+0.12$, whereas in contrast the Pearson correlation is -0.61, which shows the strong influence exerted by an outlier.

An alternative robust and resistant correlation is the biweight midcorrelation (see Section 11.2.1).

1.3.2 Autocorrelation

To determine the degree of autocorrelation in a time series, we use the autocorrelation coefficient, where a copy of the time series is shifted in time by a lag of l time intervals, and then correlated with the original time series. The lag-l autocorrelation coefficient is given by

$$\rho(l) = \frac{\sum_{i=1}^{N-l} [(x_i - \overline{x})(x_{i+l} - \overline{x})]}{\sum_{i=1}^{N} (x_i - \overline{x})^2}, \tag{1.19}$$

where $\overline{x}$ is the sample mean. There are other estimators of the autocorrelation function, besides the non-parametric estimator given here (von Storch and Zwiers, 1999, p. 252). The function $\rho(l)$, which has the value 1 at lag 0, begins to decrease as the lag increases. The lag where $\rho(l)$ first intersects the l-axis is l_0, the *first zero crossing*. A crude estimate for the *effective sample size* is $N_{\text{eff}} = N/l_0$. From symmetry, one defines $\rho(-l) = \rho(l)$. In practice, $\rho(l)$ cannot be estimated reliably when l approaches N, since the numerator of (1.19) would then involve summing over very few terms.

The autocorrelation function can be integrated to yield a *decorrelation time scale* or *integral time scale*

$$\begin{aligned} T &= \int_{-\infty}^{\infty} \rho(l)\, \mathrm{d}l \quad \text{(continuous case)} \\ &= \left(1 + 2\sum_{l=1}^{L} \rho(l)\right) \Delta t \quad \text{(discrete case)}, \end{aligned} \tag{1.20}$$

where Δt is the time increment between adjacent data values, and the maximum lag L used in the summation is usually not more than $N/3$, as $\rho(l)$ cannot be estimated reliably when l becomes large. The effective sample size is then

$$N_{\text{eff}} = N\Delta t / T, \tag{1.21}$$

with $N\Delta t$ the data record length. When the decorrelation time scale is large, $N_{\text{eff}} \ll N$.

With two time series x and y, both with N samples, the effective sample size is often estimated by

$$N_{\text{eff}} = \frac{N}{\sum_{l=-L}^{L} \left[\rho_{xx}(l)\rho_{yy}(l) + \rho_{xy}(l)\rho_{yx}(l)\right]}, \tag{1.22}$$

(Emery and Thomson, 1997), though sometimes the $\rho_{xy}\rho_{yx}$ terms are ignored (Pyper and Peterman, 1998).

1.3.3 Correlation matrix

If there are M variables, e.g. M stations reporting the air pressure, then correlations between the variables lead to a correlation matrix

$$\mathbf{C} = \begin{bmatrix} \rho_{11} & \rho_{12} & \cdots & \rho_{1M} \\ \rho_{21} & \rho_{22} & \cdots & \rho_{2M} \\ \cdots & \cdots & \cdots & \cdots \\ \rho_{M1} & \rho_{M2} & \cdots & \rho_{MM} \end{bmatrix}, \tag{1.23}$$

where ρ_{ij} is the correlation between the ith and the jth variables. The diagonal elements of the matrix satisfy $\rho_{ii} = 1$, and the matrix is symmetric, i.e. $\rho_{ij} = \rho_{ji}$. The jth column of $\mathbf{C}$ gives the correlations between the variable j and all other variables.

1.4 Regression

Regression, introduced originally by Galton (1885), is used to find a linear relation between a dependent variable y and one or more independent variables $\mathbf{x}$.

1.4.1 Linear regression

For now, consider simple linear regression where there is only one independent variable x, and the dataset contains N pairs of (x, y) measurements. The relation is

$$y_i = \tilde{y}_i + e_i = a_0 + a_1 x_i + e_i, \qquad i = 1, \ldots, N, \tag{1.24}$$

where a_0 and a_1 are the regression parameters, $\tilde{y}_i$ is the y_i predicted or described by the linear regression relation, and e_i is the error or the residual unaccounted for by the regression (Fig. 1.2). As regression is commonly used as a prediction tool (i.e. given x, use the regression relation to predict y), x is referred to as the *predictor* or independent variable, and y, the *predictand*, response or dependent variable. Curiously, the term 'predictand', widely used within the atmospheric–oceanic community, is not well known outside.

The error

$$e_i = y_i - \tilde{y}_i = y_i - a_0 - a_1 x_i. \tag{1.25}$$

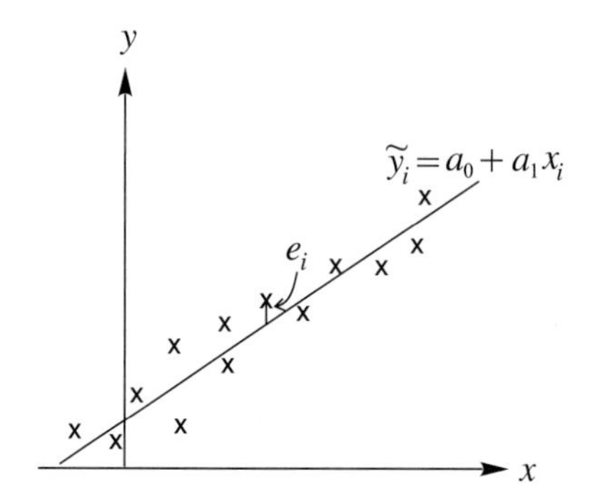

Fig. 1.2 Illustrating linear regression. A straight line $\tilde{y}_i = a_0 + a_1 x_i$ is fitted to the data, where the parameters a_0 and a_1 are determined from minimizing the sum of the square of the error e_i, which is the vertical distance between the ith data point and the line.

By finding the optimal values of the parameters a_0 and a_1, linear regression minimizes the sum of squared errors (SSE) ,

$$\mathrm{SSE} = \sum_{i=1}^{N} {e_i}^2, \tag{1.26}$$

yielding the best straight line relation between y and x. Because the SSE is minimized, this method is also referred to as the *least squares* method.

Differentiation of (1.26) by a_0 yields

$$\sum_{i=1}^{N} (y_i - a_0 - a_1 x_i) = 0. \tag{1.27}$$

Differentiation of (1.26) by a_1 gives

$$\sum_{i=1}^{N} (y_i - a_0 - a_1 x_i) x_i = 0. \tag{1.28}$$

These two equations are called the *normal equations*, from which we will obtain the optimal values of a_0 and a_1.

From (1.27), we have

$$a_0 = \frac{1}{N} \sum y_i - \frac{a_1}{N} \sum x_i, \quad \text{i.e. } a_0 = \overline{y} - a_1 \overline{x}. \tag{1.29}$$

Substituting (1.29) into (1.28) yields

$$a_1 = \frac{\sum x_i y_i - N \overline{x}\, \overline{y}}{\sum x_i^2 - N \overline{x}\, \overline{x}}. \tag{1.30}$$

Equations (1.29) and (1.30) provide the optimal values of a_0 and a_1 for minimizing the SSE, thereby yielding the best straight line fit to the data in the x-y plane. The parameter a_1 gives the slope of the regression line, while a_0 gives the y-intercept.

1.4.2 Relating regression to correlation

Since regression and correlation are two approaches to extract linear relations between two variables, one would expect the two to be related. Equation (1.30) can be rewritten as

$$a_1 = \frac{\sum(x_i - \overline{x})(y_i - \overline{y})}{\sum(x_i - \overline{x})^2}. \tag{1.31}$$

Comparing with the expression for the sample correlation, (1.15), we see that

$$a_1 = \rho_{xy}\frac{\sigma_y}{\sigma_x}, \tag{1.32}$$

i.e. the slope of the regression line is the correlation coefficient times the ratio of the standard deviation of y to that of x.

It can also be shown that

$$\sigma_e^2 = \sigma_y^2(1 - \rho_{xy}^2), \tag{1.33}$$

where $1-\rho_{xy}^2$ is the fraction of the variance of y not accounted for by the regression. For example, if $\rho_{xy} = 0.5$, then $1 - \rho_{xy}^2 = 0.75$, i.e. 75% of the variance of y is not accounted for by the regression.

1.4.3 Partitioning the variance

It can be shown that the variance, i.e. the total sum of squares (SST), can be partitioned into two: the first part is that accounted for by the regression relation, i.e. the sum of squares due to regression (SSR), and the remainder is the sum of squared errors (SSE):

$$\text{SST} = \text{SSR} + \text{SSE}, \tag{1.34}$$

where

$$\text{SST} = \sum_{i=1}^{N}(y_i - \overline{y})^2, \tag{1.35}$$

$$\text{SSR} = \sum_{i=1}^{N}(\tilde{y}_i - \overline{y})^2, \tag{1.36}$$

$$\mathrm{SSE} = \sum_{i=1}^{N} (y_i - \tilde{y}_i)^2. \tag{1.37}$$

How well the regression fitted the data can be characterized by

$$R^2 = \frac{\mathrm{SSR}}{\mathrm{SST}} = 1 - \frac{\mathrm{SSE}}{\mathrm{SST}}, \tag{1.38}$$

where R^2 approaches 1 when the fit is very good. Note that R is called the *multiple correlation coefficient*, as it can be shown that it is the correlation between $\tilde{y}$ and y (Draper and Smith, 1981, p. 46), and this holds even when there are multiple predictors in the regression, a situation to be considered in the next subsection.

1.4.4 Multiple linear regression

Often one encounters situations where there are multiple predictors x_l, ($l = 1, \ldots, k$) for the response variable y. This type of multiple linear regression (MLR) has the form

$$y_i = a_0 + \sum_{l=1}^{k} x_{il} a_l + e_i, \qquad i = 1, \ldots, N. \tag{1.39}$$

In vector form,

$$\mathbf{y} = \mathbf{X}\mathbf{a} + \mathbf{e}, \tag{1.40}$$

where

$$\mathbf{y} = \begin{bmatrix} y_1 \\ \vdots \\ y_N \end{bmatrix}, \qquad \mathbf{X} = \begin{bmatrix} 1 & x_{11} & \cdots & x_{k1} \\ \vdots & \vdots & \vdots & \vdots \\ 1 & x_{1N} & \cdots & x_{kN} \end{bmatrix}, \tag{1.41}$$

$$\mathbf{a} = \begin{bmatrix} a_0 \\ \vdots \\ a_k \end{bmatrix}, \qquad \mathbf{e} = \begin{bmatrix} e_1 \\ \vdots \\ e_N \end{bmatrix}. \tag{1.42}$$

The SSE is then

$$\mathrm{SSE} = \mathbf{e}^{\mathrm{T}}\mathbf{e} = (\mathbf{y} - \mathbf{X}\mathbf{a})^{\mathrm{T}}(\mathbf{y} - \mathbf{X}\mathbf{a}), \tag{1.43}$$

where the superscript T denotes the transpose. To minimize SSE with respect to **a**, we differentiate the SSE by **a** and set the derivatives to zero, yielding the *normal equations*,

$$\mathbf{X}^{\mathrm{T}}(\mathbf{y} - \mathbf{X}\mathbf{a}) = \mathbf{0}. \tag{1.44}$$

Thus

$$\mathbf{X}^{\mathrm{T}}\mathbf{X}\mathbf{a} = \mathbf{X}^{\mathrm{T}}\mathbf{y}, \tag{1.45}$$

and the optimal parameters are given by

$$\mathbf{a} = (\mathbf{X}^{\mathrm{T}}\mathbf{X})^{-1}\,\mathbf{X}^{\mathrm{T}}\mathbf{y}. \tag{1.46}$$

A major problem with multiple regression is that often a large number of predictors are available, though only a few of these are actually significant. If all possible predictors are used in building a MLR model, one often '*overfits*' the data, i.e. too many parameters are used in the model so that one is simply fitting to the noise in the data. While the fit to the data may appear very impressive, such overfitted MLR models generally perform poorly when used to make predictions based on new predictor data. Automatic procedures, e.g. *stepwise multiple regression* (Draper and Smith, 1981), have been devised to eliminate insignificant predictors, thereby avoiding an overfitted MLR model.

Another approach to the overfitting problem in MLR is *ridge regression*, where penalty is placed on excessive use of large regression parameters. With ridge regression, the parameters are constrained to lie within a hypersphere of certain radius to prevent use of large regression parameters, and minimization of the SSE is performed with this constraint (Draper and Smith, 1981). Penalty methods are also commonly used in neural network models to prevent overfitting (Section 6.5).

1.4.5 Perfect Prog and MOS

In many branches of environmental sciences, physical (or dynamical) prediction models have surpassed statistical models. For instance, in numerical weather forecasting, the governing equations of the atmosphere are solved by finite-element or spectral methods on supercomputers. Such dynamical models can be integrated forward in time to give weather forecasts. Nevertheless regression is commonly used to assist and improve the raw forecasts made by the dynamical models (Wilks, 1995). The reason is that the variables in the dynamical model usually have poor resolution and are sometimes too idealized. For instance, the lowest temperature level in the model may be some considerable distance above the ground. Furthermore, the local topography may be completely missed in the low resolution dynamical model. Thus it would be difficult to use the output from such a dynamical model directly to predict the ground temperature at a village located in a valley. Furthermore, some local variables such as ozone concentration or precipitation may not even be variables carried in the dynamical model.

The *Perfect Prog* (abbreviation for perfect prognosis) scheme computes a multiple regression relation from the historical data archive:

$$y(t) = \mathbf{x}(t)^{\mathrm{T}}\mathbf{a} + e(t), \tag{1.47}$$

where y is the response, $\mathbf{x}$ the predictors, and e the error. During actual forecasting, $\mathbf{x}(t)$ is provided by the forecasts from the dynamical model, and $y(t)$ is predicted by the above regression relation. The problem with this scheme is that while the regression model was developed or trained using historical data for $\mathbf{x}$, the actual forecasts used the dynamical model forecasts for $\mathbf{x}$. Hence, the systematic error between the dynamical model forecasts and real data has not been taken into account – i.e. perfect prognosis is assumed, whence the name of this scheme.

In contrast, a better approach is the *model output statistics* (MOS) scheme, where the dynamical model forecasts have been archived, so the regression was developed using $y(t)$ from the data archive and $\mathbf{x}(t)$ from the dynamical model forecast archive. Since $\mathbf{x}$ was from the dynamical model forecasts during both model training and actual forecasting, the model bias in the Perfect Prog scheme has been eliminated. While MOS is more accurate than Perfect Prog, it is considerably more difficult to implement since a slight modification of the dynamical model would require regeneration of the dynamical model forecast archive and recalculation of the regression relations.

In summary, even in areas where physical or dynamical models outperform statistical models in forecasting, post-processing of the dynamical model outputs by regression in the form of Perfect Prog or MOS can often enhance the dynamical model predictions. We shall see that machine learning methods can further improve the post-processing of dynamical model outputs (Section 12.3.4).

1.5 Bayes theorem

Bayes theorem, named after the Reverend Thomas Bayes (1702–1761), plays a central role in modern statistics (Jaynes, 2003; Le and Zidek, 2006). Historically, it had a major role in the debate around the foundations of statistics, as the traditional '*frequentist*' school and the Bayesian school disagreed on how probabilities should be assigned in applications. Frequentists assign probabilities to random events according to their frequencies of occurrence or to subsets of populations as proportions of the whole. In contrast, Bayesians describe probabilities in terms of beliefs and degrees of uncertainty, similarly to how the general public uses probability. For instance, a sports fan prior to the start of a sports tournament asserts that team A has a probability of 60% for winning the tournament. However, after a disappointing game, the fan may modify the winning probability to 30%. Bayes theorem provides the formula for modifiying prior probabilities in view of new data.

We will use a *classification* problem to illustrate the Bayes approach. Suppose a meteorologist wants to classify the approaching weather state as either storm (C_1), or non-storm (C_2). Assume there is some *a priori probability* (or simply *prior probability*) $P(C_1)$ that there is a storm, and some prior probability $P(C_2)$ that there is no storm. For instance, from the past weather records, 15% of the days were found to be stormy during this season, then the meteorologist may assign $P(C_1) = 0.15$, and $P(C_2) = 0.85$. Now suppose the meteorologist has a barometer measuring a pressure x at 6 a.m. The meteorologist would like to obtain an *a posteriori probability* (or simply *posterior probability*) $P(C_1|x)$, i.e. the conditional probability of having a storm on that day given the 6 a.m. pressure x. In essence, he would like to improve on his simple prior probability with the new information x.

The joint probability density $p(C_i, x)$ is the probability density that an event belongs to class C_i and has value x, (noting that a small p is used in this book to denote a probability density, versus P for probability). The joint probability density can be written as

$$p(C_i, x) = P(C_i|x)p(x), \tag{1.48}$$

with $p(x)$ the probability density of x. Alternatively, $p(C_i, x)$ can be written as

$$p(C_i, x) = p(x|C_i)P(C_i), \tag{1.49}$$

with $p(x|C_i)$, the conditional probability density of x, given that the event belongs to class C_i. Equating the right hand sides of these two equations, we obtain

$$P(C_i|x) = \frac{p(x|C_i)P(C_i)}{p(x)}, \tag{1.50}$$

which is *Bayes theorem*. Since $p(x)$ is the probability density of x without regard to which class, it can be decomposed into

$$p(x) = \sum_i p(x|C_i)P(C_i). \tag{1.51}$$

Substituting this for $p(x)$ in (1.50) yields

$$P(C_i|x) = \frac{p(x|C_i)P(C_i)}{\sum_i p(x|C_i)P(C_i)}, \tag{1.52}$$

where the denominator on the right hand side is seen as a normalization factor for the posterior probabilities to sum to unity. Bayes theorem says that the posterior probability $P(C_i|x)$ is simply $p(x|C_i)$ (the *likelihood* of x given the event is of class C_i) multiplied by the prior probability $P(C_i)$, and divided by a normalization factor. The advantage of Bayes theorem is that the posterior probability is now expressed in terms of quantities which can be estimated. For instance, to estimate $p(x|C_i)$, the meteorologist can divide the 6 a.m. pressure record into two classes, and estimate $p(x|C_1)$ from the pressure distribution for stormy days, and $p(x|C_2)$ from the pressure distribution for non-stormy days.

For the general situation, the scalar x is replaced by a *feature vector* $\mathbf{x}$, and the classes are $C_1, \ldots, C_k$, then Bayes theorem becomes

$$P(C_i|\mathbf{x}) = \frac{p(\mathbf{x}|C_i)P(C_i)}{\sum_i p(\mathbf{x}|C_i)P(C_i)}, \tag{1.53}$$

for $i = 1, \ldots, k$.

If instead of the discrete variable C_i, we have a continuous variable w, then Bayes theorem (1.50) takes the form

$$p(w|x) = \frac{p(x|w)p(w)}{p(x)}. \tag{1.54}$$

1.6 Discriminant functions and classification

Once the posterior probabilities $P(C_i|\mathbf{x})$ have been estimated from (1.53), we can proceed to classification: Given a feature vector $\mathbf{x}$, we choose the class C_j having the highest posterior probability, i.e.

$$P(C_j|\mathbf{x}) > P(C_i|\mathbf{x}), \quad \text{for all } i \neq j. \tag{1.55}$$

This is equivalent to

$$p(\mathbf{x}|C_j)P(C_j) > p(\mathbf{x}|C_i)P(C_i), \quad \text{for all } i \neq j. \tag{1.56}$$

In the feature space, the pattern classifier has divided the space into *decision regions* $R_1, \ldots, R_k$, so that if a feature vector lands within R_i, the classifier will assign the class C_i. The decision region R_i may be composed of several disjoint regions, all of which are assigned the class C_i. The boundaries between decision regions are called *decision boundaries* or *decision surfaces*.

To justify the decison rule (1.56), consider the probability P_{correct} of a new pattern being classified correctly:

$$P_{\text{correct}} = \sum_{j=1}^{k} P(\mathbf{x} \in R_j, C_j), \tag{1.57}$$

where $P(\mathbf{x} \in R_j, C_j)$ gives the probability that the pattern which belongs to class C_j has its feature vector falling within the decision region R_j, hence classified correctly as belonging to class C_j. Note that P_{correct} can be expressed as

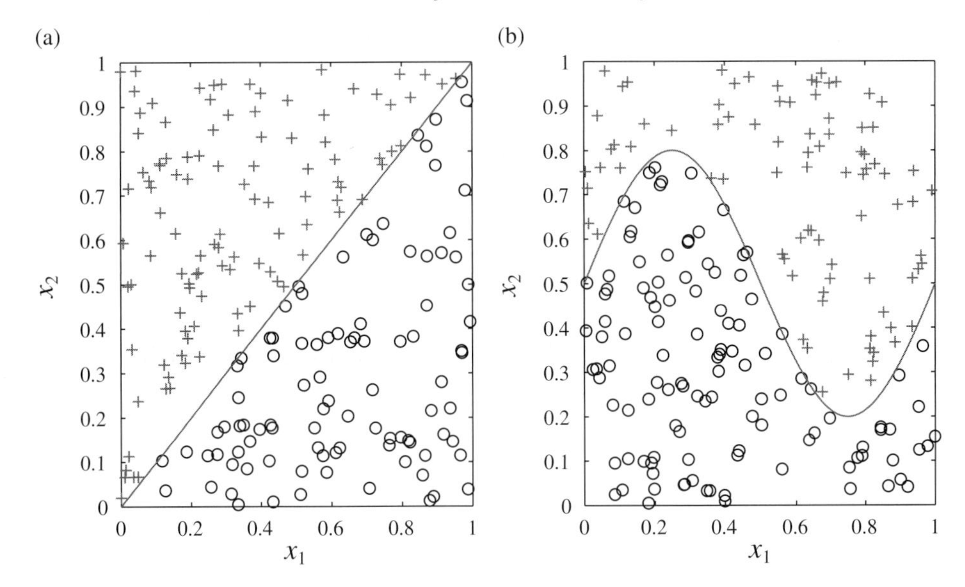

Fig. 1.3 (a) A linear decision boundary separating two classes of data denoted by circles and plus signs respectively. (b) A nonlinear decision boundary.

$$
\begin{aligned}
P_{\text{correct}} &= \sum_{j=1}^{k} P(\mathbf{x} \in R_j | C_j) P(C_j), \\
&= \sum_{j=1}^{k} \int_{R_j} p(\mathbf{x}|C_j) P(C_j) \mathrm{d}\mathbf{x}. \qquad (1.58)
\end{aligned}
$$

To maximize P_{correct}, one needs to maximize the integrand by choosing the decision regions so that $\mathbf{x}$ is assigned to the class C_j satisfying (1.56), which justifies the decision rule (1.56).

In general, classification need not be based on probability distribution functions, since in many situations, $p(\mathbf{x}|C_i)$ and $P(C_i)$ are not known. The classification procedure is then formulated in terms of *discriminant functions*, which tell us which class we should assign to the given predictor data point $\mathbf{x}$. For example, in Fig. 1.3a, $\mathbf{x} = (x_1, x_2)^{\mathrm{T}}$, and the two classes are separated by the line $x_2 = x_1$. Hence the discriminant function can be simply $y(\mathbf{x}) = -x_1 + x_2$, with C_2 assigned when $y(\mathbf{x}) > 0$, and C_1 otherwise. Hence the decision boundary is given by $y(\mathbf{x}) = 0$.

When there are more than two classes, the discriminant functions are $y_1(\mathbf{x}), \ldots,$ $y_k(\mathbf{x})$, where a feature vector $\mathbf{x}$ is assigned to class C_j if

$$
y_j(\mathbf{x}) > y_i(\mathbf{x}), \qquad \text{for all } i \neq j. \qquad (1.59)
$$

Clearly (1.55) is a special case of (1.59). An important property of a discriminant function $y_i(\mathbf{x})$ is that it can be replaced by $f(y_i(\mathbf{x}))$, for any monotonic function

f, since the classification is unchanged as the relative magnitudes of the discriminant functions are preserved by f. There are many classical *linear discriminant analysis* methods (e.g. Fisher's linear discriminant) (Duda *et al.*, 2001), where the discriminant function is a linear combination of the inputs, i.e.

$$y_i(\mathbf{x}) = \sum_l w_{il} x_l + w_{i0} \equiv \mathbf{w}_i^{\mathrm{T}} \mathbf{x} + w_{i0}, \tag{1.60}$$

with parameters $\mathbf{w}_i$ and w_{i0}. Based on (1.59), the decision boundary between class C_j and C_i is obtained from setting $y_j(\mathbf{x}) = y_i(\mathbf{x})$, yielding a hyperplane decision boundary described by

$$(\mathbf{w}_j - \mathbf{w}_i)^{\mathrm{T}} \mathbf{x} + (w_{j0} - w_{i0}) = 0. \tag{1.61}$$

Suppose $\mathbf{x}$ and $\mathbf{x}'$ both lie within the decision region R_j. Consider any point $\tilde{\mathbf{x}}$ lying on a straight line connecting $\mathbf{x}$ and $\mathbf{x}'$, i.e.

$$\tilde{\mathbf{x}} = a\mathbf{x} + (1 - a)\mathbf{x}', \tag{1.62}$$

with $0 \leq a \leq 1$. Since $\mathbf{x}$ and $\mathbf{x}'$ both lie within R_j, they satisfy $y_j(\mathbf{x}) > y_i(\mathbf{x})$ and $y_j(\mathbf{x}') > y_i(\mathbf{x}')$ for all $i \neq j$. Since the discriminant function is linear, we also have

$$y_j(\tilde{\mathbf{x}}) = a y_j(\mathbf{x}) + (1 - a) y_j(\mathbf{x}'), \tag{1.63}$$

hence $y_j(\tilde{\mathbf{x}}) > y_i(\tilde{\mathbf{x}})$ for all $i \neq j$. Thus any point on the straight line joining $\mathbf{x}$ and $\mathbf{x}'$ must also lie within R_j, meaning that the decision region R_j is simply connected and convex. As we shall see later, with neural network methods, the decision boundaries can be curved surfaces (Fig. 1.3b) instead of hyperplanes, and the decision regions need not be simply connected nor convex.

1.7 Clustering

In machine learning, there are two general approaches, *supervised learning* and *unsupervised learning*. An analogy for the former is students in a French class where the teacher demonstrates the correct French pronunciation. An analogy for the latter is students working on a team project without supervision. In unsupervised learning, the students are provided with learning rules, but must rely on self-organization to arrive at a solution, without the benefit of being able to learn from a teacher's demonstration.

When we perform regression in (1.24), the training dataset consists of pairs of predictors and predictands (x_i, y_i), $y = 1, \ldots, N$. Here y_i serves the role of the teacher or *target* for the regression model output $\tilde{y}_i$, i.e. $\tilde{y}_i$ is fitted to the given target data, similar to students trying to imitate the French accent of their teacher, hence the learning is supervised. Classification can be regarded as the discrete version of regression. For instance, with regression we are interested in predicting the air temperature in degrees Celsius, whereas with classification, we are interested in predicting whether the temperature will be 'cold', 'normal' or 'warm'. Since the classes are specified in the target data, classification is also supervised learning.

Clustering or cluster analysis is the unsupervised version of classification. The goal of clustering is to group the data into a number of subsets or 'clusters', such that the data within a cluster are more closely related to each other than data from other clusters. For instance, in the simple and widely used *K-means clustering*, one starts with initial guesses for the mean positions of the K clusters in data space

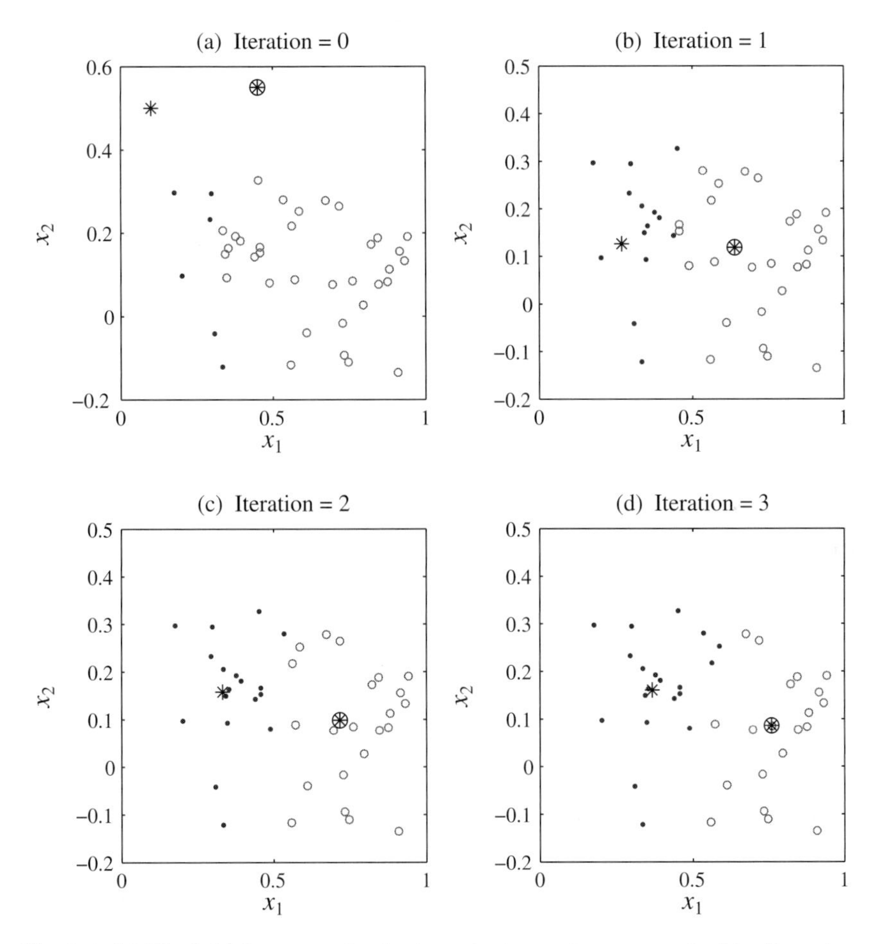

Fig. 1.4 (a) The initial guesses for the two cluster centres are marked by the asterisk and the circled asterisk. The data points closest to the asterisk are plotted as dots, while those closest to the circled asterisk are shown as circles. The location of the cluster centres and their associated clusters are shown after (b) One, (c) Two, (d) Three iterations of the K-means clustering algorithm.

(to be referred to as the cluster centres), then iterates the following two steps till convergence:

(i) For each data point, find the closest cluster centre (based on Euclidean distance).
(ii) For each cluster, reassign the cluster centre to be the mean position of all the data belonging to that cluster.

Figure 1.4 illustrates K-means clustering. As K, the number of clusters, is specified by the user, choosing a different K will lead to very different clusters.

Exercises

(1.1) A variable y is measured by two instruments placed 50 km apart in the east-west direction. Values are recorded daily for 100 days. The autocorrelation function of y shows the first zero crossing (i.e. the smallest lag at which the autocorrelation is zero) occurring at 6 days (for both stations). Furthermore, y at one station is correlated with the y at the second station, with the second time series shifted in time by various lags. The maximum correlation occurred with y from the eastern station lagging y from the western station by 2 days. Assuming a sinusoidal wave is propagating between the two stations, estimate the period, wavelength, and the speed and direction of propagation.

(1.2) Given two time series of daily observations over 42 days, the Pearson correlation was found to be -0.65. (a) Assuming the daily observations are independent, is the correlation significant? (b) If autocorrelation shows a decorrelation time scale of 6 days, is the correlation significant?

(1.3) Regression ($y = ax + b$) and correlation are related. What are the conditions when a dataset of (x, y) values yields a large correlation ρ (e.g. $\rho = 0.9$) but a small regression parameter a ($a = 0.01$)? Is it possible for ρ and a to have opposite signs?

(1.4) Using the data file provided in the book website (given in the Preface), compare the Pearson correlation with the Spearman rank correlation for the time series x and y (each with 40 observations). Repeat the comparison for the time series x_2 and y_2 (from the same data file as above), where x_2 and y_2 are the same as x and y, except that the fifth data point in y is replaced by an outlier in y_2. Repeat the comparison for the time series x_3 and y_3, where x_3 and y_3 are the same as x and y, except that the fifth data point in x and y is replaced by an outlier in x_3 and y_3. Make scatterplots of the data points in the x–y space, the x_2–y_2 space and the x_3–y_3 space. Also plot the linear regression line in the scatterplots.

(1.5) Using the data file provided in the book website, perform multiple linear regression with predictors x_1, x_2 and x_3, and the response variable y. Rank the importance of the predictors in their influence on y.

(1.6) Suppose a test for the presence of a toxic chemical in a lake gives the following results: if a lake has the toxin, the test returns a positive result 99% of the time; if a lake does not have the toxin, the test still returns a positive result 2% of the time. Suppose only 5% of the lakes contain the toxin. According to Bayes theorem, what is the probability that a positive test result for a lake turns out to be a false positive?

2

Linear multivariate statistical analysis

As one often encounters datasets with more than a few variables, multivariate statistical techniques are needed to extract the information contained in these datasets effectively. In the environmental sciences, examples of multivariate datasets are ubiquitous – the air temperatures recorded by all the weather stations around the globe, the satellite infrared images composed of numerous small pixels, the gridded output from a general circulation model, etc. The number of variables or time series from these datasets ranges from thousands to millions. Without a mastery of multivariate techniques, one is overwhelmed by these gigantic datasets. In this chapter, we review the principal component analysis method and its many variants, and the canonical correlation analysis method. These methods, using standard matrix techniques such as singular value decomposition, are relatively easy to use, but suffer from being *linear*, a limitation which will be lifted with neural network and kernel methods in later chapters.

2.1 Principal component analysis (PCA)

2.1.1 Geometric approach to PCA

We have a dataset with variables $y_1, \ldots, y_m$. These variables have been sampled n times. In many situations, the m variables are m time series each containing n observations in time. For instance, one may have a dataset containing the monthly air temperature measured at m stations over n months. If m is a large number, we would like to capture the essence of $y_1, \ldots, y_m$ by a smaller set of variables $z_1, \ldots, z_k$ (i.e. $k < m$; and hopefully $k \ll m$, for truly large m). This is the objective of *principal component analysis* (PCA), also called *empirical orthogonal function* (EOF) analysis in meteorology and oceanography. We first begin with a geometric approach, which is more intuitive than the standard eigenvector approach to PCA.

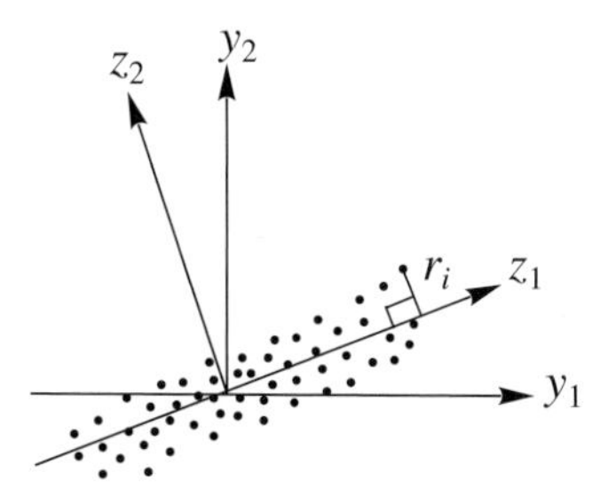

Fig. 2.1 The PCA problem formulated as a minimization of the sum of r_i^2, where r_i is the shortest distance from the ith data point to the first PCA axis z_1.

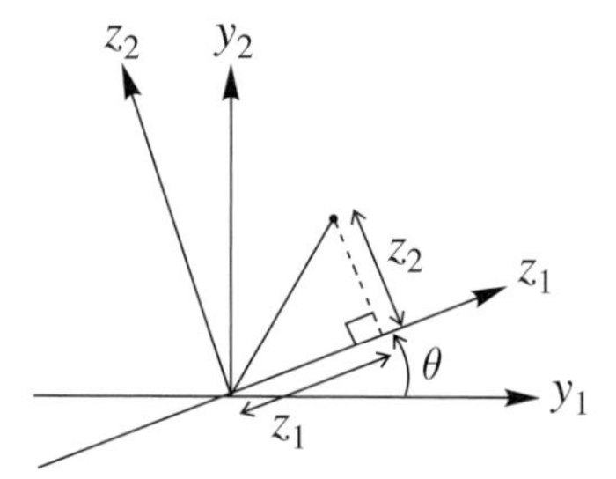

Fig. 2.2 Rotation of coordinate axes by an angle θ in a 2-dimensional space.

Let us start with only two variables, y_1 and y_2, as illustrated in Fig. 2.1. Clearly the bulk of the variance is along the axis z_1. If r_i is the distance between the ith data point and the axis z_1, then the optimal z_1 is found by minimizing $\sum_{i=1}^{n} r_i^2$. This type of geometric approach to PCA was first proposed by Pearson (1901). Note that PCA treats all variables equally, whereas regression divides variables into independent and dependent variables (cf. Fig. 2.1 and Fig. 1.2), hence the straight line described by z_1 is in general different from the regression line.

In 3-D, z_1 is the best 1-D line fit to the data, while z_1 and z_2 span a 2-D plane giving the best plane fit to the data. In general, with an m-dimensional dataset, we want to find the k-dimensional hyperplane giving the best fit.

2.1.2 Eigenvector approach to PCA

The more systematic eigenvector approach to PCA is due to Hotelling (1933). Here again with the 2-D example, a data point is transformed from its old coordinates (y_1, y_2) to new coordinates (z_1, z_2) via a rotation of the coordinate system (Fig. 2.2):

$$z_1 = y_1 \cos\theta + y_2 \sin\theta,$$
$$z_2 = -y_1 \sin\theta + y_2 \cos\theta. \tag{2.1}$$

In the general m-dimensional problem, we want to introduce new coordinates

$$z_j = \sum_{l=1}^{m} e_{jl} y_l, \qquad j = 1, \ldots, m. \tag{2.2}$$

The objective is to find

$$\mathbf{e}_1 = [e_{11}, \ldots, e_{1m}]^{\mathrm{T}}, \tag{2.3}$$

which maximizes $\mathrm{var}(z_1)$, i.e. find the coordinate transformation such that the variance of the dataset along the direction of the z_1 axis is maximized.

With

$$z_1 = \sum_{l=1}^{m} e_{1l} y_l = \mathbf{e}_1^{\mathrm{T}} \mathbf{y}, \qquad \mathbf{y} = [y_1, \ldots, y_m]^{\mathrm{T}}, \tag{2.4}$$

i.e. projecting the data point $\mathbf{y}$ onto the vector $\mathbf{e}_1$ gives a distance of z_1 along the $\mathbf{e}_1$ direction, we have

$$\mathrm{var}(z_1) = \mathrm{E}\,[(z_1 - \bar{z}_1)(z_1 - \bar{z}_1)] = \mathrm{E}\,[\mathbf{e}_1^{\mathrm{T}}(\mathbf{y} - \bar{\mathbf{y}})(\mathbf{y} - \bar{\mathbf{y}})^{\mathrm{T}} \mathbf{e}_1], \tag{2.5}$$

where we have used the vector property $\mathbf{a}^{\mathrm{T}}\mathbf{b} = \mathbf{b}^{\mathrm{T}}\mathbf{a}$. Thus,

$$\mathrm{var}(z_1) = \mathbf{e}_1^{\mathrm{T}}\,\mathrm{E}\,[(\mathbf{y} - \bar{\mathbf{y}})(\mathbf{y} - \bar{\mathbf{y}})^{\mathrm{T}}]\,\mathbf{e}_1 = \mathbf{e}_1^{\mathrm{T}} \mathbf{C} \mathbf{e}_1, \tag{2.6}$$

where the *covariance matrix* $\mathbf{C}$ is given by

$$\mathbf{C} = \mathrm{E}\,[(\mathbf{y} - \bar{\mathbf{y}})(\mathbf{y} - \bar{\mathbf{y}})^{\mathrm{T}}]. \tag{2.7}$$

Clearly, the larger is the vector norm $\|\mathbf{e}_1\|$, the larger $\mathrm{var}(z_1)$ will be. Hence, we need to place a constraint on $\|\mathbf{e}_1\|$ while we try to maximize $\mathrm{var}(z_1)$. Let us impose a normalization constraint $\|\mathbf{e}_1\| = 1$, i.e.

$$\mathbf{e}_1^{\mathrm{T}} \mathbf{e}_1 = 1. \tag{2.8}$$

Thus our optimization problem is to find $\mathbf{e}_1$ which maximizes $\mathbf{e}_1^{\mathrm{T}} \mathbf{C} \mathbf{e}_1$, subject to the constraint

$$\mathbf{e}_1^{\mathrm{T}} \mathbf{e}_1 - 1 = 0. \tag{2.9}$$

The method of Lagrange multipliers is commonly used to tackle optimization under constraints (see Appendix B). Instead of finding stationary points of $\mathbf{e}_1^{\mathrm{T}} \mathbf{C} \mathbf{e}_1$, we search for the stationary points of the Lagrange function L,

$$L = \mathbf{e}_1^{\mathrm{T}} \mathbf{C} \mathbf{e}_1 - \lambda(\mathbf{e}_1^{\mathrm{T}} \mathbf{e}_1 - 1), \tag{2.10}$$

where λ is a Lagrange multiplier. Differentiating L by the elements of $\mathbf{e}_1$, and setting the derivatives to zero, we obtain

$$\mathbf{C}\mathbf{e}_1 - \lambda \mathbf{e}_1 = 0, \tag{2.11}$$

which says that λ is an eigenvalue of the covariance matrix $\mathbf{C}$, with $\mathbf{e}_1$ the eigenvector. Multiplying this equation by $\mathbf{e}_1^{\mathrm{T}}$ on the left, we obtain

$$\lambda = \mathbf{e}_1^{\mathrm{T}}\mathbf{C}\mathbf{e}_1 = \mathrm{var}(z_1). \tag{2.12}$$

Since $\mathbf{e}_1^{\mathrm{T}}\mathbf{C}\mathbf{e}_1$ is maximized, so are $\mathrm{var}(z_1)$ and λ. The new coordinate z_1, called the *principal component* (PC), is found from (2.4).

Next, we want to find z_2 – our task is to find $\mathbf{e}_2$ which maximizes $\mathrm{var}(z_2) = \mathbf{e}_2^{\mathrm{T}}\mathbf{C}\mathbf{e}_2$, subject to the constraint $\mathbf{e}_2^{\mathrm{T}}\mathbf{e}_2 = 1$, and the constraint that z_2 be uncorrelated with z_1, i.e. the covariance between z_2 and z_1 be zero,

$$\mathrm{cov}(z_1, z_2) = 0. \tag{2.13}$$

As $\mathbf{C} = \mathbf{C}^{\mathrm{T}}$, we can write

$$\begin{aligned} 0 = \mathrm{cov}(z_1, z_2) &= \mathrm{cov}(\mathbf{e}_1^{\mathrm{T}}\mathbf{y}, \mathbf{e}_2^{\mathrm{T}}\mathbf{y}) \\ &= \mathrm{E}[\mathbf{e}_1^{\mathrm{T}}(\mathbf{y}-\bar{\mathbf{y}})(\mathbf{y}-\bar{\mathbf{y}})^{\mathrm{T}}\mathbf{e}_2] = \mathbf{e}_1^{\mathrm{T}}\mathrm{E}[(\mathbf{y}-\bar{\mathbf{y}})(\mathbf{y}-\bar{\mathbf{y}})^{\mathrm{T}}]\mathbf{e}_2 \\ &= \mathbf{e}_1^{\mathrm{T}}\mathbf{C}\mathbf{e}_2 = \mathbf{e}_2^{\mathrm{T}}\mathbf{C}\mathbf{e}_1 = \mathbf{e}_2^{\mathrm{T}}\lambda_1\mathbf{e}_1 = \lambda_1\mathbf{e}_2^{\mathrm{T}}\mathbf{e}_1 = \lambda_1\mathbf{e}_1^{\mathrm{T}}\mathbf{e}_2\,. \end{aligned} \tag{2.14}$$

The orthogonality condition

$$\mathbf{e}_2^{\mathrm{T}}\mathbf{e}_1 = 0, \tag{2.15}$$

can be used as a constraint in place of (2.13).

Upon introducing another Lagrange multiplier γ, we want to find an $\mathbf{e}_2$ which gives a stationary point of the Lagrange function L,

$$L = \mathbf{e}_2^{\mathrm{T}}\mathbf{C}\mathbf{e}_2 - \lambda(\mathbf{e}_2^{\mathrm{T}}\mathbf{e}_2 - 1) - \gamma\,\mathbf{e}_2^{\mathrm{T}}\mathbf{e}_1. \tag{2.16}$$

Differentiating L by the elements of $\mathbf{e}_2$, and setting the derivatives to zero, we obtain

$$\mathbf{C}\mathbf{e}_2 - \lambda\mathbf{e}_2 - \gamma\mathbf{e}_1 = 0. \tag{2.17}$$

Left multiplying this equation by $\mathbf{e}_1^{\mathrm{T}}$ yields

$$\mathbf{e}_1^{\mathrm{T}}\mathbf{C}\mathbf{e}_2 - \lambda\mathbf{e}_1^{\mathrm{T}}\mathbf{e}_2 - \gamma\,\mathbf{e}_1^{\mathrm{T}}\mathbf{e}_1 = 0. \tag{2.18}$$

On the left hand side, the first two terms are both zero from (2.14) while the third term is simply γ, so we have $\gamma = 0$, and (2.17) reduces to

$$\mathbf{C}\mathbf{e}_2 - \lambda\mathbf{e}_2 = 0. \tag{2.19}$$

Once again λ is an eigenvalue of $\mathbf{C}$, with $\mathbf{e}_2$ the eigenvector. As

$$\lambda = \mathbf{e}_2^{\mathrm{T}}\mathbf{C}\mathbf{e}_2 = \mathrm{var}(z_2), \tag{2.20}$$

which is maximized, this $\lambda = \lambda_2$ is as large as possible with $\lambda_2 < \lambda_1$. (The case $\lambda_2 = \lambda_1$ is degenerate and will be discussed later.) Hence, λ_2 is the second largest eigenvalue of $\mathbf{C}$, with $\lambda_2 = \text{var}(z_2)$. This process can be repeated for $z_3, z_4, \ldots$

How do we reconcile the geometric approach of the previous subsection and the present eigenvector approach? First we subtract the mean $\overline{\mathbf{y}}$ from $\mathbf{y}$, so the transformed data are centred around the origin with $\overline{\mathbf{y}} = 0$. In the geometric approach, we minimize the distance between the data points and the new axis. If the unit vector $\mathbf{e}_1$ gives the direction of the new axis, then the projection of a data point (described by the vector $\mathbf{y}$) onto $\mathbf{e}_1$ is $(\mathbf{e}_1^{\mathrm{T}}\mathbf{y})\mathbf{e}_1$. The component of $\mathbf{y}$ normal to $\mathbf{e}_1$ is $\mathbf{y} - (\mathbf{e}_1^{\mathrm{T}}\mathbf{y})\mathbf{e}_1$. Thus, minimizing the distance between the data points and the new axis amounts to minimizing

$$\epsilon = \mathrm{E}[\, \| \mathbf{y} - (\mathbf{e}_1^{\mathrm{T}}\mathbf{y})\mathbf{e}_1 \|^2 \,]. \tag{2.21}$$

Simplifying this yields

$$\epsilon = \mathrm{E}[\, \|\mathbf{y}\|^2 - (\mathbf{e}_1^{\mathrm{T}}\mathbf{y})\mathbf{y}^{\mathrm{T}}\mathbf{e}_1 \,] = \text{var}(\mathbf{y}) - \text{var}(\mathbf{e}_1^{\mathrm{T}}\mathbf{y}), \tag{2.22}$$

where $\text{var}(\mathbf{y}) \equiv \mathrm{E}[\, \|\mathbf{y}\|^2]$, with $\overline{\mathbf{y}}$ assumed to be zero. Since $\text{var}(\mathbf{y})$ is constant, minimizing ϵ is equivalent to maximizing $\text{var}(\mathbf{e}_1^{\mathrm{T}}\mathbf{y})$, which is equivalent to maximizing $\text{var}(z_1)$. Hence the geometric approach of minimizing the distance between the data points and the new axis is equivalent to the eigenvector approach in finding the largest eigenvalue λ, which is simply $\max[\text{var}(z_1)]$.

So far, $\mathbf{C}$ is the data covariance matrix, but it can also be the data correlation matrix, if one prefers correlation over covariance. In *combined PCA*, where two or more variables with different units are combined into one large data matrix for PCA – e.g. finding the PCA modes of the combined sea surface temperature data and the sea level pressure data – then one needs to normalize the variables, so that $\mathbf{C}$ is the correlation matrix.

2.1.3 Real and complex data

In general, for $\mathbf{y}$ real,

$$\mathbf{C} \equiv \mathrm{E}[(\mathbf{y} - \overline{\mathbf{y}})(\mathbf{y} - \overline{\mathbf{y}})^{\mathrm{T}}], \tag{2.23}$$

implies that $\mathbf{C}^{\mathrm{T}} = \mathbf{C}$, i.e. $\mathbf{C}$ is a real, symmetric matrix. A *positive semi-definite matrix* $\mathbf{A}$ is defined by the property that for any $\mathbf{v} \neq \mathbf{0}$, it follows that $\mathbf{v}^{\mathrm{T}}\mathbf{A}\mathbf{v} \geq 0$ (Strang, 2005). From the definition of $\mathbf{C}$ (2.23), it is clear that $\mathbf{v}^{\mathrm{T}}\mathbf{C}\mathbf{v} \geq 0$ is satisfied. Hence $\mathbf{C}$ is a real, symmetric, positive semi-definite matrix.

If $\mathbf{y}$ is complex, then

$$\mathbf{C} \equiv \mathrm{E}[(\mathbf{y} - \overline{\mathbf{y}})(\mathbf{y} - \overline{\mathbf{y}})^{\mathrm{T}*}], \tag{2.24}$$

with complex conjugation denoted by the superscript asterisk. As $\mathbf{C}^{\mathrm{T}*} = \mathbf{C}$, $\mathbf{C}$ is a *Hermitian matrix*. It is also a positive semi-definite matrix. Theorems on Hermitian, positive semi-definite matrices tell us that $\mathbf{C}$ has real eigenvalues

$$\lambda_1 \geq \lambda_2 \geq \cdots \geq \lambda_m \geq 0, \qquad \sum_{j=1}^{m} \lambda_j = \mathrm{var}(\mathbf{y}), \tag{2.25}$$

with corresponding orthonormal eigenvectors, $\mathbf{e}_1, \ldots, \mathbf{e}_m$, and that the k eigenvectors corresponding to $\lambda_1, \ldots, \lambda_k$ minimize

$$\epsilon_k = \mathrm{E}[\,\|(\mathbf{y} - \overline{\mathbf{y}}) - \sum_{j=1}^{k} (\mathbf{e}_j^{\mathrm{T}}(\mathbf{y} - \overline{\mathbf{y}}))\mathbf{e}_j\|^2\,], \tag{2.26}$$

which can be expressed as

$$\epsilon_k = \mathrm{var}(\mathbf{y}) - \sum_{j=1}^{k} \lambda_j. \tag{2.27}$$

2.1.4 Orthogonality relations

Thus PCA amounts to finding the eigenvectors and eigenvalues of $\mathbf{C}$. The orthonormal eigenvectors then provide a basis, i.e. the data $\mathbf{y}$ can be expanded in terms of the eigenvectors $\mathbf{e}_j$:

$$\mathbf{y} - \overline{\mathbf{y}} = \sum_{j=1}^{m} a_j(t)\mathbf{e}_j, \tag{2.28}$$

where $a_j(t)$ are the expansion coefficients. To obtain $a_j(t)$, left multiply the above equation by $\mathbf{e}_i^{\mathrm{T}}$, and use the orthonormal relation of the eigenvectors,

$$\mathbf{e}_i^{\mathrm{T}}\mathbf{e}_j = \delta_{ij}, \tag{2.29}$$

(with δ_{ij} denoting the Kronecker delta function, which equals 1 if $i = j$, and 0 otherwise) to get

$$a_j(t) = \mathbf{e}_j^{\mathrm{T}}(\mathbf{y} - \overline{\mathbf{y}}), \tag{2.30}$$

i.e. $a_j(t)$ is obtained by projection of the data vector $\mathbf{y} - \overline{\mathbf{y}}$ onto the eigenvector $\mathbf{e}_j$, as the right hand side of this equation is simply a dot product between the two vectors. The nomenclature varies considerably in the literature: a_j are called principal components, scores, temporal coefficients and amplitudes; while the eigenvectors $\mathbf{e}_j$ are also referred to as principal vectors, loadings, spatial patterns and EOFs (Empirical Orthogonal Functions). In this book, we prefer calling a_j *principal components* (PCs), $\mathbf{e}_j$ eigenvectors or EOFs (and the elements e_{ji} *loadings*), and j the mode number. Note that for time series, a_j is a function of time while $\mathbf{e}_j$ is a function of space, hence the names temporal coefficients and spatial patterns describe them well. However, in many cases, the dataset may not consist of time series. For instance, the dataset could be plankton collected from various oceanographic stations – t then becomes the label for a station, while 'space' here could represent the various plankton species, and the data $\mathbf{y}(t) = [y_1(t), \ldots, y_m(t)]^{\mathrm{T}}$ could be the amount of species $1, \ldots, m$ found in station t. Another example comes from the multi-channel satellite image data, where images of the Earth's surface have been collected at several frequency channels. Here t becomes the location label for a pixel in an image, and 'space' indicates the various frequency channels.

There are two important properties of PCAs. The expansion $\sum_{j=1}^{k} a_j(t)\mathbf{e}_j(\mathbf{x})$, with $k \leq m$, explains more of the variance of the data than any other linear combination $\sum_{j=1}^{k} b_j(t)\mathbf{f}_j(\mathbf{x})$. Thus PCA provides the most efficient way to compress data, using k eigenvectors $\mathbf{e}_j$ and corresponding time series a_j.

The second important property is that the time series in the set $\{a_j\}$ are uncorrelated. We can write

$$a_j(t) = \mathbf{e}_j^{\mathrm{T}}(\mathbf{y} - \overline{\mathbf{y}}) = (\mathbf{y} - \overline{\mathbf{y}})^{\mathrm{T}}\mathbf{e}_j. \tag{2.31}$$

For $i \neq j$,

$$\begin{aligned}\mathrm{cov}(a_i, a_j) &= \mathrm{E}[\mathbf{e}_i^{\mathrm{T}}(\mathbf{y} - \overline{\mathbf{y}})(\mathbf{y} - \overline{\mathbf{y}})^{\mathrm{T}}\mathbf{e}_j] = \mathbf{e}_i^{\mathrm{T}}\mathrm{E}[(\mathbf{y} - \overline{\mathbf{y}})(\mathbf{y} - \overline{\mathbf{y}})^{\mathrm{T}}]\mathbf{e}_j \\ &= \mathbf{e}_i^{\mathrm{T}}\mathbf{C}\mathbf{e}_j = \mathbf{e}_i^{\mathrm{T}}\lambda_j\mathbf{e}_j = \lambda_j\mathbf{e}_i^{\mathrm{T}}\mathbf{e}_j = 0,\end{aligned} \tag{2.32}$$

implying zero correlation between $a_i(t)$ and $a_j(t)$. Hence PCA extracts the uncorrelated modes of variability of the data field. Note that no correlation between $a_i(t)$ and $a_j(t)$ only means no linear relation between the two, there may still be nonlinear relations between them, which can be extracted by nonlinear PCA methods (Chapter 10).

When applying PCA to gridded data over the globe, one should take into account the decrease in the area of a grid cell with latitude. By scaling the variance from each grid cell by the area of the cell (which is proportional to the cosine of the latitude ϕ), one can avoid having the anomalies in the higher latitudes overweighted in the PCA (North *et al.*, 1982). This scaling of the variance can be accomplished simply by multiplying the anomaly $y_l - \overline{y_l}$ at the lth grid cell by the factor $(\cos\phi)^{1/2}$ for that cell.

2.1.5 PCA of the tropical Pacific climate variability

Let us illustrate the PCA technique with data from the tropical Pacific, a region renowned for the *El Niño* phenomenon (Philander, 1990). Every 2–10 years, a sudden warming of the coastal waters occurs off Peru. As the maximum warming occurs around Christmas, the local fishermen called this warming 'El Niño' (the Child in Spanish), after the Christ child. During normal times, the easterly equatorial winds drive surface waters offshore, forcing the cool, nutrient-rich, sub-surface waters to upwell, thereby replenishing the nutrients in the surface waters, hence the rich biological productivity and fisheries in the Peruvian waters. During an El Niño, upwelling suddenly stops and the Peruvian fisheries crash. A major El Niño can bring a maximum warming of 5 °C or more to the surface waters off Peru. Sometimes the opposite of an El Niño develops, i.e. anomalously cool waters appear in the equatorial Pacific, and this has been named the '*La Niña*' (the girl in Spanish) (also called 'El Viejo', the old man, by some researchers). Unlike El Niño episodes, which were documented as far back as 1726, La Niña episodes were not noticed until recent decades, because its cool sea surface temperature (SST) anomalies are located much further offshore than the El Niño warm anomalies, and La Niña does not harm the Peruvian fisheries. The SST averaged over some regions (Niño 3, Niño 3.4, Niño 4, etc.) of the equatorial Pacific (Fig. 2.3) are shown in Fig. 2.4, where El Niño warm episodes and La Niña cool episodes can easily be seen.

Let us study the monthly tropical Pacific SST from NOAA (Reynolds and Smith, 1994; Smith *et al.*, 1996) for the period January 1950 to August 2000, (where the original 2° by 2° resolution data had been combined into 4° by 4° gridded data; and each grid point had its climatological seasonal cycle removed, and smoothed by a 3 month moving average). The SST field has two spatial dimensions, but can easily be rearranged into the form of $\mathbf{y}(t)$ for the analysis with PCA. The first six PCA modes account for 51.8%, 10.1%, 7.3%, 4.3%, 3.5% and 3.1%, respectively, of the total SST variance. The spatial patterns (i.e. the eigenvectors) for the first three

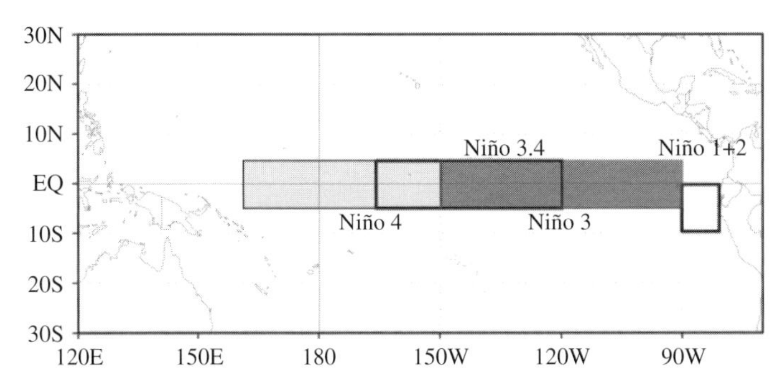

Fig. 2.3 Regions of interest in the tropical Pacific: Niño 1+2 (0°-10°S, 80°W-90°W), Niño 3 (5°S-5°N, 150°W-90°W, shaded in dark grey), Niño 3.4 (5°S-5°N, 170°W-120°W) and Niño 4 (5°S-5°N, 160°E-150°W, shaded in light grey). SST anomalies averaged over each of these regions are used as indices. The Niño 3 and Niño 3.4 SST anomalies are commonly used as indices for El Niño/La Niña episodes. (Figure downloaded from the website of Climate Prediction Center, NOAA.)

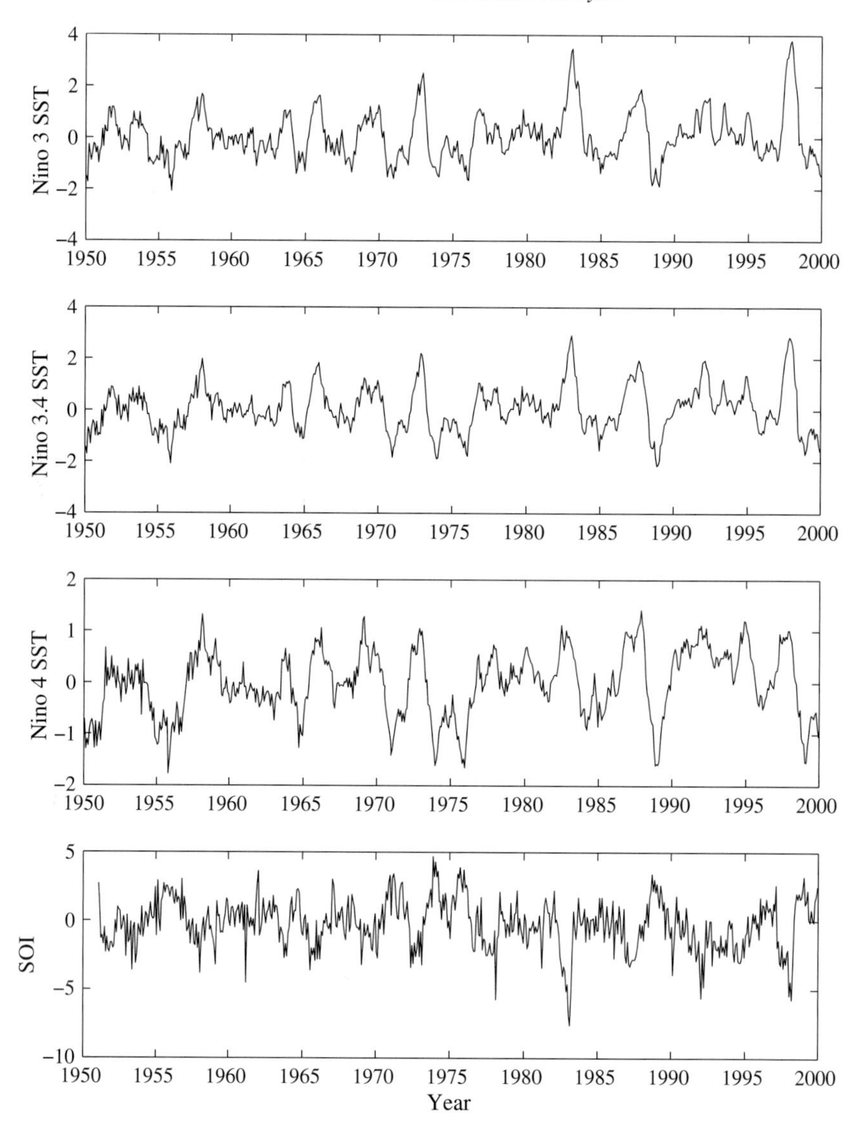

Fig. 2.4 The monthly SST anomalies in Niño 3, Niño 3.4 and Niño 4 (in °C), and the monthly Southern Oscillation Index, SOI [i.e. standardized Tahiti sea level pressure (SLP) minus standardized Darwin SLP]. During El Niño episodes, the SST rises in Niño 3 and Niño 3.4 while the SOI drops. The reverse occurs during a La Niña episode. The grid mark for a year marks the January of that year.

modes are shown in Fig. 2.5, and the associated PC time series in Fig. 2.6. All three modes have their most intense variability located close to the equator (Fig. 2.5). Only until the fourth mode and beyond (not shown), do we find modes where their most intense variability occurs off equator. It turns out that the first three modes are

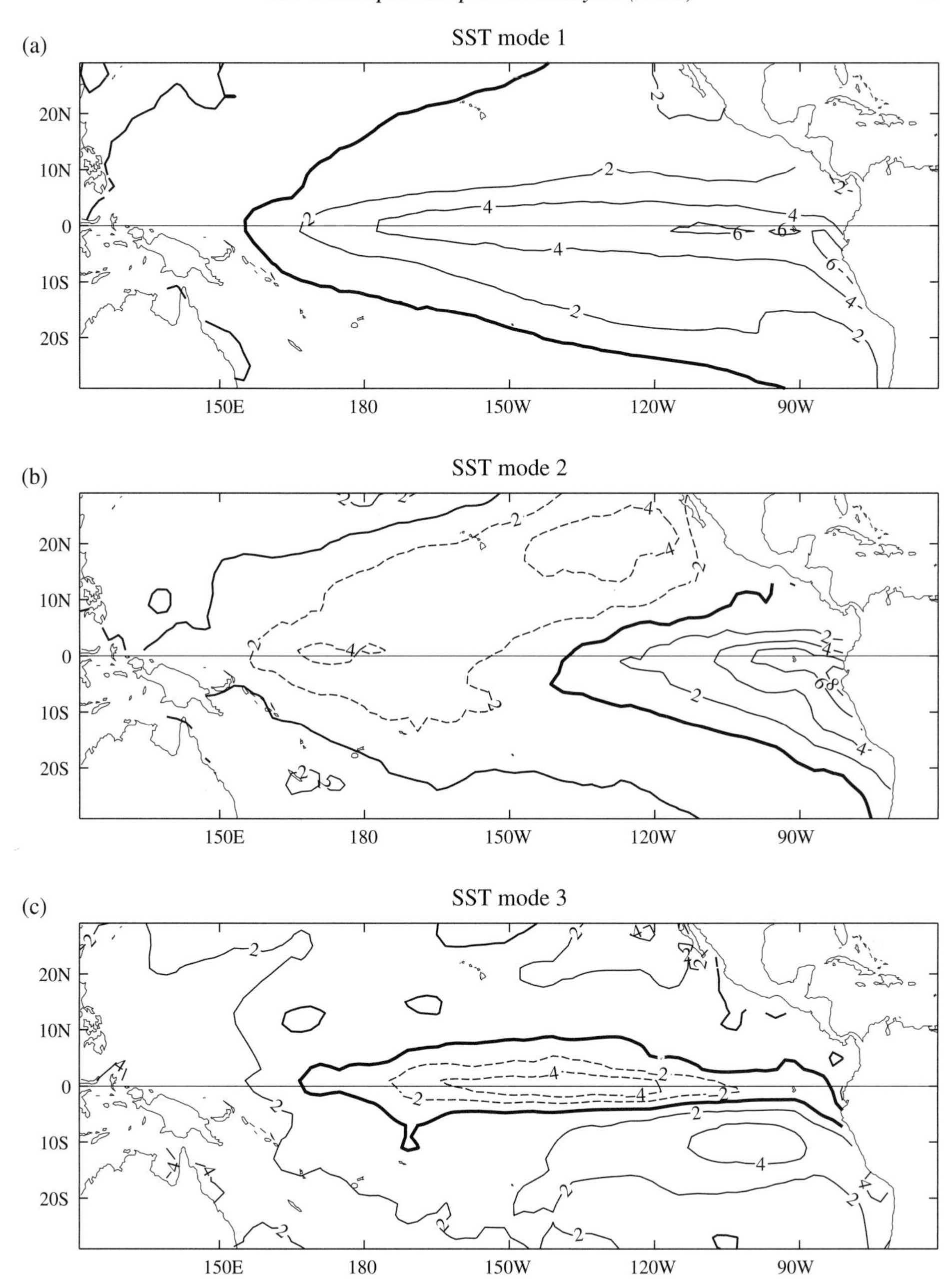

Fig. 2.5 The spatial patterns (i.e. eigenvectors or EOFs) of PCA modes (a) 1, (b) 2 and (c) 3 for the SST. Positive contours are indicated by the solid curves, negative contours by dashed curves, and the zero contour by the thick solid curve. The contour unit is 0.01 °C. The eigenvectors have been normalized to unit norm.

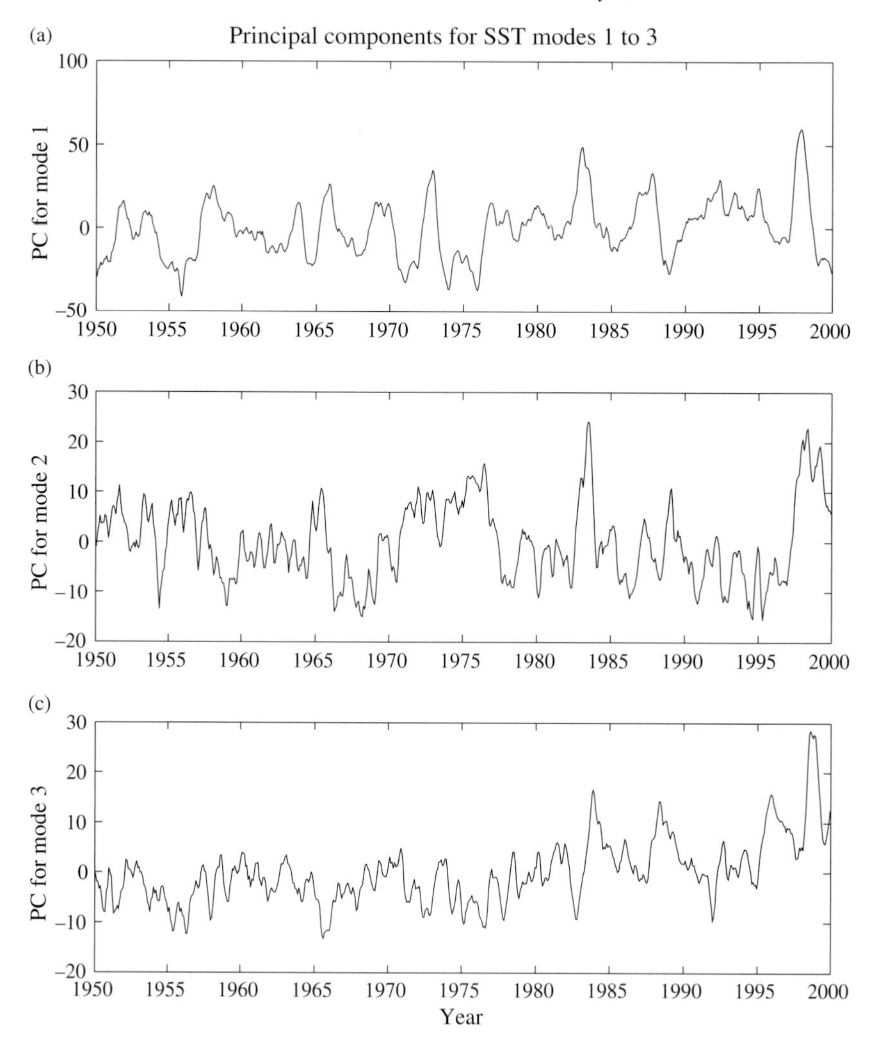

Fig. 2.6 The principal component time series for the SST modes (a) 1, (b) 2 and (c) 3.

all related to the El Niño/La Niña phenomenon. That it takes at least three modes to represent this phenomenon accurately is a result of the limitations of a linear approach like PCA – later we will see how a single nonlinear PCA mode by neural network (NN) modelling can accurately represent the El Niño/La Niña oscillation (Section 10.1.2).

Mode 1 (Fig. 2.5a) shows the largest SST anomalies occurring in the eastern and central equatorial Pacific. The mode 1 PC (Fig. 2.6) closely resembles the Niño3.4 SST (Fig. 2.4), which is commonly used as an index for El Niño/La Niña. The SST anomalies associated with mode 1 at a given time are the product of PC at that time and the spatial EOF pattern.

Mode 2 (Fig. 2.5b) has, along the equator, positive anomalies near the east and negative anomalies further west. Its PC (Fig. 2.6) shows positive values during both El Niño and La Niña episodes. Mode 3 (Fig. 2.5c) shows the largest anomaly occurring in the central equatorial Pacific, and the PC (Fig. 2.6) shows a rising trend after the mid 1970s.

Since the late nineteenth century, it has been known that the normal high air pressure (at sea level) in the eastern equatorial Pacific and the low pressure in the western equatorial Pacific and Indian Ocean may undergo a see-saw oscillation once very 2–10 years. The '*Southern Oscillation*' (termed by Sir Gilbert Walker in the 1920s) is the east-west seesaw oscillation in the sea level pressure (SLP) centred in the equatorial Pacific. The SLP of Tahiti (in the eastern equatorial Pacific) minus that of Darwin (in the western equatorial Pacific/Indian Ocean domain) is commonly called the *Southern Oscillation Index* (SOI). Clearly the SOI is negatively correlated with the Niño3.4 SST index (Fig. 2.4), i.e. when an El Niño occurs, the SLP of the eastern equatorial Pacific drops relative to the SLP of the western equatorial Pacific/Indian Ocean domain. By the mid 1960s, El Niño, the oceanographic phenomenon, has been found to be connected to the Southern Oscillation, the atmospheric phenomenon, and the coupled phenomenon named the *El Niño-Southern Oscillation* (ENSO).

Let us also consider the tropical Pacific monthly SLP data from COADS (Comprehensive Ocean-Atmosphere Data Set) (Woodruff *et al.*, 1987) during January 1950 to August 2000. The 2° by 2° resolution data were combined into 10° longitude by 4° latitude gridded data, with climatological seasonal cycle removed, and smoothed by a 3 month running average. PCA of the data resulted in the first six modes accounting for 29.5%, 16.5%, 4.8%, 3.4%, 2.5% and 2.2%, respectively, of the total variance. The first three spatial modes (Fig. 2.7) and their associated PCs (Fig. 2.8) are also shown. The first mode describes the east-west seesaw in the SLP associated with the Southern Oscillation (Fig. 2.7a).

2.1.6 Scaling the PCs and eigenvectors

There are various options for scaling of the PCs $\{a_j(t)\}$ and the eigenvectors $\{\mathbf{e}_j\}$. One can introduce an arbitrary scale factor α,

$$a'_j = \frac{1}{\alpha} a_j, \quad \mathbf{e}'_j = \alpha \mathbf{e}_j, \tag{2.33}$$

so that

$$\mathbf{y} - \bar{\mathbf{y}} = \sum_j a'_j \mathbf{e}'_j. \tag{2.34}$$

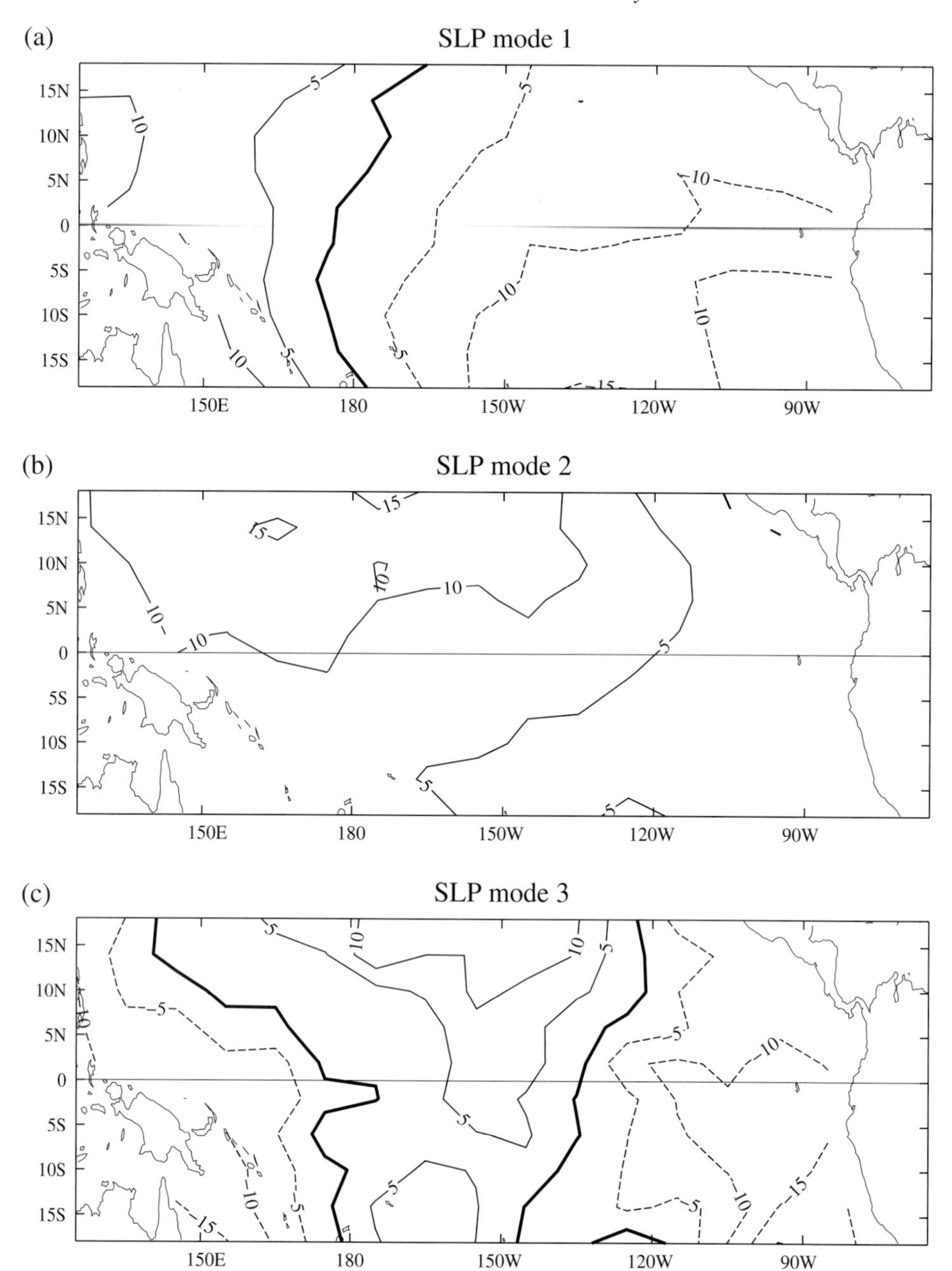

Fig. 2.7 The spatial patterns of PCA modes (a) 1, (b) 2 and (c) 3 for the SLP. The contour unit is 0.01 mb. Positive contours are indicated by the solid curves, negative contours by dashed curves, and the zero contour by the thick solid curve.

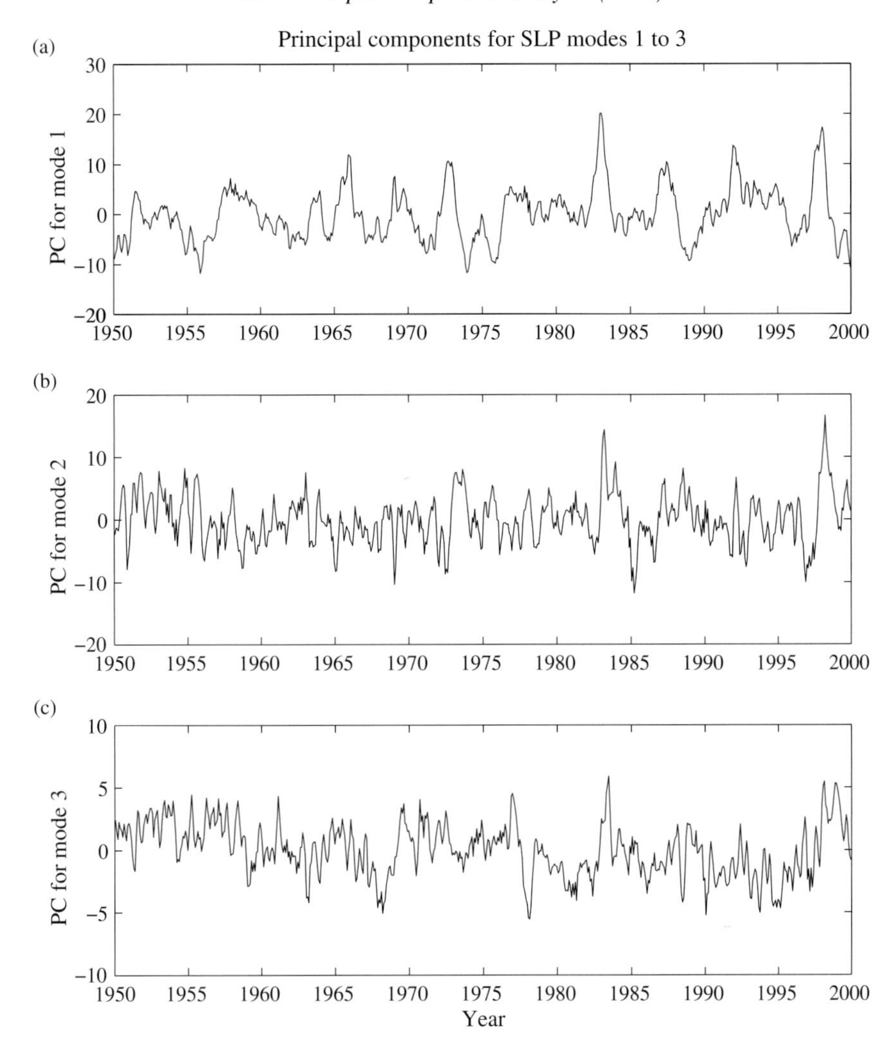

Fig. 2.8 The principal component time series for the SLP modes (a) 1, (b) 2 and (c) 3. PC1 is clearly well correlated with the SST PC1 in Fig. 2.6.

Thus $a_j(t)$ and $\mathbf{e}_j$ are defined only up to an arbitrary scale factor. With $\alpha = -1$, one reverses the sign of both $a_j(t)$ and $\mathbf{e}_j$, which is often done to make them more interpretable.

Our choice for the scaling has so far been

$$\mathbf{e}_i^{\mathrm{T}}\mathbf{e}_j = \delta_{ij}, \tag{2.35}$$

which was the choice of Lorenz (1956). The variance of the original data $\mathbf{y}$ is then contained in $\{a_j(t)\}$, with

$$\mathrm{var}(\mathbf{y}) = \mathrm{E}\left[\sum_{j=1}^{m} a_j^2\right]. \tag{2.36}$$

Another common choice is Hotelling's original choice

$$a'_j = \frac{1}{\sqrt{\lambda_j}} a_j, \quad \mathbf{e}'_j = \sqrt{\lambda_j}\mathbf{e}_j, \tag{2.37}$$

whence

$$\mathrm{var}(\mathbf{y}) = \sum_{j=1}^{m} \lambda_j = \sum_{j=1}^{m} \|\mathbf{e}'_j\|^2, \tag{2.38}$$

$$\mathrm{cov}(a'_i, a'_j) = \delta_{ij}. \tag{2.39}$$

The variance of the original data is now contained in $\{\mathbf{e}_j(t)\}$ instead. In sum, regardless of the arbitrary scale factor, the PCA eigenvectors are orthogonal and the PCs are uncorrelated.

If PCA is performed on the standardized variables $\tilde{y}_l$, i.e. y_l with mean removed and normalized by standard deviation, then one can show that the correlation

$$\rho(a'_j(t), \tilde{y}_l(t)) = e'_{jl}, \tag{2.40}$$

the lth element of $\mathbf{e}'_j$ (Jolliffe, 2002, p. 25). Hence the lth element of $\mathbf{e}'_j$ conveniently provides the correlation between the PC a'_j and the standardized variable $\tilde{y}_l$, which is a reason why Hotelling's scaling (2.37) is also widely used.

2.1.7 Degeneracy of eigenvalues

A degenerate case arises when $\lambda_i = \lambda_j$, $(i \neq j)$. When two eigenvalues are equal, their eigenspace is 2-D, i.e. a plane in which any two orthogonal vectors can be chosen as the eigenvectors – hence the eigenvectors are not unique. If l eigenvalues are equal, l non-unique orthogonal vectors can be chosen in the l-dimensional eigenspace.

A simple example of degeneracy is illustrated by a propagating plane wave,

$$h(x, y, t) = A\cos(ky - \omega t), \tag{2.41}$$

which can be expressed in terms of two standing waves:

$$h = A\cos(ky)\cos(\omega t) + A\sin(ky)\sin(\omega t). \tag{2.42}$$

If we perform PCA on $h(x, y, t)$, we get two modes with equal eigenvalues. To see this, note that in the x-y plane, $\cos(ky)$ and $\sin(ky)$ are orthogonal, while $\cos(\omega t)$ and $\sin(\omega t)$ are uncorrelated, so (2.42) satisfies the properties of PCA modes in that the eigenvectors are orthogonal and the PCs are uncorrelated. Equation (2.42) is a PCA decomposition, with the two modes having the same amplitude A, hence the eigenvalues $\lambda_1 = \lambda_2$, and the case is degenerate. Thus propagating waves in the data lead to degeneracy in the eigenvalues. If one finds eigenvalues of very similar magnitudes from a PCA analysis, that implies near degeneracy and there may be propagating waves in the data. In reality, noise in the data usually precludes $\lambda_1 = \lambda_2$ exactly. Nevertheless, when $\lambda_1 \approx \lambda_2$, the near degeneracy causes the eigenvectors to be rather poorly defined (i.e. very sensitive to noise in the data) (North *et al.*, 1982).

2.1.8 A smaller covariance matrix

Let the data matrix be

$$\mathbf{Y} = \begin{bmatrix} y_{11} & \cdots & y_{1n} \\ \cdots & \cdots & \cdots \\ y_{m1} & \cdots & y_{mn} \end{bmatrix}, \tag{2.43}$$

where m is the number of spatial points and n the number of time points. The columns of this matrix are simply the vectors $\mathbf{y}(t_1), \mathbf{y}(t_2), \ldots, \mathbf{y}(t_n)$. Assuming

$$\frac{1}{n}\sum_{i=1}^{n} y_{ji} = 0, \tag{2.44}$$

i.e. the temporal mean has been removed, then

$$\mathbf{C} = \frac{1}{n}\mathbf{Y}\mathbf{Y}^{\mathrm{T}} \tag{2.45}$$

is an $m \times m$ matrix. The theory of singular value decomposition (SVD) (see Section 2.1.10) tells us that the non-zero eigenvalues of $\mathbf{Y}\mathbf{Y}^{\mathrm{T}}$ (an $m \times m$ matrix) are exactly the non-zero eigenvalues of $\mathbf{Y}^{\mathrm{T}}\mathbf{Y}$ (an $n \times n$ matrix).

In most problems, the size of the two matrices is very different. For instance, for global $5° \times 5°$ monthly sea level pressure data collected over 50 years, the total number of spatial grid points is $m = 2592$ while the number of time points is $n = 600$. Obviously, it will be much easier to solve the eigen problem for the 600×600 matrix than that for the 2592×2592 matrix.

Hence, when $n < m$, considerable computational savings can be made by first finding the eigenvalues $\{\lambda_j\}$ and eigenvectors $\{\mathbf{v}_j\}$ for the alternative covariance matrix

$$\mathbf{C}' = \frac{1}{n}\mathbf{Y}^{\mathrm{T}}\mathbf{Y}, \tag{2.46}$$

i.e.

$$\frac{1}{n}\mathbf{Y}^{\mathrm{T}}\mathbf{Y}\mathbf{v}_j = \lambda_j \mathbf{v}_j. \tag{2.47}$$

Since

$$\lambda_j \mathbf{Y}\mathbf{v}_j = \mathbf{Y}\lambda_j \mathbf{v}_j = \mathbf{Y}\frac{1}{n}\mathbf{Y}^{\mathrm{T}}\mathbf{Y}\mathbf{v}_j,$$

$$\left(\frac{1}{n}\mathbf{Y}\mathbf{Y}^{\mathrm{T}}\right)(\mathbf{Y}\mathbf{v}_j) = \lambda_j(\mathbf{Y}\mathbf{v}_j). \tag{2.48}$$

This equation is easily seen to be of the form

$$\mathbf{C}\mathbf{e}_j = \lambda_j \mathbf{e}_j, \tag{2.49}$$

with

$$\mathbf{e}_j = \mathbf{Y}\mathbf{v}_j, \tag{2.50}$$

which means $\mathbf{e}_j$ is an eigenvector for $\mathbf{C}$. In summary, solving the eigen problem for the smaller matrix $\mathbf{C}'$ yields the eigenvalues $\{\lambda_j\}$ and eigenvectors $\{\mathbf{v}_j\}$. The eigenvectors $\{\mathbf{e}_j\}$ for the bigger matrix $\mathbf{C}$ are then obtained from (2.50).

2.1.9 Temporal and spatial mean removal

Given a data matrix $\mathbf{Y}$ as in (2.43), what type of mean are we trying to remove from the data? So far, we have removed the temporal mean, i.e. the average of the jth row, from each datum y_{ji}. We could instead have removed the spatial mean, i.e. the average of the ith column, from each datum y_{ji}.

Which type of mean should be removed is very dependent on the type of data one has. For most applications, one removes the temporal mean. However, for satellite sensed sea surface temperature data, the precision is much better than the accuracy. Also, the subsequent satellite image may be collected by a different satellite, which would have different systematic errors. So it is more appropriate to subtract the spatial mean of an image from each pixel (as was done in Fang and Hsieh, 1993).

It is also possible to remove both the temporal and spatial means, by subtracting the average of the jth row and then the average of the ith column from each datum y_{ji}.

2.1.10 Singular value decomposition

Instead of solving the eigen problem of the data covariance matrix $\mathbf{C}$, a computationally more efficient way to perform PCA is via *singular value decomposition* (SVD) of the $m \times n$ data matrix $\mathbf{Y}$ given by (2.43) (Kelly, 1988). Without loss of generality, we can assume $m \geq n$, then the SVD theorem (Strang, 2005) says that

$$
\mathbf{Y} = \mathbf{ESF}^{\mathrm{T}} = \underbrace{\left[\begin{array}{c:c} \underset{m \times n}{\mathbf{E}'} & 0 \end{array}\right]}_{m \times m} \underbrace{\left[\begin{array}{c} \underset{n \times n}{\mathbf{S}'} \\ \hdashline 0 \end{array}\right]}_{m \times n} \underbrace{\left[\quad\right]}_{n \times n} . \tag{2.51}
$$

The $m \times m$ matrix $\mathbf{E}$ contains an $m \times n$ sub-matrix $\mathbf{E}'$ – and if $m > n$, some zero column vectors. The $m \times n$ matrix $\mathbf{S}$ contains the diagonal $n \times n$ sub-matrix $\mathbf{S}'$, and possibly some zero row vectors. $\mathbf{F}^{\mathrm{T}}$ is an $n \times n$ matrix. (If $m < n$, one can apply the above arguments to the transpose of the data matrix.)

$\mathbf{E}$ and $\mathbf{F}$ are *orthonormal matrices*, i.e. they satisfy

$$
\mathbf{E}^{\mathrm{T}}\mathbf{E} = \mathbf{I}, \qquad \mathbf{F}^{\mathrm{T}}\mathbf{F} = \mathbf{I}, \tag{2.52}
$$

where $\mathbf{I}$ is the identity matrix. The leftmost n columns of $\mathbf{E}$ contain the n *left singular vectors*, and the columns of $\mathbf{F}$ the n *right singular vectors*, while the diagonal elements of $\mathbf{S}'$ are the *singular values*.

The covariance matrix $\mathbf{C}$ can be rewritten as

$$
\mathbf{C} = \frac{1}{n}\mathbf{Y}\mathbf{Y}^{\mathrm{T}} = \frac{1}{n}\mathbf{ESS}^{\mathrm{T}}\mathbf{E}^{\mathrm{T}}, \tag{2.53}
$$

where (2.51) and (2.52) have been invoked. The matrix

$$
\mathbf{SS}^{\mathrm{T}} \equiv \mathbf{\Lambda}, \tag{2.54}
$$

is diagonal and zero everywhere, except in the upper left $n \times n$ corner, containing $\mathbf{S}'^2$.

Right multiplying (2.53) by $n\mathbf{E}$ gives

$$
n\mathbf{CE} = \mathbf{E\Lambda}, \tag{2.55}
$$

where (2.54) and (2.52) have been invoked, and $\mathbf{\Lambda}$ contains the eigenvalues for the matrix $n\mathbf{C}$. Instead of solving the eigen problem (2.55), we use SVD to get $\mathbf{E}$ from $\mathbf{Y}$ by (2.51). Equation (2.55) implies that there are only n eigenvalues in $\mathbf{\Lambda}$ from $\mathbf{S}'^2$, and the eigenvalues $=$ (singular values)^2. As (2.55) and (2.49) are equivalent except for the constant n, the eigenvalues in $\mathbf{\Lambda}$ are simply $n\lambda_j$, with λ_j the eigenvalues from (2.49).

Similarly, for the other covariance matrix

$$
\mathbf{C}' = \frac{1}{n}\mathbf{Y}^{\mathrm{T}}\mathbf{Y}, \tag{2.56}
$$

we can rewrite it as

$$
\mathbf{C}' = \frac{1}{n}\mathbf{FS}'^2\mathbf{F}^{\mathrm{T}}, \tag{2.57}
$$

and ultimately,

$$n\mathbf{C}'\mathbf{F} = \mathbf{F}\mathbf{S}'^2 . \tag{2.58}$$

Hence the eigen problem (2.58) has the same eigenvalues as (2.55).

The PCA decomposition

$$\mathbf{y}(t) = \sum_j \mathbf{e}_j a_j(t) , \tag{2.59}$$

is equivalent to the matrix form

$$\mathbf{Y} = \mathbf{E}\mathbf{A}^{\mathrm{T}} = \sum_j \mathbf{e}_j \mathbf{a}_j^{\mathrm{T}} , \tag{2.60}$$

where the eigenvector $\mathbf{e}_j$ is the jth column in the matrix $\mathbf{E}$, and the PC $a_j(t)$ is the vector $\mathbf{a}_j$, the jth column in the matrix $\mathbf{A}$. Equations (2.51) and (2.60) yield

$$\mathbf{A}^{\mathrm{T}} = \mathbf{S}\mathbf{F}^{\mathrm{T}}. \tag{2.61}$$

Hence by the SVD (2.51), we obtain the eigenvectors $\mathbf{e}_j$ from $\mathbf{E}$, and the PCs $a_j(t)$ from $\mathbf{A}$ in (2.61). We can also left multiply (2.60) by $\mathbf{E}^{\mathrm{T}}$, and invoke (2.52) to get

$$\mathbf{A}^{\mathrm{T}} = \mathbf{E}^{\mathrm{T}}\mathbf{Y}. \tag{2.62}$$

There are many computing packages (e.g. MATLAB) with standard codes for performing SVD. Kelly (1988) pointed out that the SVD approach to PCA is at least twice as fast as the eigen approach, as SVD requires $\mathrm{O}(mn^2)$ operations to compute, while the eigen approach requires $\mathrm{O}(mn^2)$ to compute the smaller of $\mathbf{C}$ or $\mathbf{C}'$, then $\mathrm{O}(n^3)$ to solve the eigen problem, and then $\mathrm{O}(mn^2)$ to get the PCs.

2.1.11 Missing data

Missing data produce gaps in data records. If the gaps are small, one can interpolate the missing values using neighbouring data. If the gaps are not small, then instead of

$$\mathbf{C} = \frac{1}{n}\mathbf{Y}\mathbf{Y}^{\mathrm{T}}, \tag{2.63}$$

(assuming that the means have been removed from the data), one computes

$$c_{kl} = \frac{1}{n'}\sum_i{}' y_{ki}\, y_{il} \tag{2.64}$$

where the prime denotes that the summation is only over i with neither y_{ki} nor y_{il} missing – with a total of n' terms in the summation. The eigenvectors $\mathbf{e}_j$ can then be obtained from this new covariance matrix. The principal components a_j cannot be computed from

$$a_j(t_l) = \sum_i e_{ji}\, y_{il}, \tag{2.65}$$

as some values of y_{il} are missing. Instead a_j is estimated (von Storch and Zwiers, 1999, Section 13.2.8) as a least squares solution to minimizing $\mathrm{E}[\|\mathbf{y} - \sum a_j \mathbf{e}_j\|^2]$, i.e.

$$a_j(t_l) = \frac{\sum_i' e_{ji}\, y_{il}}{\sum_i' |e_{ji}|^2}, \tag{2.66}$$

where for a given value of l, the superscript prime means that the summations are only over i for which y_{il} is not missing.

PCA is also used to supply missing data. With climate data, one often finds the earlier records to be sparse. Suppose the data record can be divided into two parts, $\mathbf{Y}$ which contains no missing values, and $\tilde{\mathbf{Y}}$ which contains missing values. From (2.60), PCA applied to $\mathbf{Y}$ yields $\mathbf{E}$, which contains the eigenvectors $\mathbf{e}_j$. The PCs for $\tilde{\mathbf{Y}}$ are then computed from (2.66)

$$\tilde{a}_j(t_l) = \frac{\sum_i' e_{ji}\, \tilde{y}_{il}}{\sum_i' |e_{ji}|^2}. \tag{2.67}$$

The missing values in $\tilde{\mathbf{Y}}$ are filled in $\tilde{\mathbf{Y}}'$, where

$$\tilde{\mathbf{Y}}' = \mathbf{E}\tilde{\mathbf{A}}^{\mathrm{T}}, \tag{2.68}$$

where the jth column of $\tilde{\mathbf{A}}$ is given by $\tilde{a}_j(t_l)$. More sophisticated interpolation of missing data by PCA is described by Kaplan *et al.* (2000).

2.1.12 Significance tests

In practice, the higher PCA modes, which basically contain noise, are rejected. How does one decide how many modes to retain? There are some 'rules of thumb'. One of the simplest approaches is to plot the eigenvalues λ_j as a function of the mode number j. Hopefully, from the plot, one finds an abrupt transition from large eigenvalues to small eigenvalues around mode number k. One can then retain the first k modes. Alternatively, the Kaiser test rejects the modes with eigenvalues λ less than the mean value $\overline{\lambda}$.

Computationally more involved is the Monte Carlo test (Preisendorfer, 1988), which involves setting up random data matrices $\mathbf{R}_l$ ($l = 1, \ldots, L$), of the same size as the data matrix $\mathbf{Y}$. The random elements are normally distributed, with the variance of the random data matching the variance of the actual data. Principal component analysis is performed on each of the random matrices, yielding eigenvalues $\lambda_j^{(l)}$. Assume for each l, the set of eigenvalues is sorted in descending order. For each j, one examines the distribution of the L values of $\lambda_j^{(l)}$, and finds the level $\lambda_{0.05}$, which is exceeded by only 5% of the $\lambda_j^{(l)}$ values. The eigenvalues λ_j from $\mathbf{Y}$ which failed to rise above this $\lambda_{0.05}$ level are then rejected. If the data have strong autocorrelation, then the dimension of $\mathbf{R}_l$ should be reduced, with the effective sample size n_{eff} replacing the sample size n.

Since the Monte Carlo method performs PCA on L matrices and L is typically about 100–1000, it can be costly for large data matrices. Hence asymptotic methods based on the central limit theorem are often used in the case of large data matrices (Mardia *et al.*, 1979, pp. 230–237; Preisendorfer, 1988, pp. 204–206).

2.2 Rotated PCA

In PCA, the linear mode which accounts for the most variance of the dataset is sought. However, as illustrated in Fig. 2.9, the resulting eigenvectors may not align close to local data clusters, so the eigenvectors may not represent actual physical states well. Rotated PCA (RPCA) methods rotate the PCA eigenvectors, so that they point closer to the local clusters of data points. Thus the rotated eigenvectors

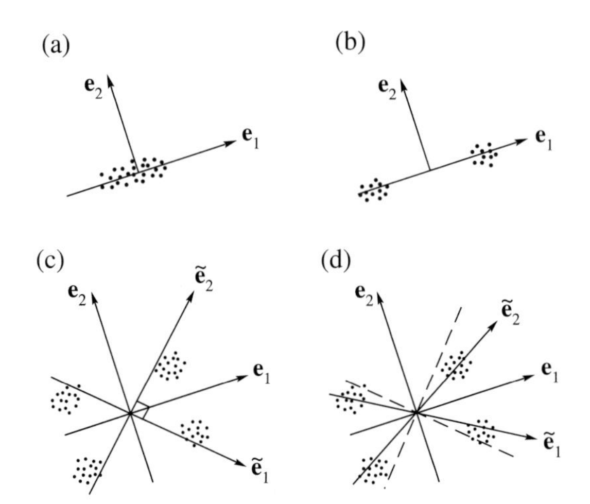

Fig. 2.9 The case of PCA applied to a dataset composed of (a) a single cluster, (b) two clusters, (c) and (d) four clusters. In (c), an orthonormal rotation has yielded rotated eigenvectors $\tilde{\mathbf{e}}_j$, ($j = 1, 2$) which pass much closer to the data clusters than the unrotated eigenvectors $\mathbf{e}_j$. In (d), an oblique rotation is used instead of an orthonormal rotation to spear through the data clusters. The dashed lines indicate the orthonormally rotated eigenvectors. Eigenvectors which failed to approach any data clusters generally bear little resemblance to physical states. (Based on Preisendorfer (1988).)

may bear greater resemblance to actual physical states (though they account for less variance) than the unrotated eigenvectors. Rotated PCA, also called rotated EOF analysis, and in statistics, PC factor analysis, is a more general but also more subjective technique than PCA. This dichotomy between PCA and RPCA methods arises because it is generally impossible to have a linear solution simultaneously (a) explaining maximum global variance of the data, and (b) approaching local data clusters. Thus PCA excels in (a) while RPCA excels in (b). However, in the nonlinear PCA (NLPCA) approach using neural networks (Chapter 10), both objectives (a) and (b) can be attained together, thereby unifying the PCA and RPCA approaches.

Let us quickly review the rotation of vectors and matrices. Given a matrix $\mathbf{P}$ composed of the column vectors $\mathbf{p}_1, \ldots, \mathbf{p}_m$, and a matrix $\mathbf{Q}$ containing the column vectors $\mathbf{q}_1, \ldots, \mathbf{q}_m$, $\mathbf{P}$ can be transformed into $\mathbf{Q}$ by $\mathbf{Q} = \mathbf{PR}$, i.e.

$$q_{il} = \sum_j p_{ij} r_{jl}, \tag{2.69}$$

where $\mathbf{R}$ is a rotation matrix with elements r_{jl}. When $\mathbf{R}$ is orthonormal, i.e.

$$\mathbf{R}^{\mathrm{T}}\mathbf{R} = \mathbf{I}, \tag{2.70}$$

the rotation is called an *orthonormal rotation*. Clearly,

$$\mathbf{R}^{-1} = \mathbf{R}^{\mathrm{T}}, \tag{2.71}$$

for an orthonormal rotation. If $\mathbf{R}$ is not orthonormal, the rotation is an *oblique rotation*.

Given the data matrix $\mathbf{Y}$,

$$\mathbf{Y} = (y_{il}) = \left(\sum_{j=1}^{m} e_{ij} a_{jl} \right) = \sum_j \mathbf{e}_j \mathbf{a}_j^{\mathrm{T}} = \mathbf{E}\mathbf{A}^{\mathrm{T}}, \tag{2.72}$$

we rewrite it as

$$\mathbf{Y} = \mathbf{E}\mathbf{R}\mathbf{R}^{-1}\mathbf{A}^{\mathrm{T}} = \tilde{\mathbf{E}}\tilde{\mathbf{A}}^{\mathrm{T}}, \tag{2.73}$$

with

$$\tilde{\mathbf{E}} = \mathbf{E}\mathbf{R} \tag{2.74}$$

and

$$\tilde{\mathbf{A}}^{\mathrm{T}} = \mathbf{R}^{-1}\mathbf{A}^{\mathrm{T}}. \tag{2.75}$$

Note that $\mathbf{E}$ has been rotated into $\tilde{\mathbf{E}}$, and $\mathbf{A}$ into $\tilde{\mathbf{A}}$.

If $\mathbf{R}$ is orthonormal, (2.71) and (2.75) yield

$$\tilde{\mathbf{A}} = \mathbf{A}\mathbf{R}. \tag{2.76}$$

To see the orthogonality properties of the rotated eigenvectors, we note that

$$\tilde{\mathbf{E}}^{\mathrm{T}}\tilde{\mathbf{E}} = \mathbf{R}^{\mathrm{T}}\mathbf{E}^{\mathrm{T}}\mathbf{E}\mathbf{R} = \mathbf{R}^{\mathrm{T}}\mathbf{D}\mathbf{R}, \tag{2.77}$$

where the diagonal matrix $\mathbf{D}$ is

$$\mathbf{D} = \mathrm{diag}(\mathbf{e}_1^{\mathrm{T}}\mathbf{e}_1, \ldots, \mathbf{e}_m^{\mathrm{T}}\mathbf{e}_m). \tag{2.78}$$

If $\mathbf{e}_j^{\mathrm{T}}\mathbf{e}_j = 1$, for all j, then $\mathbf{D} = \mathbf{I}$ and (2.77) reduces to

$$\tilde{\mathbf{E}}^{\mathrm{T}}\tilde{\mathbf{E}} = \mathbf{R}^{\mathrm{T}}\mathbf{R} = \mathbf{I}, \tag{2.79}$$

which means the $\{\tilde{\mathbf{e}}_j\}$ are orthonormal. Hence the rotated eigenvectors $\{\tilde{\mathbf{e}}_j\}$ are orthonormal only if the original eigenvectors $\{\mathbf{e}_j\}$ are orthonormal. If $\{\mathbf{e}_j\}$ are orthogonal but not orthonormal, then $\{\tilde{\mathbf{e}}_j\}$ are in general not orthogonal.

From (2.32), the PCs $\{a_j(t_l)\}$ are uncorrelated, i.e. the covariance matrix

$$\mathbf{C}_{\mathbf{AA}} = \mathrm{diag}(\alpha_1^2, \ldots, \alpha_m^2), \tag{2.80}$$

where

$$\mathbf{a}_j^{\mathrm{T}}\mathbf{a}_j = \alpha_j^2. \tag{2.81}$$

With the rotated PCs, their covariance matrix is

$$\mathbf{C}_{\tilde{\mathbf{A}}\tilde{\mathbf{A}}} = \mathrm{cov}(\mathbf{R}^{\mathrm{T}}\mathbf{A}^{\mathrm{T}}, \mathbf{A}\mathbf{R}) = \mathbf{R}^{\mathrm{T}}\,\mathrm{cov}(\mathbf{A}^{\mathrm{T}}, \mathbf{A})\,\mathbf{R} = \mathbf{R}^{\mathrm{T}}\,\mathbf{C}_{\mathbf{AA}}\,\mathbf{R}. \tag{2.82}$$

Hence $\mathbf{C}_{\tilde{\mathbf{A}}\tilde{\mathbf{A}}}$ is diagonal only if $\mathbf{C}_{\mathbf{AA}} = \mathbf{I}$, i.e. $\mathbf{a}_j^{\mathrm{T}}\mathbf{a}_j = 1$, for all j.

There are now two cases.

Case a: If we choose $\mathbf{e}_j^{\mathrm{T}}\mathbf{e}_j = 1$, for all j, then we cannot have $\mathbf{a}_j^{\mathrm{T}}\mathbf{a}_j = 1$, for all j. This implies that $\{\tilde{\mathbf{a}}_j\}$ are not uncorrelated, but $\{\tilde{\mathbf{e}}_j\}$ are orthonormal.

Case b: If we choose $\mathbf{a}_j^{\mathrm{T}}\mathbf{a}_j = 1$, for all j, then we cannot have $\mathbf{e}_j^{\mathrm{T}}\mathbf{e}_j = 1$, for all j. This implies that $\{\tilde{\mathbf{a}}_j\}$ are uncorrelated, but $\{\tilde{\mathbf{e}}_j\}$ are not orthonormal.

Thus there is a notable difference between PCA and RPCA: PCA can have both $\{\mathbf{e}_j\}$ orthonormal and $\{\mathbf{a}_j\}$ uncorrelated, but RPCA can only possess one of these two properties.

In general, out of a total of m PCA modes, only the k leading ones are selected for rotation, while the higher modes are discarded as noise. As there are many possible criteria for rotation, there are many RPCA schemes – Richman (1986) listed 5 orthogonal and 14 oblique rotation schemes. The *varimax* scheme proposed by Kaiser (1958) is the most popular among orthogonal rotation schemes. For illustration, suppose that only the first two eigenvectors are chosen for rotation. The data are first projected onto the two PCA eigenvectors $\mathbf{e}_j$ $(j = 1, 2)$ to get the first two PCs

$$a_j(t_l) = \sum_i e_{ji}\, y_{il}. \tag{2.83}$$

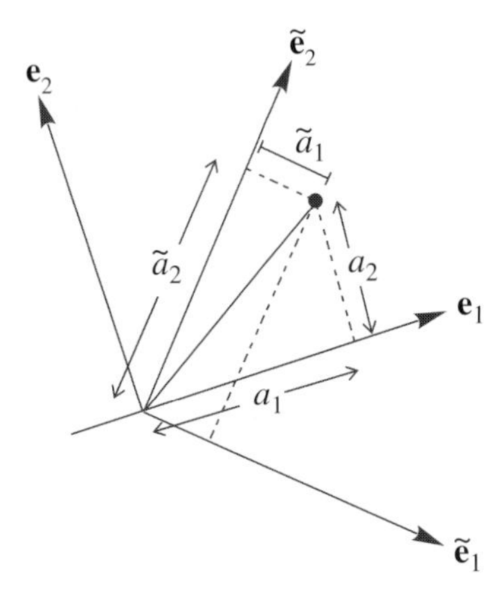

Fig. 2.10 The coordinates of a data point in the original coordinate system (a_1, a_2), and in the rotated coordinate system ($\tilde{a}_1$, $\tilde{a}_2$).

With the rotated eigenvectors $\tilde{\mathbf{e}}_j$, the rotated PCs are

$$\tilde{a}_j(t_l) = \sum_i \tilde{e}_{ji} y_{il}. \tag{2.84}$$

A common objective in rotation is to make $\tilde{a}_j^2(t_l)$ either as large as possible, or as close to zero as possible, i.e. to maximize the variance of the square of the rotated PCs. Figure 2.10 illustrates a rotation which has yielded $|\tilde{a}_1| < |a_1|$, $|a_2| < |\tilde{a}_2|$, i.e. instead of intermediate magnitudes for a_1, a_2, the rotated PCs have either larger or smaller magnitudes. Geometrically, this means the rotated axes (i.e. the eigenvectors) point closer to actual data points than the unrotated axes, hence the rotated eigenvectors have closer resemblance to observed states than the unrotated ones. In the event the rotated vector $\tilde{\mathbf{e}}_2$ actually passes through the data point in Fig. 2.10, then $|\tilde{a}_1|$ is zero, while $|\tilde{a}_2|$ assumes its largest possible value.

The varimax criterion is to maximize $f(\tilde{\mathbf{A}}) = \sum_{j=1}^k \mathrm{var}(\tilde{a}_j^2)$, i.e.

$$f(\tilde{\mathbf{A}}) = \sum_{j=1}^{k} \left\{ \frac{1}{n} \sum_{l=1}^{n} [\tilde{a}_j^2(t_l)]^2 - \left[\frac{1}{n} \sum_{l=1}^{n} \tilde{a}_j^2(t_l) \right]^2 \right\}. \tag{2.85}$$

Kaiser (1958) found an iterative algorithm for finding the rotation matrix **R** (see also Preisendorfer, 1988, pp. 273–277).

If $\{\tilde{\mathbf{e}}_j\}$ are not required to be orthogonal, one can concentrate on the eigenvectors individually, so as to best ‘spear’ through local data clusters (Fig. 2.9d), as done in oblique rotations, e.g. the *procrustes* method (Preisendorfer, 1988, pp. 278–282).

In the above varimax criterion, (2.85) maximizes the variance of the rotated squared PCs. An alternative (which is actually the one used in traditional factor analysis) is to maximize the variance of the rotated squared loadings $\tilde{e}_{ji}^2$, i.e. maximize

$$f(\tilde{\mathbf{E}}) = \sum_{j=1}^{k} \left\{ \frac{1}{m} \sum_{i=1}^{m} [\tilde{e}_{ji}^2]^2 - \left[\frac{1}{m} \sum_{i=1}^{m} \tilde{e}_{ji}^2 \right]^2 \right\}, \tag{2.86}$$

(with some user choices on how the loadings are to be normalized described in von Storch and Zwiers, 1999). That the squared loadings are made as large as possible or as close to zero as possible means that many of the loadings are essentially set to zero, yielding loading patterns which have more localized features than the unrotated patterns. For instance, Horel (1981) showed both the rotated and unrotated loading patterns for the 500 mb height data of winter months in the Northern Hemisphere, with the rotated patterns showing more regionalized anomalies than the unrotated patterns, where the anomalies were spread all over the Northern Hemisphere.

These two ways of performing rotation can be seen as working with either the data matrix or the transpose of the data matrix. In PCA, using the transpose of the data matrix does not change the results (but can be exploited to save considerable computional time by working with the smaller data covariance matrix as seen in Section 2.1.8). In RPCA, taking the transpose reverses the role of the PCs and the loadings, thereby changing from a rotational criterion on the loadings to one on the PCs. Richman (1986, Section 6) mentioned that applying a rotational criterion on the loadings yielded loading patterns with far fewer anomaly centres than observed in typical 700 mb height anomaly maps of the Northern Hemisphere, whereas applying a rotational criterion on PCs yielded loading patterns in good agreement with commonly observed patterns.

To summarize, we list the disadvantages of the PCA and those of the RPCA. There are four main disadvantages with PCA.

(1) Domain shape dependence: often the PCA spatial modes (i.e. eigenvectors) are related simply to the spatial harmonics rather than to physical states.
(2) Subdomain instability: if the domain is divided into two parts, then the PCA mode 1 spatial patterns for the subdomains may not be similar to the spatial mode calculated for the whole domain, as illustrated in Fig. 2.11.
(3) Degeneracy: if $\lambda_i \approx \lambda_j$, the near degeneracy of eigenvalues means that the eigenvectors $\mathbf{e}_i$ and $\mathbf{e}_j$ cannot be estimated accurately by PCA.
(4) Neglect of regional correlated patterns: small regional correlated patterns tend to be ignored by PCA, as PCA spatial modes tend to be related to the dominant spatial harmonics.

RPCA improves on all 1–4 above.

There are also four disadvantages with RPCA.

(1) Many possible choices for the rotation criterion: Richman (1986) listed 19 types of rotation scheme. Critics complain that rotation is too subjective. Furthermore, the rotational criterion can be applied to the loadings or to the PCs, yielding different results.

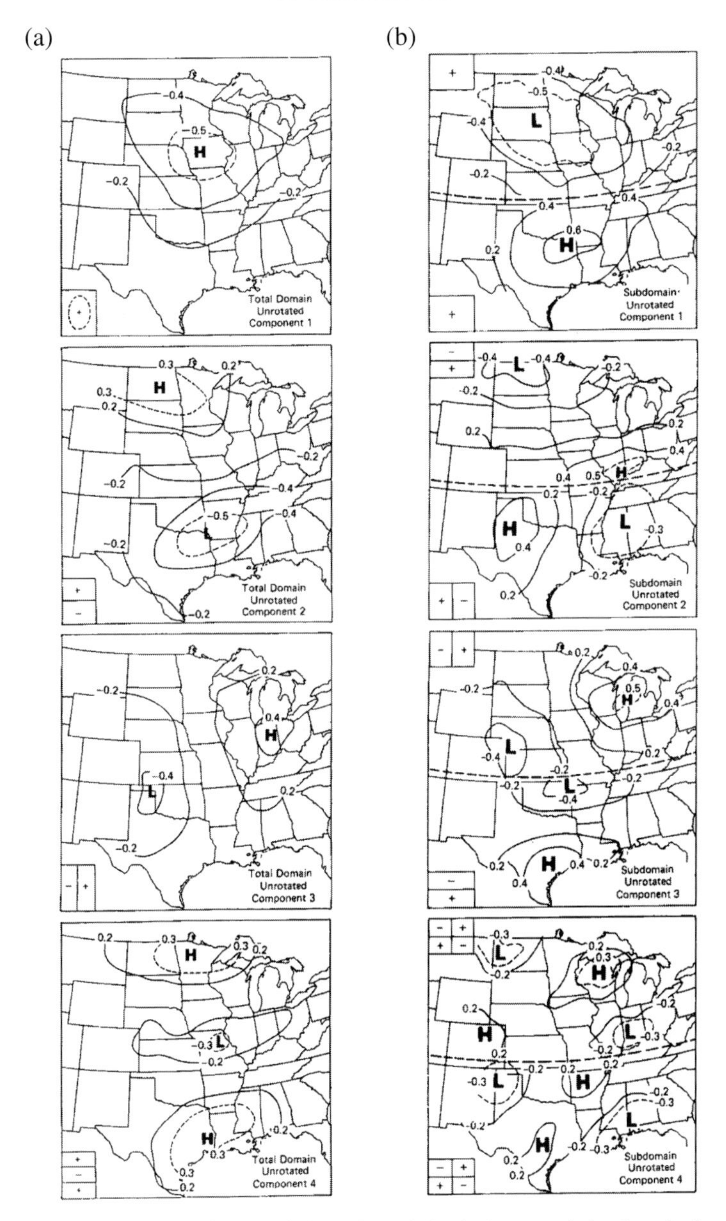

Fig. 2.11 The first four PCA spatial modes of the 3-day precipitation during May to August over the central USA. (a) The left panels show the four modes computed for the whole domain. (b) The right panels show the modes computed separately for the northern and southern halves of the full domain. The dashed lines in (b) indicate the boundary of the two halves. The insets show the basic harmonic patterns found by the modes. (Reproduced from Richman (1986, Fig. 2) with permission of the Royal Meteorological Society.)

(2) Dependence on k, the number of PCA modes chosen for rotation: if the first k PCA modes are selected for rotation, changing k can lead to large changes in the RPCAs. For instance, in RPCA, if one first chooses $k = 3$, then one chooses $k = 4$, the first three RPCAs are changed. In contrast, in PCA, if one first chooses $k = 3$, then $k = 4$, the first three PCAs are unchanged.

(3) Dependence on how the PCA eigenvectors and PCs are normalized before rotation is performed.

(4) Less variance explained: the variance of the data accounted for by the first k RPCA modes is $\leq$ the variance explained by the first k PCA modes.

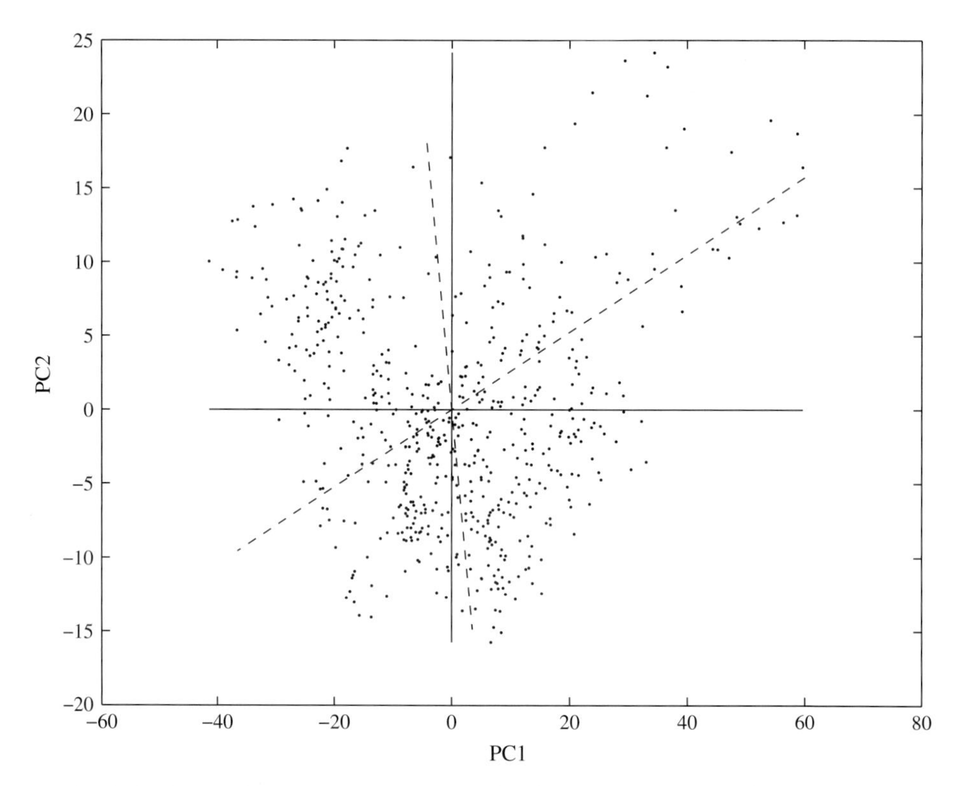

Fig. 2.12 PCA with and without rotation for the tropical Pacific SST. In the PC1-PC2 plane of the scatter plot, where the monthly data are shown as dots, the cool La Niña states lie in the upper left corner, while the warm El Niño states lie in the upper right corner. The first PCA eigenvector lies along the horizontal line, and the second PCA, along the vertical line. A varimax rotation is performed on the first three PCA eigenvectors. The first RPCA eigenvector, shown as a dashed line, spears through the cluster of El Niño states in the upper right corner, thereby yielding a more accurate description of the SST anomalies during an El Niño. The second dashed line shows the second RPCA eigenvector, which is orthogonal to the first RPCA eigenvector, (though it may not seem so in this 2-dimensional projection of 3-dimensional vectors).

Let us return to the tropical Pacific SST PCA modes. The PC1-PC2 values are shown as dots in a scatter plot (Fig. 2.12), where the cool La Niña states lie in the upper left corner, and the warm El Niño states in the upper right corner. The first PCA eigenvector lies along the horizontal line, and the second PCA along the vertical line, neither of which would come close to the El Niño nor the La Niña states. Using the varimax criterion on the squared PCs, a rotation is performed on the first three PCA eigenvectors. The first RPCA eigenvector, shown as a dashed line, spears through the cluster of El Niño states in the upper right corner, thereby yielding a more accurate description of the SST anomalies during El Niño (Fig. 2.13a) than the first PCA mode (Fig. 2.5a), which did not fully represent the intense warming of Peruvian waters during El Niño. In terms of variance explained, the first RPCA mode explained only 91.7% as much variance as the first PCA mode. The second RPCA eigenvector, also shown as a dashed line in Fig. 2.12, did not improve

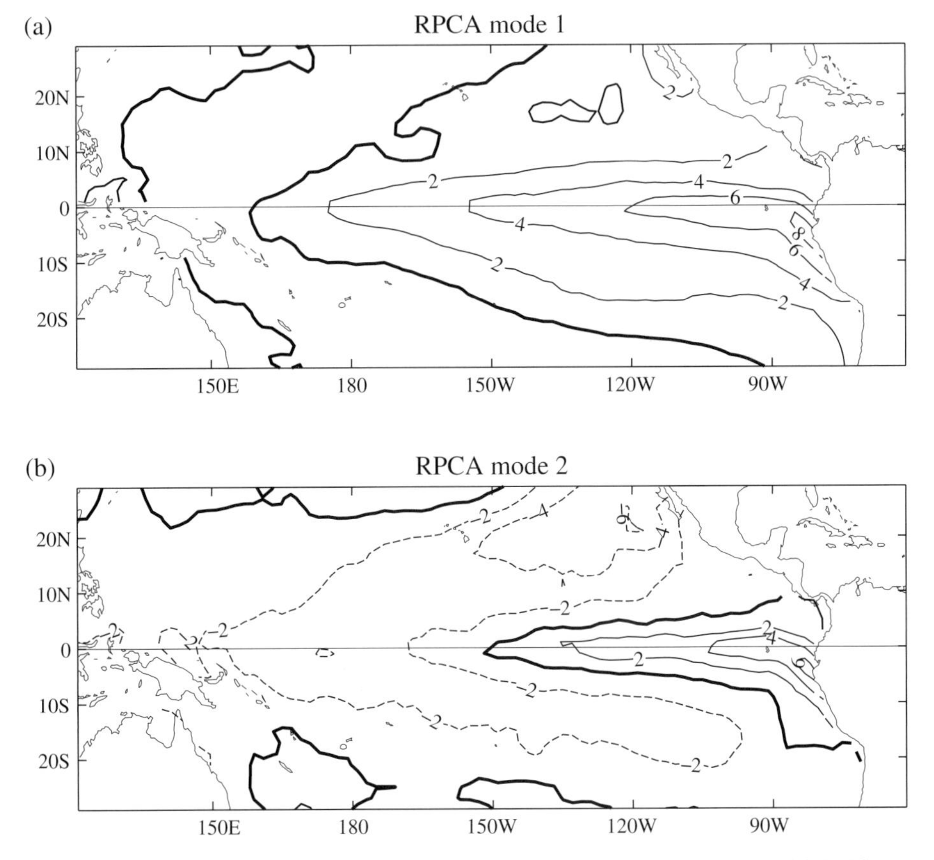

Fig. 2.13 The varimax RPCA spatial modes (a) 1 and (b) 2 for the SST. The contour unit is 0.01 °C. More intense SST anomalies are found in the equatorial waters off Peru in the RPCA mode 1 than in the PCA mode 1.

much on the second PCA mode, with the RPCA spatial pattern shown in Fig. 2.13b (cf. Fig. 2.5b).

2.3 PCA for vectors

When one has vector variables, e.g. wind velocity (u, v), there are several options for performing PCA. (a) One can simply apply PCA to the u field and to the v field separately. (b) One can do a *combined PCA*, i.e. treat the v variables as though they were extra u variables, so the data matrix becomes

$$\mathbf{Y} = \begin{bmatrix} u_{11} & \cdots & u_{1n} \\ \cdots & \cdots & \cdots \\ u_{m1} & \cdots & u_{mn} \\ v_{11} & \cdots & v_{1n} \\ \cdots & \cdots & \cdots \\ v_{m1} & \cdots & v_{mn} \end{bmatrix}, \tag{2.87}$$

where m is the number of spatial points and n the number of time points. In cases (a) and (b), the vector can of course be generalized to more than two dimensions. If the vector is two-dimensional, then one has option (c) as well, namely one can combine u and v into a complex variable, and perform a complex PCA (Hardy, 1977; Hardy and Walton, 1978).

Let

$$w = u + \mathrm{i}v. \tag{2.88}$$

Applying PCA to w allows the data matrix to be expressed as

$$\mathbf{Y} = \sum_j \mathbf{e}_j \mathbf{a}_j^{*\mathrm{T}}, \tag{2.89}$$

where the superscript $^{*\mathrm{T}}$ denotes the complex conjugate transpose. Since the covariance matrix is Hermitian and positive semi-definite (see Section 2.1.3), the eigenvalues of $\mathbf{C}$ are real and non-negative, though $\mathbf{e}_j$ and $\mathbf{a}_j$ are in general complex.

If we write the lth component of $\mathbf{a}_j$ as

$$a_{lj} = |a_{lj}|\,\mathrm{e}^{\mathrm{i}\theta_{lj}}, \tag{2.90}$$

then

$$Y_{il} = \sum_j e_{ij} \mathrm{e}^{-\mathrm{i}\theta_{jl}} |a_{jl}|. \tag{2.91}$$

One can interpret $e_{ij}e^{-i\theta_{jl}}$ as each complex element of $\mathbf{e}_j$ being rotated by the same angle θ_{jl} during the lth time interval. Similarly, each element of $\mathbf{e}_j$ is amplified by the same factor $|a_{jl}|$.

When PCA is applied to real variables, the real $\mathbf{e}_j$ and $\mathbf{a}_j$ can both be multiplied by -1. When PCA is applied to complex variables, an arbitrary phase ϕ_j can be attached to the complex $\mathbf{e}_j$ and $\mathbf{a}_j$, as follows

$$\mathbf{Y} = \sum_j (\mathbf{e}_j e^{i\phi_j})(e^{-i\phi_j}\mathbf{a}_j^{*T}). \tag{2.92}$$

Often the arbitrary phase is chosen to make interpretation of the modes easier. For instance, in analysis of the tropical Pacific wind field, Legler (1983) chose ϕ_j so that $e^{-i\phi_j}\mathbf{a}_j^{*T}$ lies mainly along the real axis. In the ocean, dynamical theory predicts the near-surface wind-driven current to spiral and diminish with depth, in the shape of an 'Ekman spiral'. This fascinating spiral shape was detected by the complex PCA (Stacey *et al.*, 1986).

2.4 Canonical correlation analysis (CCA)

Given a set of variables $\{y_j\}$, PCA finds the linear modes accounting for the maximum amount of variance in the dataset. When there are two sets of variables $\{x_i\}$ and $\{y_j\}$, *canonical correlation analysis* (CCA), first introduced by Hotelling (1936), finds the modes of maximum correlation between $\{x_i\}$ and $\{y_j\}$, rendering CCA a standard tool for discovering linear relations between two fields. CCA is a generalization of the Pearson correlation between two variables x and y to two sets of variables $\{x_i\}$ and $\{y_j\}$. Thus CCA can be viewed as a 'double-barrelled PCA'. A variant of the CCA method finds the modes of maximum *covariance* between $\{x_i\}$ and $\{y_j\}$ – this variant is called the *maximum covariance analysis* (MCA) by von Storch and Zwiers (1999), and because it uses the SVD matrix technique, it is also simply called the SVD (singular value decomposition) method by other researchers, though this name is confusing as it is used to denote both a matrix technique and a multivariate statistical technique.

In PCA, one finds a linear combination of the y_j variables, i.e. $\mathbf{e}_1^{T}\mathbf{y}$, which has the largest variance (subject to $\|\mathbf{e}_1\| = 1$). Next, one finds $\mathbf{e}_2^{T}\mathbf{y}$ with the largest variance, but with $\mathbf{e}_2^{T}\mathbf{y}$ uncorrelated with $\mathbf{e}_1^{T}\mathbf{y}$, and similarly for the higher modes.

In CCA, one finds $\mathbf{f}_1$ and $\mathbf{g}_1$, so that the correlation between $\mathbf{f}_1^{T}\mathbf{x}$ and $\mathbf{g}_1^{T}\mathbf{y}$ is maximized. Next find $\mathbf{f}_2$ and $\mathbf{g}_2$ so that the correlation between $\mathbf{f}_2^{T}\mathbf{x}$ and $\mathbf{g}_2^{T}\mathbf{y}$ is maximized, with $\mathbf{f}_2^{T}\mathbf{x}$ and $\mathbf{g}_2^{T}\mathbf{y}$ uncorrelated with both $\mathbf{f}_1^{T}\mathbf{x}$ and $\mathbf{g}_1^{T}\mathbf{y}$, and so forth for the higher modes.

2.4.1 CCA theory

Consider two datasets

$$\mathbf{x}(t) = x_{il}, \qquad i = 1, \ldots n_x, \quad l = 1, \ldots, n_t, \tag{2.93}$$

and

$$\mathbf{y}(t) = y_{jl}, \qquad j = 1, \ldots, n_y, \quad l = 1, \ldots, n_t, \tag{2.94}$$

i.e. $\mathbf{x}$ and $\mathbf{y}$ need not have the same spatial dimensions, but need the same time dimension n_t. Assume $\mathbf{x}$ and $\mathbf{y}$ have zero means. Let

$$u = \mathbf{f}^{\mathrm{T}}\mathbf{x}, \quad v = \mathbf{g}^{\mathrm{T}}\mathbf{y}. \tag{2.95}$$

The correlation

$$\rho = \frac{\mathrm{cov}(u, v)}{\sqrt{\mathrm{var}(u)\,\mathrm{var}(v)}} = \frac{\mathrm{cov}(\mathbf{f}^{\mathrm{T}}\mathbf{x}, \mathbf{g}^{\mathrm{T}}\mathbf{y})}{\sqrt{\mathrm{var}(u)\,\mathrm{var}(v)}} = \frac{\mathbf{f}^{\mathrm{T}}\mathrm{cov}(\mathbf{x}, \mathbf{y})\mathbf{g}}{\sqrt{\mathrm{var}(\mathbf{f}^{\mathrm{T}}\mathbf{x})\,\mathrm{var}(\mathbf{g}^{\mathrm{T}}\mathbf{y})}}, \tag{2.96}$$

where we have invoked

$$\mathrm{cov}(\mathbf{f}^{\mathrm{T}}\mathbf{x}, \mathbf{g}^{\mathrm{T}}\mathbf{y}) = \mathrm{E}[\mathbf{f}^{\mathrm{T}}\mathbf{x}(\mathbf{g}^{\mathrm{T}}\mathbf{y})^{\mathrm{T}}] = \mathrm{E}[\mathbf{f}^{\mathrm{T}}\mathbf{x}\mathbf{y}^{\mathrm{T}}\mathbf{g}] = \mathbf{f}^{\mathrm{T}}\mathrm{E}[\mathbf{x}\mathbf{y}^{\mathrm{T}}]\mathbf{g}. \tag{2.97}$$

We want u and v, the two *canonical variates* or *canonical correlation coordinates*, to have maximum correlation between them, i.e. $\mathbf{f}$ and $\mathbf{g}$ are chosen to maximize ρ. We are of course free to normalize $\mathbf{f}$ and $\mathbf{g}$ as we like, because if $\mathbf{f}$ and $\mathbf{g}$ maximize ρ, so will $\alpha\mathbf{f}$ and $\beta\mathbf{g}$, for any positive α and β. We choose the normalization condition

$$\mathrm{var}(\mathbf{f}^{\mathrm{T}}\mathbf{x}) = 1 = \mathrm{var}(\mathbf{g}^{\mathrm{T}}\mathbf{y}). \tag{2.98}$$

Since

$$\mathrm{var}(\mathbf{f}^{\mathrm{T}}\mathbf{x}) = \mathrm{cov}(\mathbf{f}^{\mathrm{T}}\mathbf{x}, \mathbf{f}^{\mathrm{T}}\mathbf{x}) = \mathbf{f}^{\mathrm{T}}\mathrm{cov}(\mathbf{x}, \mathbf{x})\mathbf{f} \equiv \mathbf{f}^{\mathrm{T}}\mathbf{C}_{xx}\mathbf{f}, \tag{2.99}$$

and

$$\mathrm{var}(\mathbf{g}^{\mathrm{T}}\mathbf{y}) = \mathbf{g}^{\mathrm{T}}\mathbf{C}_{yy}\mathbf{g}, \tag{2.100}$$

equation (2.98) implies

$$\mathbf{f}^{\mathrm{T}}\mathbf{C}_{xx}\mathbf{f} = 1, \quad \mathbf{g}^{\mathrm{T}}\mathbf{C}_{yy}\mathbf{g} = 1. \tag{2.101}$$

With (2.98), (2.96) reduces to

$$\rho = \mathbf{f}^{\mathrm{T}}\mathbf{C}_{xy}\mathbf{g}, \tag{2.102}$$

where $\mathbf{C}_{xy} = \mathrm{cov}(\mathbf{x}, \mathbf{y})$.

The problem is to maximize (2.102) subject to constraints (2.101). We will again use the method of Lagrange multipliers (Appendix B), where we incoroporate the constraints into the Lagrange function L,

$$L = \mathbf{f}^{\mathrm{T}}\mathbf{C}_{xy}\mathbf{g} + \alpha(\mathbf{f}^{\mathrm{T}}\mathbf{C}_{xx}\mathbf{f} - 1) + \beta(\mathbf{g}^{\mathrm{T}}\mathbf{C}_{yy}\mathbf{g} - 1), \tag{2.103}$$

where α and β are the unknown Lagrange multipliers. To find the stationary points of L, we need

$$\frac{\partial L}{\partial \mathbf{f}} = \mathbf{C}_{xy}\mathbf{g} + 2\alpha\mathbf{C}_{xx}\mathbf{f} = 0, \tag{2.104}$$

and

$$\frac{\partial L}{\partial \mathbf{g}} = \mathbf{C}_{xy}^{\mathrm{T}}\mathbf{f} + 2\beta\mathbf{C}_{yy}\mathbf{g} = 0. \tag{2.105}$$

Hence

$$\mathbf{C}_{xx}^{-1}\mathbf{C}_{xy}\mathbf{g} = -2\alpha\mathbf{f}, \tag{2.106}$$

and

$$\mathbf{C}_{yy}^{-1}\mathbf{C}_{xy}^{\mathrm{T}}\mathbf{f} = -2\beta\mathbf{g}. \tag{2.107}$$

Substituting (2.107) into (2.106) yields

$$\mathbf{C}_{xx}^{-1}\mathbf{C}_{xy}\mathbf{C}_{yy}^{-1}\mathbf{C}_{xy}^{\mathrm{T}}\,\mathbf{f} \equiv \mathbf{M}_f\mathbf{f} = \lambda\mathbf{f}, \tag{2.108}$$

with $\lambda = 4\alpha\beta$. Similarly, substituting (2.106) into (2.107) gives

$$\mathbf{C}_{yy}^{-1}\mathbf{C}_{xy}^{\mathrm{T}}\mathbf{C}_{xx}^{-1}\mathbf{C}_{xy}\,\mathbf{g} \equiv \mathbf{M}_g\mathbf{g} = \lambda\mathbf{g}. \tag{2.109}$$

Both these equations can be viewed as eigenvalue equations, with $\mathbf{M}_f$ and $\mathbf{M}_g$ sharing the same non-zero eigenvalues λ. As $\mathbf{M}_f$ and $\mathbf{M}_g$ are known from the data, $\mathbf{f}$ can be found by solving the eigenvalue problem (2.108). Then $\beta\mathbf{g}$ can be obtained from (2.107). Since β is unknown, the magnitude of $\mathbf{g}$ is unknown, and the normalization conditions (2.101) are used to determine the magnitude of $\mathbf{g}$ and $\mathbf{f}$. Alternatively, one can use (2.109) to solve for $\mathbf{g}$ first, then obtain $\mathbf{f}$ from (2.106) and the normalization condition (2.101). The matrix $\mathbf{M}_f$ is of dimension $n_x \times n_x$, while $\mathbf{M}_g$ is $n_y \times n_y$, so one usually picks the smaller of the two to solve the eigenvalue problem.

From (2.102),

$$\rho^2 = \mathbf{f}^{\mathrm{T}}\mathbf{C}_{xy}\mathbf{g}\,\mathbf{g}^{\mathrm{T}}\mathbf{C}_{xy}^{\mathrm{T}}\mathbf{f} = 4\alpha\beta\,(\mathbf{f}^{\mathrm{T}}\mathbf{C}_{xx}\mathbf{f})\,(\mathbf{g}^{\mathrm{T}}\mathbf{C}_{yy}\mathbf{g}), \tag{2.110}$$

where (2.104) and (2.105) have been invoked. From (2.101), (2.110) reduces to

$$\rho^2 = \lambda. \tag{2.111}$$

The eigenvalue problems (2.108) and (2.109) yield n number of λs, with $n = \min(n_x, n_y)$. Assuming the λs to be all distinct and non-zero, we have for each λ_j ($j = 1, \ldots, n$), a pair of eigenvectors, $\mathbf{f}_j$ and $\mathbf{g}_j$, and a pair of canonical variates, u_j and v_j, with correlation $\rho_j = \sqrt{\lambda_j}$ between the two. It can also be shown that

$$\mathrm{cov}(u_j, u_k) = \mathrm{cov}(v_j, v_k) = \delta_{jk}, \quad \text{and} \quad \mathrm{cov}(u_j, v_k) = 0 \quad \text{if} \quad j \neq k. \tag{2.112}$$

Let us write the forward mappings from the variables $\mathbf{x}(t)$ and $\mathbf{y}(t)$ to the canonical variates $\mathbf{u}(t) = [u_1(t), \ldots, u_n(t)]^\mathrm{T}$ and $\mathbf{v}(t) = [v_1(t), \ldots, v_n(t)]^\mathrm{T}$ as

$$\mathbf{u} = [\mathbf{f}_1^\mathrm{T}\mathbf{x}, \ldots, \mathbf{f}_n^\mathrm{T}\mathbf{x}]^\mathrm{T} = \mathcal{F}^\mathrm{T}\mathbf{x}, \quad \mathbf{v} = \mathcal{G}^\mathrm{T}\mathbf{y}. \tag{2.113}$$

Next, we need to find the inverse mapping from $\mathbf{u} = [u_1, \ldots, u_n]^\mathrm{T}$ and $\mathbf{v} = [v_1, \ldots, v_n]^\mathrm{T}$ to the original variables $\mathbf{x}$ and $\mathbf{y}$. Let

$$\mathbf{x} = \mathbf{F}\mathbf{u}, \quad \mathbf{y} = \mathbf{G}\mathbf{v}. \tag{2.114}$$

We note that

$$\mathrm{cov}(\mathbf{x}, \mathbf{u}) = \mathrm{cov}(\mathbf{x}, \mathcal{F}^\mathrm{T}\mathbf{x}) = \mathrm{E}[\mathbf{x}(\mathcal{F}^\mathrm{T}\mathbf{x})^\mathrm{T}] = \mathrm{E}[\mathbf{x}\mathbf{x}^\mathrm{T}\mathcal{F}] = \mathbf{C}_{xx}\mathcal{F}, \tag{2.115}$$

and

$$\mathrm{cov}(\mathbf{x}, \mathbf{u}) = \mathrm{cov}(\mathbf{F}\mathbf{u}, \mathbf{u}) = \mathbf{F}\,\mathrm{cov}(\mathbf{u}, \mathbf{u}) = \mathbf{F}. \tag{2.116}$$

Equations (2.115) and (2.116) imply

$$\mathbf{F} = \mathbf{C}_{xx}\mathcal{F}. \tag{2.117}$$

Similarly,

$$\mathbf{G} = \mathbf{C}_{yy}\mathcal{G}. \tag{2.118}$$

Hence the inverse mappings $\mathbf{F}$ and $\mathbf{G}$ (from the canonical variates to $\mathbf{x}$ and $\mathbf{y}$) can be calculated from the forward mappings $\mathcal{F}^T$ and $\mathcal{G}^T$. The matrix $\mathbf{F}$ is composed of column vectors $\mathbf{F}_j$, and $\mathbf{G}$, of column vectors $\mathbf{G}_j$. Note that $\mathbf{F}_j$ and $\mathbf{G}_j$ are the *canonical correlation patterns* associated with u_j and v_j, the canonical variates. In general, orthogonality of vectors within a set is not satisfied by any of the four sets $\{\mathbf{F}_j\}$, $\{\mathbf{G}_j\}$, $\{\mathbf{f}_j\}$ and $\{\mathbf{g}_j\}$, while

$$\mathrm{cov}(u_i, u_j) = \mathrm{cov}(v_i, v_j) = \mathrm{cov}(u_i, v_j) = 0, \quad \text{for } i \neq j. \tag{2.119}$$

Figure 2.14 schematically illustrates the canonical correlation patterns.

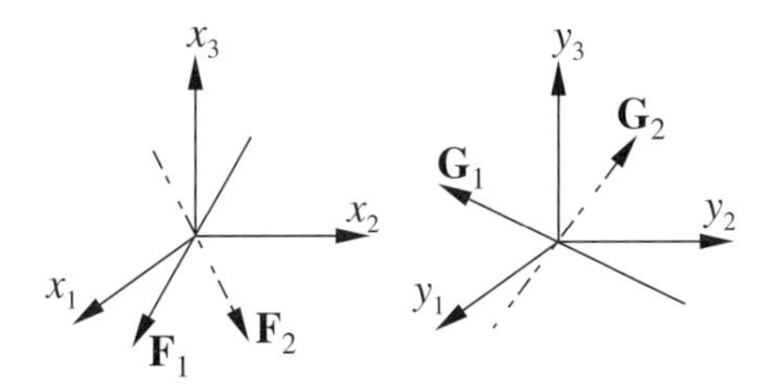

Fig. 2.14 Illustrating the CCA solution in the **x** and **y** spaces. The vectors $\mathbf{F}_1$ and $\mathbf{G}_1$ are the canonical correlation patterns for mode 1, and $u_1(t)$ is the amplitude of the 'oscillation' along $\mathbf{F}_1$, and $v_1(t)$, the amplitude along $\mathbf{G}_1$. The vectors $\mathbf{F}_1$ and $\mathbf{G}_1$ have been chosen so that the correlation between u_1 and v_1 is maximized. Next $\mathbf{F}_2$ and $\mathbf{G}_2$ are found, with $u_2(t)$ the amplitude of the 'oscillation' along $\mathbf{F}_2$, and $v_2(t)$ along $\mathbf{G}_2$. The correlation between u_2 and v_2 is again maximized, but with $\text{cov}(u_1, u_2) = \text{cov}(v_1, v_2) = \text{cov}(u_1, v_2) = \text{cov}(v_1, u_2) = 0$. In general, $\mathbf{F}_2$ is not orthogonal to $\mathbf{F}_1$, and $\mathbf{G}_2$ not orthogonal to $\mathbf{G}_1$. Unlike PCA, $\mathbf{F}_1$ and $\mathbf{G}_1$ need not be oriented in the direction of maximum variance. Solving for $\mathbf{F}_1$ and $\mathbf{G}_1$ is analogous to performing rotated PCA in the **x** and **y** spaces separately, with the rotations determined from maximizing the correlation between u_1 and v_1.

2.4.2 Pre-filter with PCA

When **x** and **y** contain many variables, it is common to use PCA to pre-filter the data to reduce the dimensions of the datasets, i.e. apply PCA to **x** and **y** separately, extract the leading PCs, then apply CCA to the leading PCs of **x** and **y**.

Using Hotelling's choice of scaling for the PCAs (Eq. 2.37), we express the PCA expansions as

$$\mathbf{x} = \sum_j a'_j \mathbf{e}'_j, \quad \mathbf{y} = \sum_j a''_j \mathbf{e}''_j. \tag{2.120}$$

CCA is then applied to

$$\tilde{\mathbf{x}} = [a'_1, \ldots, a'_{m_x}]^{\mathrm{T}}, \quad \tilde{\mathbf{y}} = [a''_1, \ldots, a''_{m_y}]^{\mathrm{T}}, \tag{2.121}$$

where only the first m_x and m_y modes are used. Another reason for using the PCA pre-filtering is that when the number of variables is not small relative to the sample size, the CCA method may become unstable (Bretherton *et al.*, 1992). The reason is that in the relatively high-dimensional **x** and **y** spaces, among the many dimensions and using correlations calculated with relatively small samples, CCA can often find directions of high correlation but with little variance, thereby extracting a spurious leading CCA mode, as illustrated in Fig. 2.15. This problem can be avoided by pre-filtering using PCA, as this avoids applying CCA directly to high-dimensional input spaces (Barnett and Preisendorfer, 1987).

With Hotelling's scaling,

$$\text{cov}(a'_j, a'_k) = \delta_{jk}, \quad \text{cov}(a''_j, a''_k) = \delta_{jk}, \tag{2.122}$$

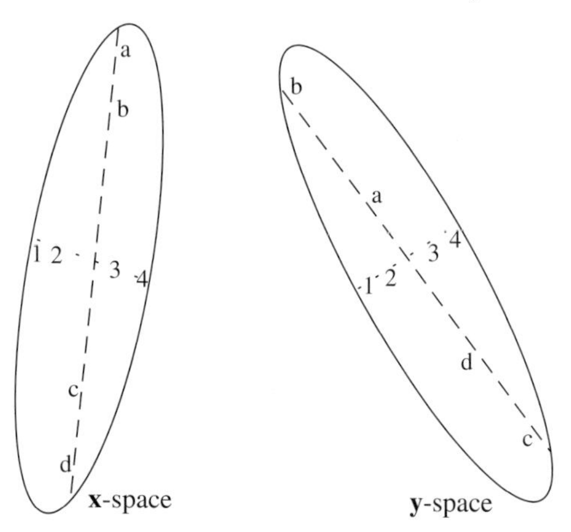

Fig. 2.15 Illustrating how CCA may end up extracting a spurious leading mode when working with relatively high-dimensional input spaces. With the ellipses denoting the data clouds in the two input spaces, the dotted lines illustrate directions with little variance but by chance with high correlation (as illustrated by the perfect order in which the data points 1, 2, 3 and 4 are arranged in the **x** and **y** spaces). Since CCA finds the correlation of the data points along the dotted lines to be higher than that along the dashed lines (where the data points a, b, c and d in the **x**-space are ordered as b, a, d and c in the **y**-space), the dotted lines are chosen as the first CCA mode. Maximum covariance analysis (MCA), which looks for modes of maximum covariance instead of maximum correlation, would select the dashed lines over the dotted lines since the lengths of the lines do count in the covariance but not in the correlation, hence MCA is stable even without pre-filtering by PCA.

leading to

$$\mathbf{C}_{\tilde{x}\tilde{x}} = \mathbf{C}_{\tilde{y}\tilde{y}} = \mathbf{I}. \tag{2.123}$$

Equations (2.108) and (2.109) simplify to

$$\mathbf{C}_{\tilde{x}\tilde{y}}\mathbf{C}^{\mathrm{T}}_{\tilde{x}\tilde{y}}\,\mathbf{f} \equiv \mathbf{M}_f\mathbf{f} = \lambda\mathbf{f}, \tag{2.124}$$

$$\mathbf{C}^{\mathrm{T}}_{\tilde{x}\tilde{y}}\mathbf{C}_{\tilde{x}\tilde{y}}\,\mathbf{g} \equiv \mathbf{M}_g\mathbf{g} = \lambda\mathbf{g}. \tag{2.125}$$

As $\mathbf{M}_f$ and $\mathbf{M}_g$ are positive semi-definite symmetric matrices, the eigenvectors $\{\mathbf{f}_j\}$ $\{\mathbf{g}_j\}$ are now sets of orthogonal vectors. Equations (2.117) and (2.118) simplify to

$$\mathbf{F} = \mathcal{F}, \quad \mathbf{G} = \mathcal{G}. \tag{2.126}$$

Hence $\{\mathbf{F}_j\}$ and $\{\mathbf{G}_j\}$ are also two sets of orthogonal vectors, and are identical to $\{\mathbf{f}_j\}$ and $\{\mathbf{g}_j\}$, respectively. Because of these nice properties, pre-filtering by PCA (with the Hotelling scaling) is recommended when **x** and **y** have many variables (relative to the sample size). However, the orthogonality holds only in the reduced

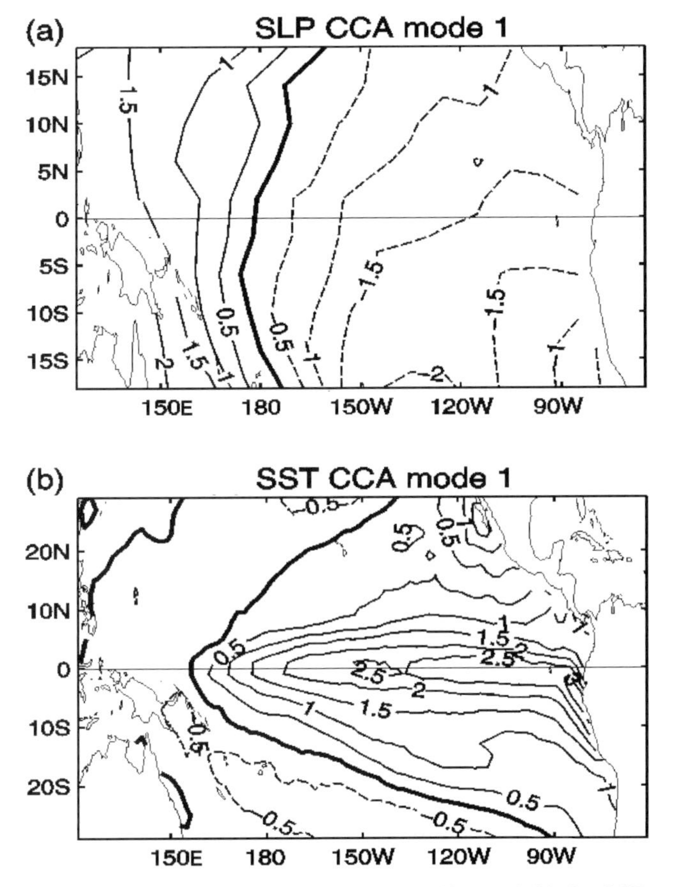

Fig. 2.16 The CCA mode 1 for (a) the SLP anomalies and (b) the SST anomalies of the tropical Pacific. As $u_1(t)$ and $v_1(t)$ fluctuate together from one extreme to the other as time progresses, the SLP and SST anomaly fields, oscillating as standing wave patterns, evolve from an El Niño to a La Niña state. The pattern in (a) is scaled by $\tilde{u}_1 = [\max(u_1) - \min(u_1)]/2$, and (b) by $\tilde{v}_1 = [\max(v_1) - \min(v_1)]/2$. Contour interval is 0.5 hPa in (a) and 0.5 °C in (b).

dimensional spaces, $\tilde{\mathbf{x}}$ and $\tilde{\mathbf{y}}$. If transformed into the original space $\mathbf{x}$ and $\mathbf{y}$, $\{\mathbf{F}_j\}$ and $\{\mathbf{G}_j\}$ are in general not two sets of orthogonal vectors.

Figure 2.16 shows the mode 1 CCA of the tropical Pacific sea level pressure (SLP) field and the SST field, showing clearly the Southern Oscillation pattern in the SLP and the El Niño/La Niña pattern in the SST. The canonical variates u and v (not shown) fluctuate with time, both attaining high values during El Niño, low values during La Niña, and neutral values around zero during normal conditions. Note that CCA is widely used for seasonal climate prediction (Barnston and Ropelewski, 1992; Shabbar and Barnston, 1996).

2.4.3 Singular value decomposition and maximum covariance analysis

Instead of maximizing the correlation as in CCA, one can maximize the covariance between two datasets. This alternative method is often called *singular value decomposition* (SVD). However, von Storch and Zwiers (1999) proposed the name *maximum covariance analysis* (MCA) as being more appropriate. The reason is that the name SVD is already used to denote a matrix technique (Strang, 2005), so there is potential confusion in using it to denote a statistical technique. In this book, we will follow the suggestion of von Storch and Zwiers, and refer to the statistical method as MCA, retaining the name SVD for the matrix technique.

Note that MCA is identical to CCA except that it maximizes the covariance instead of the correlation. As mentioned in the previous subsection, CCA can be unstable when working with a relatively large number of variables, in that directions with high correlation but negligible variance may be selected by CCA, hence the recommended pre-filtering of data by PCA before applying CCA. By using covariance instead of correlation, MCA does not have the unstable nature of CCA (Fig. 2.15), and does not need pre-filtering by PCA. Bretherton *et al.* (1992) compared the MCA ('SVD') and the CCA.

In MCA, one simply performs SVD (see Section 2.1.10) on the data covariance matrix $\mathbf{C}_{xy}$,

$$\mathbf{C}_{xy} = \mathbf{USV}^{\mathrm{T}}, \tag{2.127}$$

where the matrix $\mathbf{U}$ contains the left singular vectors $\mathbf{f}_i$, $\mathbf{V}$ the right singular vectors $\mathbf{g}_i$, and $\mathbf{S}$ the singular values. Maximum covariance between u_i and v_i is attained (Bretherton *et al.*, 1992) with

$$u_i = \mathbf{f}_i^{\mathrm{T}}\mathbf{x}, \quad v_i = \mathbf{g}_i^{\mathrm{T}}\mathbf{y}. \tag{2.128}$$

The inverse transform is given by

$$\mathbf{x} = \sum_i u_i \mathbf{f}_i, \quad \mathbf{y} = \sum_i v_i \mathbf{g}_i. \tag{2.129}$$

For most applications, MCA yields rather similar results to CCA (with PCA pre-filtering) (Bretherton *et al.*, 1992; Wallace *et al.*, 1992).

The matrix technique SVD can also be used to solve the CCA problem: instead of solving the eigenequations (2.108) and (2.109), simply perform SVD on $(\mathbf{C}_{xx}^{-1/2}\mathbf{C}_{xy}\mathbf{C}_{yy}^{-1/2})$. When the data have been prefiltered by PCA, then instead of solving eigenequations (2.124) and (2.125), simply perform SVD on $\mathbf{C}_{\tilde{x}\tilde{y}}$ (Bretherton *et al.*, 1992). Conversely, eigenequations can be used instead of SVD to solve the MCA problem (von Storch and Zwiers, 1999).

Exercises

(2.1) Suppose principal component analysis (PCA) allows one to write the data matrix $\mathbf{Y} = \mathbf{E}\,\mathbf{A}^{\mathrm{T}}$, where

$$\mathbf{E} = \begin{bmatrix} 0.7071 & -0.7071 \\ 0.7071 & 0.7071 \end{bmatrix}, \qquad \mathbf{A}^{\mathrm{T}} = \begin{bmatrix} 1 & 3 & -2 & -1 \\ 2 & -1 & 1 & -3 \end{bmatrix},$$

with $\mathbf{E}$ the matrix of eigenvectors, and $\mathbf{A}$ the principal components. Write down the 2×2 rotation matrix $\mathbf{R}$ for rotating eigenvectors anticlockwise by 30°. Derive the rotated matrix of eigenvectors and the rotated principal components.

(2.2) In principal component analysis (PCA) under the Hotelling (1933) scaling (Section 2.1.6), one has $\mathrm{cov}(a'_i, a'_j) = \delta_{ij}$, where a'_i is the ith principal component. Suppose each variable $y'_l(t)$ in a dataset has its mean removed and has been normalized by its standard deviation. Prove that upon applying PCA (with Hotelling scaling) to this normalized dataset, the lth element of the jth eigenvector (e'_{jl}) is simply the correlation between $a'_j(t)$ and $y'_l(t)$, i.e. $\rho(a'_j(t), y'_l(t)) = e'_{jl}$.

(2.3) Canonical correlation analysis (CCA) was used to analyze a set of $\mathbf{x}$ variables consisting of air pressure observations from three stations, and $\mathbf{y}$ variables consisting of precipitation observations from two stations. For the first CCA mode, the canonical variates $u(t) = 4.5, -5.0, -1.0, 5.5$, and $v(t) = 2.5, -4.0, 1.5, 4.0$, for $t = 1, 2, 3, 4$, and the corresponding canonical patterns are $\mathbf{F}^{\mathrm{T}} = [0.4, -0.5, 0.7]$, and $\mathbf{G}^{\mathrm{T}} = [0.8, 0.6]$, respectively. (a) What is the canonical correlation coefficient ρ between u and v? (b) If at $t = 5$, air pressure data are available, yielding $u = 3.6$, but the precipitation sensors are not working, estimate the precipitation at the two stations at $t = 5$.

(2.4) Analyze the dataset provided in the book website using principal component analysis (PCA). The dataset contains four time series x_1, x_2, x_3 and x_4.

(2.5) Using the data file provided in the book website, perform canonical correlation analysis between the two groups of time series, x_1, x_2, x_3 and y_1, y_2, y_3.

3

Basic time series analysis

In the previous chapter, we surveyed linear multivariate techniques for extracting features from, or recognizing patterns in, a dataset or two datasets, without considering time as an explicit factor. In the real world, the variables may be governed by dynamical equations, and the manifested patterns evolve with time. Classical Fourier spectral analysis is a first step in analyzing the temporal behaviour in a time series. Basic notions of windows and filters are introduced. More modern linear techniques, singular spectrum analysis (SSA), principal oscillation patterns (POP), and complex spectral PCA are also presented to analyze temporal–spatial patterns in a dataset.

3.1 Spectrum

With time series data, an alternative way to view the data is via the frequency representation first developed by J. B. Fourier in the early nineteenth century. Given a function $y(t)$ defined on the interval $[0, T]$, the Fourier series representation for $y(t)$ is

$$\hat{y}(t) = \frac{a_0}{2} + \sum_{m=1}^{\infty} [a_m \cos(\omega_m t) + b_m \sin(\omega_m t)], \tag{3.1}$$

with the (angular) frequency ω_m given by

$$\omega_m = \frac{2\pi m}{T}, \qquad m = 1, 2, \ldots \tag{3.2}$$

and the Fourier coefficients a_m and b_m given by

$$a_m = \frac{2}{T} \int_0^T y(t) \cos(\omega_m t)\, \mathrm{d}t, \quad m = 0, 1, 2, \ldots \tag{3.3}$$

$$b_m = \frac{2}{T} \int_0^T y(t) \sin(\omega_m t)\, \mathrm{d}t, \quad m = 1, 2, \ldots \tag{3.4}$$

With

$$a_0 = \frac{2}{T}\int_0^T y(t)\,\mathrm{d}t, \tag{3.5}$$

we see that

$$a_0/2 = \overline{y}, \tag{3.6}$$

the mean of y. If $y(t)$ is a continuous function, then (3.1) has $\hat{y}(t) \to y(t)$. If y is discontinuous at t, then $\hat{y}(t) \to [y(t+) + y(t-)]/2$.

For a discrete time series, $y(t)$ is replaced by $y(t_n) \equiv y_n$, $n = 1, \cdots, N$. With a sampling interval $\Delta t = T/N$, the observations are made at time $t_n = n\Delta t$. The discrete Fourier series representation is then given by

$$y_n = \frac{a_0}{2} + \sum_{m=1}^{M} [a_m \cos(\omega_m t_n) + b_m \sin(\omega_m t_n)], \tag{3.7}$$

where M is the largest integer $\leq N/2$, with the Fourier coefficients given by

$$a_m = \frac{2}{N}\sum_{n=1}^{N} y_n \cos(\omega_m t_n), \quad m = 0, 1, 2, \ldots, M, \tag{3.8}$$

$$b_m = \frac{2}{N}\sum_{n=1}^{N} y_n \sin(\omega_m t_n), \quad m = 1, 2, \ldots, M. \tag{3.9}$$

For N even, $b_M = 0$, so the number of non-trivial Fourier coefficients is N.

The cosine and sine functions have orthogonality properties:

$$\begin{aligned} \sum_{n=1}^{N} \cos(\omega_l t_n)\cos(\omega_m t_n) &= \frac{N}{2}\delta_{lm}, \\ \sum_{n} \sin(\omega_l t_n)\sin(\omega_m t_n) &= \frac{N}{2}\delta_{lm}, \\ \sum_{n} \cos(\omega_l t_n)\sin(\omega_m t_n) &= 0, \end{aligned} \tag{3.10}$$

where δ_{lm} is the Kronecker delta function. The original data y_n can be recovered by substituting (3.8) and (3.9) into the right side of (3.7), and invoking the orthogonal relations (3.10).

3.1.1 Autospectrum

The variance of the time series y can be written in terms of its Fourier coefficients:

$$\begin{aligned} \mathrm{var}(y) &= \frac{1}{N}\sum_{n=1}^{N}(y_n - \overline{y}_n)^2 = \frac{1}{N}\sum_{n}\left(y_n - \frac{a_0}{2}\right)^2 \\ &= \frac{1}{N}\sum_{n}\left[\sum_{m}(a_m\cos(\omega_m t_n) + b_m \sin(\omega_m t_n))\right]^2. \end{aligned} \tag{3.11}$$

Using (3.10), var(y) can be expressed in terms of the Fourier coefficients,

$$\mathrm{var}(y) = \frac{1}{2}\sum_m (a_m^2 + b_m^2). \tag{3.12}$$

The *autospectrum*, also called the *spectrum*, the *power spectrum* or the *periodogram*, is defined as

$$S_m = \frac{N\Delta t}{4\pi}(a_m^2 + b_m^2). \tag{3.13}$$

Thus (3.12) can be expressed as

$$\mathrm{var}(y) = \sum_m S_m \Delta\omega, \tag{3.14}$$

with

$$\Delta\omega = \frac{2\pi}{T} = \frac{2\pi}{N\Delta t}. \tag{3.15}$$

Hence, the spectrum S_m can be viewed as the variance or 'energy' in the ω_m frequency band (with bandwidth $\Delta\omega$), and the total variance var(y) can be computed by integrating S_m over all frequency bands. When S_m is plotted as a function of the frequency, peaks in S_m reveal the frequencies where the energy is relatively high.

The lowest frequency in the spectrum, known as the *fundamental frequency*, is

$$\omega_1 = \frac{2\pi}{T}, \quad \text{or} \quad f_1 = \frac{1}{T}. \tag{3.16}$$

Often a time series displays a *trend*, i.e. a positive or negative slope in the data over the time record. For instance, the Canadian prairie wheat yield (Fig. 3.1) shows a positive trend with time, largely due to the gradual improvement in agricultural technology. The frequency associated with a trend is lower than the fundamental frequency, thus energy from the trend will leak to other low frequency spectral bands, thereby distorting the low frequency part of the spectrum. By subtracting the linear regression line from the data time series, trends can easily be removed prior to spectral analysis.

The highest resolvable frequency from (3.2) is $\omega = 2\pi\, M/T$, but with $M \approx N/2$, we have $M/T \approx 1/(2\Delta t)$. Hence the highest resolvable frequency, called the *Nyquist frequency*, is

$$\omega_N = \frac{\pi}{\Delta T}, \quad \text{or} \quad f_N = \frac{1}{2\Delta T}. \tag{3.17}$$

To resolve a wave of period τ, we need at least two data points to cover the period τ, i.e. $\tau = 2\Delta t = 1/f_N$. *Aliasing* arises when Δt is too large to resolve the highest frequency oscillations in the data. Figure 3.2 illustrates a signal measured too infrequently, resulting in an incorrect inference aliased signal of much lower

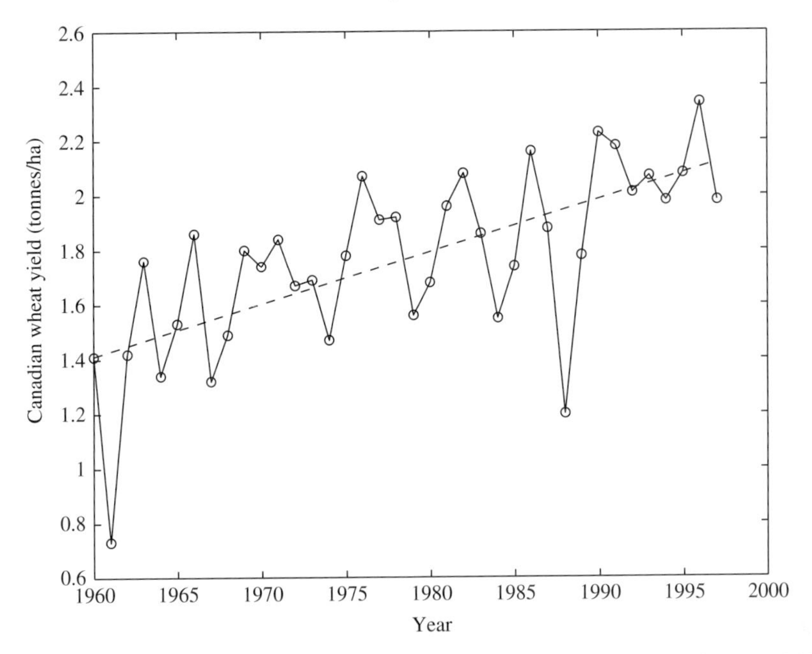

Fig. 3.1 Annual Canadian prairie hard red spring wheat yield from 1960 to 1997, with the dashed line indicating the linear trend. The upward trend results from the gradual advance in agricultural technology. (Reproduced from Hsieh *et al.* (1999) with permission of Elsevier.)

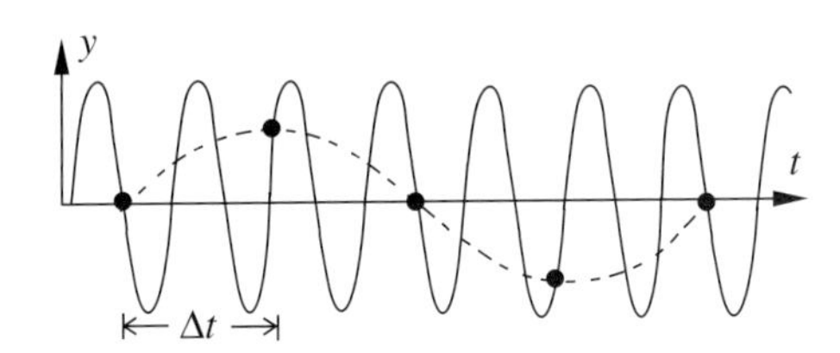

Fig. 3.2 Illustrating the phenomenon of aliasing. The sampling time interval is Δt, but the signal (solid curve) is oscillating too quickly to be resolved by the sampling. From the observations (dots), an incorrect signal (dashed curve) of much lower frequency is inferred by the observer.

frequency. In a spectrum, signals with frequency above the Nyquist frequency are reflected across the Nyquist frequency into the frequency bands below the Nyquist frequency – resulting in a distortion of the high frequency part of the spectrum. In movies, one often notices aliasing, where fast spinning wheels and propellors appear to rotate backwards.

From (3.2), the frequency $\Delta\omega$ between adjacent frequency bands is

$$\Delta\omega = \frac{2\pi}{T}. \tag{3.18}$$

The ability to resolve neighbouring spectral peaks is controlled by $\Delta\omega$, which is proportional to $1/T$. Hence a longer record T will yield sharper spectral peaks, thereby allowing resolution of two signals with close-by frequencies as distinct peaks in the spectrum. A shorter record will blur the two signals into a single spectral peak.

The raw spectrum S_m calculated from (3.13) is often very noisy in appearance. There are two common methods for smoothing the spectrum: (a) band-averaging (the Daniell estimator) and (b) ensemble averaging. In (a), a moving average (or running mean) is applied to the raw spectrum

$$\tilde{S}_m = \frac{1}{(2K+1)} \sum_{k=-K}^{K} S_{m+k}, \tag{3.19}$$

where $\tilde{S}_m$ is the smoothed spectrum resulting from averaging the raw spectrum over $2K + 1$ frequency bands.

In method (b), the data record is divided into J blocks of equal length $L = T/J$. We compute the periodogram for each block to get $S_m^{(j)}$ $(j = 1, \ldots, J)$. The spectrum S_m is then the ensemble average over the J periodograms:

$$S_m = \frac{1}{J} \sum_{j=1}^{J} S_m^{(j)}. \tag{3.20}$$

Method (b) has an advantage over (a) when there are data gaps – in (b), the data gaps do not pose a serious problem since the data record is to be chopped into J blocks anyway, whereas in (a), the data gaps may have to be filled with interpolated values or zeros. The disadvantage of (b) is that the lowest resolvable frequency is $f_1 = 1/L = J/T$, hence there is a loss of low frequency information when the record is chopped up.

There is a trade-off between the variance of the spectrum S and the band width. Increasing the band width (by increasing K or J) leads to a less noisy S, but spectral peaks are broadened, so that nearby spectral peaks may merge together, resulting in a loss of resolution.

There is an important relation between the spectrum and the auto-covariance function. In complex notation, the Fourier transform

$$\hat{y}_m = \frac{2}{N} \sum_{n=1}^{N} y_n \, e^{-i\omega_m t_n} \tag{3.21}$$

$$= a_m - ib_m, \tag{3.22}$$

where (3.8) and (3.9) have been invoked. Equation (3.13) can be written as

$$S_m = \frac{N\,\Delta t}{4\pi} |\hat{y}_m|^2. \tag{3.23}$$

Assume $\{y_n\}$ is stationary and the mean $\overline{y}$ has been subtracted from the data, then

$$S_m = \frac{\Delta t}{N\pi} \left[\sum_n y_n\, \mathrm{e}^{-\mathrm{i}\omega_m t_n} \right] \left[\sum_j y_j\, \mathrm{e}^{\mathrm{i}\omega_m t_j} \right] \tag{3.24}$$

$$= \frac{\Delta t}{\pi} \sum_{l=-(N-1)}^{N-1} \left[\frac{1}{N} \left(\sum_{j-n=l} y_n y_j \right) \right] \mathrm{e}^{\mathrm{i}\omega_m t_l}. \tag{3.25}$$

In general, the *auto-covariance* function with lag l is defined as

$$C_l = \frac{1}{N} \sum_{n=1}^{N-l} (y_n - \overline{y})(y_{n+l} - \overline{y}). \tag{3.26}$$

Here (with $\overline{y} = 0$), we have the important relation

$$S_m = \frac{\Delta t}{\pi} \sum_{l=-(N-1)}^{N-1} C(l)\, \mathrm{e}^{\mathrm{i}\omega_m t_l}, \tag{3.27}$$

i.e. the spectrum is related to the auto-covariance function by a Fourier transform.

3.1.2 Cross-spectrum

Now let us consider two time series, $x_1, \ldots, x_N$ and $y_1, \ldots, y_N$, with respective Fourier transforms $\hat{x}_m$ and $\hat{y}_m$, which are in general complex numbers. The *cross-spectrum*

$$C_m = \frac{N\,\Delta t}{4\pi} \hat{x}_m^* \hat{y}_m, \tag{3.28}$$

where the asterisk denotes complex conjugation, so C_m is in general complex. If $\hat{y}_m = \hat{x}_m$, C_m reduces to S_m, which is real.

C_m can be split into a real part and an imaginary part,

$$C_m = R_m + \mathrm{i}\, I_m, \tag{3.29}$$

where R_m is the *co-spectrum* and I_m is the *quadrature spectrum*. Note that C_m can also be expressed in polar form,

$$C_m = A_m\, \mathrm{e}^{\mathrm{i}\theta_m}, \tag{3.30}$$

where A_m is the *amplitude spectrum* and θ_m, the *phase spectrum*, with

$$A_m = [\, R_m^2 + I_m^2 \,]^{\frac{1}{2}} \quad \text{and} \quad \theta_m = \tan^{-1}(I_m \,/\, R_m). \tag{3.31}$$

A useful quantity is the *squared coherency spectrum* (where the word 'squared' is often omitted for brevity):

$$r_m^2 = \frac{A_m^2}{S_m^{(x)}\, S_m^{(y)}}, \tag{3.32}$$

where $S_m^{(x)}$, $S_m^{(y)}$ are the autospectrum for the x and y time series, respectively. One can interpret r_m^2 as the magnitude of the correlation between x and y in the mth frequency band. However, if one does not perform band averaging or ensemble averaging, then $r_m^2 = 1$, i.e. perfect correlation for all m! To see this, let

$$\hat{x}_m = a_m\, \mathrm{e}^{\mathrm{i}\alpha_m} \quad \text{and} \quad \hat{y}_m = b_m\, \mathrm{e}^{\mathrm{i}\beta_m}. \tag{3.33}$$

Equation (3.28) becomes

$$C_m = \frac{N\Delta t}{4\pi}\, a_m\, b_m\, \mathrm{e}^{\mathrm{i}(\alpha_m - \beta_m)}. \tag{3.34}$$

Thus

$$A_m = \frac{N\Delta t}{4\pi}\, a_m\, b_m \quad \text{and} \quad \theta_m = \alpha_m - \beta_m. \tag{3.35}$$

Also,

$$S_m^{(x)} = \frac{N\Delta t}{4\pi}\, a_m^2, \quad \text{and} \quad S_m^{(y)} = \frac{N\Delta t}{4\pi}\, b_m^2. \tag{3.36}$$

Substituting these into (3.32) yields $r_m^2 = 1$. The reason is that in a single frequency band, the x and y signals are simply sinusoidals of the same frequency, which are perfectly correlated (other than a possible phase shift between the two).

Suppose there is no real relation between $\hat{x}_m$ and $\hat{y}_m$, then the phase $\alpha_m - \beta_m$ tends to be random. Consider ensemble averaging, with

$$C_m = \frac{1}{J} \sum_{j=1}^{J} C_m^{(j)}. \tag{3.37}$$

With random phase, the $C_m^{(j)}$ vectors are randomly oriented in the complex plane, so summing of the $C_m^{(j)}$ vectors tends not to produce a C_m vector with large magnitude A_m. In general for large J, $A_m^2 \ll S_m^{(x)}\, S_m^{(y)}$, resulting in a small value for r_m^2, as desired. Thus some form of ensemble averaging or band averaging is essential for computing the squared coherency spectrum – without the averaging, even random noise has r_m^2 equal to unity.

It can also be shown that the cross-spectrum is the Fourier transform of the *cross-covariance* γ, where

$$\gamma = \frac{1}{N} \sum_{n=1}^{N-l} (x_n - \overline{x})(y_{n+l} - \overline{y}). \tag{3.38}$$

3.2 Windows

The data record for a variable y is always of finite duration T, commencing at $t = 0$, and ending at $t = T$. When applying the Fourier transform to a finite record, periodicity is assumed for y, which presents a potentially serious problem. Unless $y(0)$ and $y(T)$ are of the same value, the periodicity assumption creates a step discontinuity at $y(T)$ (Fig. 3.3). The Fourier representation of a step discontinuity requires use of many spectral components, i.e. spurious energy is leaked to many frequency bands.

Another way to view the situation is to consider the true time series $Y(t)$ as extending from $-\infty$ to $+\infty$. It is multiplied by a *window* function

$$w(t) = \begin{cases} 1 & \text{for } -T/2 \leq t \leq T/2, \\ 0 & \text{elsewhere} \end{cases} \tag{3.39}$$

to yield the finite data record $y(t)$ of duration T (for convenience, y is now defined for $-T/2 \leq t \leq T/2$). Thus the data record can be regarded as the product between the true time series and the window function. If $\hat{w}$ and $\hat{Y}$ are the Fourier transforms of $w(t)$ and $Y(t)$ over $(-\infty, \infty)$, then the Fourier transform of the product wY is the convolution of $\hat{w}$ and $\hat{Y}$. For the rectangular window, $\hat{w}$ has many significant side-lobes, thus the convolution of $\hat{w}$ and $\hat{Y}$ leads to spurious energy leakage into other frequency bands (Jenkins and Watts, 1968, p.49).

If the ends of the window are tapered (e.g. by a cosine-shaped taper) to avoid the abrupt ends, the size of the side-lobes can be greatly reduced, thereby reducing the spurious energy leakage. In effect, a tapered window tapers the ends of the data record, so that the two ends continuously approach $y(-T/2) = y(T/2) = 0$, avoiding the original step discontinuity. As there are many possible windows, finding an optimal one can be quite involved (see e.g. Emery and Thomson, 1997).

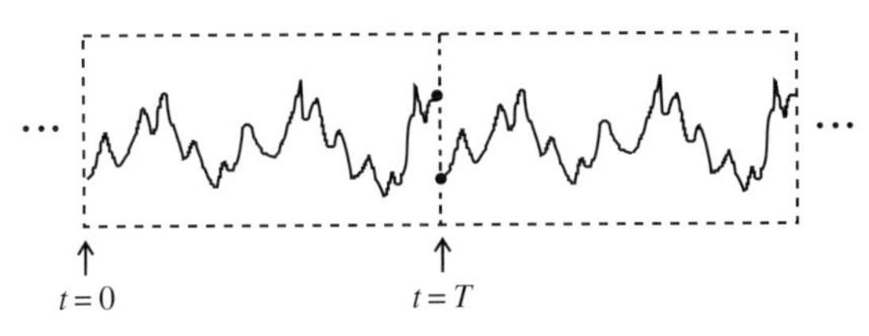

Fig. 3.3 A troublesome consequence of the periodicity assumption in Fourier spectral analysis. Here the observed record is from $t = 0$ to $t = T$. For Fourier spectral analysis, the data record is assumed to repeat itself periodically outside the interval of $[0, T]$. As illustrated here, the periodicity assumption leads to a sharp jump at $t = T$. Such a step discontinuity causes trouble for Fourier transforms in that spurious energy is leaked to many other frequency bands. To avoid this, the data record is often tapered, so the values at the two ends, $t = 0$ and $t = T$, do not give a major step.

3.3 Filters

One often would like to perform digital filtering on the raw data. For instance, one may want a smoother data field, or want to concentrate on the low-frequency or high-frequency signals in the time series. Let us express a time series $x(t)$ in terms of its complex Fourier components $X(\omega)$:

$$x(t) = \sum_{\omega} X(\omega)\, e^{i\omega t}, \tag{3.40}$$

where it is understood that ω and t denote the discrete variables ω_m and t_n. A filtered time series is given by

$$\tilde{x}(t) = \sum_{\omega} f(\omega) X(\omega)\, e^{i\omega t}, \tag{3.41}$$

where $f(\omega)$ is the filter 'response' function. Figure 3.4 illustrates several ideal filters for low-pass, high-pass and band-pass filtering. In these ideal filters, the step discontinuity at the cut-off frequency ω_c produces 'ringing' (i.e. oscillations) in the filtered time series (especially at the two ends) (Emery and Thomson, 1997, Fig. 5.10.19). This problem of a step discontinuity in the frequency domain leading to ringing in the time domain mirrors the one mentioned in the previous section, where a time series truncated by a rectangular window leads to energy leakage in the frequency domain. Thus in practice, $f(\omega)$ needs to be tapered at ω_c to suppress ringing in the filtered time series.

To perform filtering in the frequency domain, the steps are: (i) Fourier transform $x(t)$ to $X(\omega)$, (ii) multiply $X(\omega)$ by $f(\omega)$, and (iii) inverse transform $f(\omega)X(\omega)$ to get $\tilde{x}(t)$, the filtered time series.

Alternatively, filtering can be performed in the time domain as well. In fact, prior to the invention of fast Fourier transform algorithms, filtering in the frequency domain was prohibitively expensive, although nowadays, the tasks are trivial in

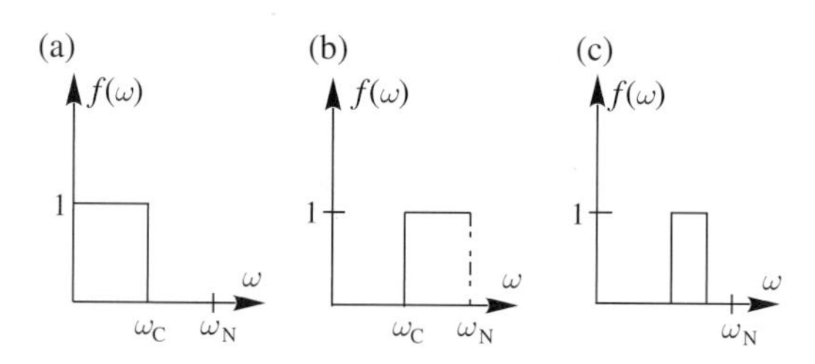

Fig. 3.4 Ideal filters: (a) Low-pass, (b) High-pass, and (c) Band-pass, where $f(\omega)$ is the filter 'response function', and ω_N is the Nyquist frequency. In (a), the low-frequency part of the signal is allowed to pass through the filter, while the high-frequency part is removed. In (b), the situation is reversed, while in (c), only frequencies within a selected band are allowed to pass through the filter. In reality, one tends to use filters where $f(\omega)$ does not exhibit abrupt behaviour (e.g. at the cut-off frequency ω_c).

computing costs, and filtering can be performed in either the frequency or the time domain.

A commonly used time domain filter is the 3-point moving average (or running mean) filter

$$\tilde{x}_n = \frac{1}{3}x_{n-1} + \frac{1}{3}x_n + \frac{1}{3}x_{n+1}, \tag{3.42}$$

i.e. average over the immediate neighbours. More generally, a filtered time series is given by

$$\tilde{x}_n = \sum_{l=-L}^{L} w_l \, x_{n+l}, \tag{3.43}$$

where w_l are the weights of the filter.

Suppose the filtered time series has the Fourier decomposition

$$\tilde{x}_n = \sum_{\omega} \tilde{X}(\omega) \, \mathrm{e}^{\mathrm{i}\omega t_n}. \tag{3.44}$$

Comparing with (3.41), one sees that

$$\tilde{X}(\omega) = f(\omega) X(\omega). \tag{3.45}$$

Thus

$$f(\omega) = \tilde{X}(\omega)/X(\omega) = \frac{\sum_l w_l \, \mathrm{e}^{\mathrm{i}\omega l \Delta t} X(\omega)}{X(\omega)}, \tag{3.46}$$

where we have used the fact that the Fourier transform is linear, so $\tilde{X}(\omega)$ is simply a linear combination of w_l times the Fourier transform of x_{n+l}. Hence, with $t_l = l\Delta t$,

$$f(\omega) = \sum_{l=-L}^{L} w_l \, \mathrm{e}^{\mathrm{i}\omega t_l}, \tag{3.47}$$

which allows us to calculate the filter response function $f(\omega)$ from the given weights of a time domain filter.

For example, moving average filters have the general form

$$\tilde{x}_n = \sum_{l=-L}^{L} \left(\frac{1}{2L+1}\right) x_{n+l}. \tag{3.48}$$

Another commonly used filter is the 3-point triangular filter,

$$\tilde{x}_n = \frac{1}{4}x_{n-1} + \frac{1}{2}x_n + \frac{1}{4}x_{n+1}, \tag{3.49}$$

which is better than the 3-point moving average filter in removing grid-scale noise (Emery and Thomson, 1997).

One often encounters time series containing strong periodic signals, e.g. the seasonal cycle or tidal cycles. While these periodic signals are important, it is often the non-periodic signals which have the most impact on humans, as they produce the unexpected events. One would often remove the strong periodic signals from the time series first.

Suppose one has monthly data for a variable x, and one would like to extract the seasonal cycle. Average all x values in January to get $\overline{x}_{\text{jan}}$, and similarly for the other months. The *climatological seasonal cycle* is then given by

$$\overline{x}_{\text{seasonal}} = [\overline{x}_{\text{jan}}, \dots, \overline{x}_{\text{dec}}]. \tag{3.50}$$

The filtered time series is obtained by subtracting this climatological seasonal cycle from the raw data – i.e. all January values of x will have $\overline{x}_{\text{jan}}$ subtracted, and similarly for the other months.

For tidal cycles, *harmonic analysis* is commonly used to extract the tidal cycles from a record of duration T. The tidal frequencies ω_n are known from astronomy, and one assumes the tidal signals are sinusoidal functions of amplitude A_n and phase θ_n. The best fit of the tidal cycle to the data is obtained by a least squares fit, i.e. minimize

$$\int_0^T [x(t) - \sum_n A_n \cos(\omega_n t + \theta_n)]^2 \, \mathrm{d}t, \tag{3.51}$$

by finding the optimal values of A_n and θ_n. If T is short, then tidal components with closely related frequencies cannot be separately resolved. A time series with the tides filtered is given by

$$\tilde{x}(t) = x(t) - \sum_n A_n \cos(\omega_n t + \theta_n). \tag{3.52}$$

3.4 Singular spectrum analysis

So far, our PCA involves finding eigenvectors containing spatial information. It is possible to use the PCA approach to incorporate time information into the eigenvectors. This method is known as *singular spectrum analysis* (SSA), or *time-PCA* (T-PCA) (Elsner and Tsonis, 1996; Ghil *et al.*, 2002).

Given a time series $y_j = y(t_j)$ $(j = 1, \dots, n)$, lagged copies of the time series are stacked to form the *augmented data matrix* $\mathbf{Y}$,

$$\mathbf{Y} = \begin{bmatrix} y_1 & y_2 & \cdots & y_{n-L+1} \\ y_2 & y_3 & \cdots & y_{n-L+2} \\ \vdots & \vdots & \vdots & \vdots \\ y_L & y_{L+1} & \cdots & y_n \end{bmatrix}. \tag{3.53}$$

This matrix has the same form as the data matrix produced by L variables, each being a time series of length $n - L + 1$. $\mathbf{Y}$ can also be viewed as composed of its column vectors $\mathbf{y}^{(l)}$, i.e.

$$\mathbf{Y} \equiv \left[\mathbf{y}^{(1)}|\mathbf{y}^{(2)}|\cdots|\mathbf{y}^{(n-L+1)}\right], \tag{3.54}$$

where

$$\mathbf{y}^{(l)} = \begin{bmatrix} y_l \\ y_{l+1} \\ \vdots \\ y_{l+L-1} \end{bmatrix}. \tag{3.55}$$

The vector space spanned by $\mathbf{y}^{(l)}$ is called the *delay coordinate space*. The number of lags L is usually taken to be at most $1/4$ of the total record length.

The standard PCA can be performed on $\mathbf{Y}$, resulting in

$$\mathbf{y}^{(l)} = \mathbf{y}(t_l) = \sum_j a_j(t_l)\,\mathbf{e}_j\,, \tag{3.56}$$

where a_j is the jth principal component (PC), a time series of length $n - L + 1$, and $\mathbf{e}_j$ is the jth eigenvector (or loading vector) of length L. Together, a_j and $\mathbf{e}_j$, represent the jth SSA mode. This method is called singular spectrum analysis, as it studies the ordered set (spectrum) of singular values (the square roots of the eigenvalues).

SSA has become popular in the field of dynamical systems (including chaos theory), where delay coordinates are commonly used. By lagging a time series, one is providing information on the first-order differencing of the discrete time series – with the first-order difference on discrete variables being the counterpart of the derivative. In the delay coordinate approach, repeated lags are supplied, thus information on the time series and its higher-order differences (hence derivatives) are provided.

The first SSA reconstructed component (RC) is the approximation of the original time series $y(t)$ by the first SSA mode. As the eigenvector $\mathbf{e}_1$ contains the loading over a range of lags, the first SSA mode, i.e. $a_1(t_l)\,\mathbf{e}_1$, provides an estimate for the y values over a range of lags starting from the time t_l. For instance, at time t_L, estimates of $y(t_L)$ can be obtained from any one of the delay coordinate vectors $\mathbf{y}^{(1)}, \ldots, \mathbf{y}^{(L)}$. Hence, each value in the reconstructed RC time series $\tilde{y}$ at time t_i involves averaging over the contributions at t_i from the L delay coordinate vectors which provide estimates of y at time t_i. Near the beginning or end of the record, the averaging will involve fewer than L terms. The RCs for higher modes are similarly defined.

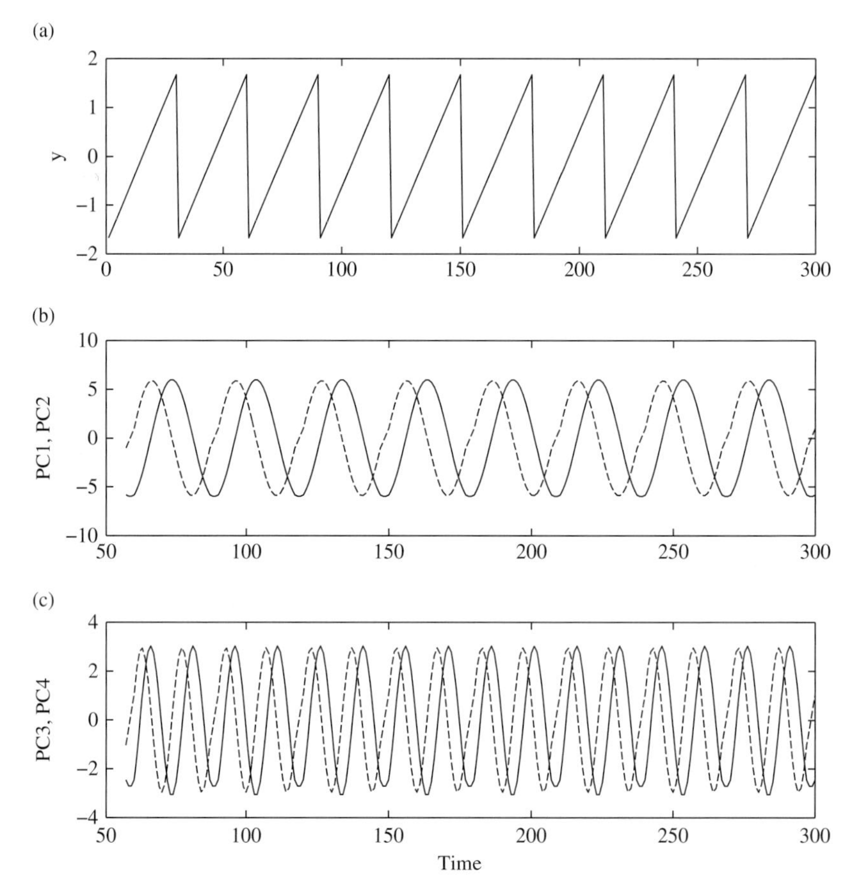

Fig. 3.5 (a) Sawtooth wave signal used for the SSA analysis. (b) The SSA PC1 (solid) and PC2 (dashed). (c) The SSA PC3 (solid) and PC4 (dashed).

A comparison with the Fourier spectral analysis is in order: unlike Fourier spectral analysis, SSA does not in general assume the time series to be periodic; hence, there is no need to taper the ends of the time series as commonly done in Fourier spectral analysis. As the wave forms extracted from the SSA eigenvectors are not restricted to sinusoidal shapes, the SSA can in principle capture an anharmonic wave more efficiently than the Fourier method. However, in many cases, the SSA eigenvectors may turn out to be not very different from sinusoidal-shaped functions. For instance, let us analyze the sawtooth wave in Fig. 3.5. The first four SSA PCs are shown in Fig. 3.5, and the corresponding SSA eigenvectors in Fig. 3.6. The PCs and the eigenvectors are paired, where the paired members have very similar amplitudes, but are phase shifted by about 90° (i.e. in quadrature). That the modes naturally occur in quadratured pairs should not be too surprising, since even in Fourier decomposition, each frequency band in general requires a sine

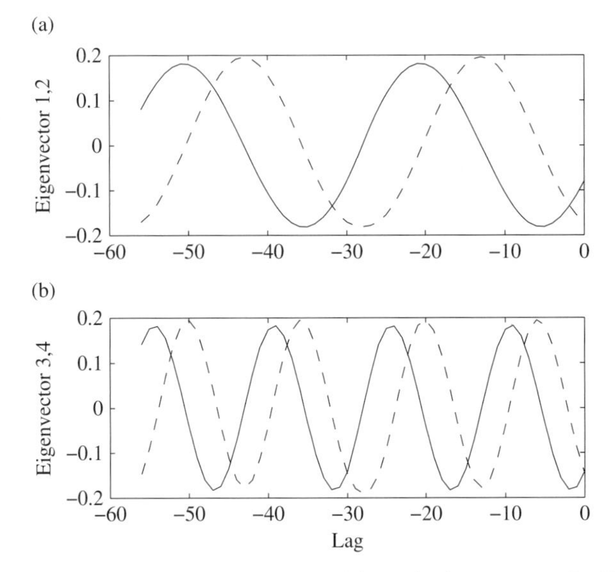

Fig. 3.6 (a) The SSA eigenvector 1 (solid) and eigenvector 2 (dashed). (b) Eigenvector 3 (solid) and 4 (dashed).

and a cosine function. One might have expected the eigenvectors to have approximately sawtooth shape, but they are visually indistinguishable from sinuosidal functions. The second pair of modes has double the frequency of the first pair, corresponding to the first harmonic above the fundamental frequency. The first pair of SSA modes captured 61.3% of the variance, while the second pair captured 15.4%. When Fourier spectral analysis was used instead, the fundamental frequency band captured 61.1%, while the first harmonic band captured 15.4% of the variance. Hence in this example, the SSA modes are nearly identical to the Fourier modes.

Next, Gaussian random noise is added to the sawtooth wave, with the standard deviation of the noise set to be the same as the standard deviation of the sawtooth wave. The SSA modes extracted (not shown) are similar to those in Fig. 3.5 and 3.6, but the variance accounted for is obviously reduced. For the first pair of SSA modes, the variance accounted for was 34.0%, versus 32.7% from the Fourier approach.

Thus for a single time series, while SSA can in principle extract more variance in its first pair of modes than the Fourier fundamental frequency band, as SSA is not restricted to using sinusoidal shaped functions, the advantage of SSA over the Fourier represention may turn out to be minor, and the SSA eigenvectors may turn out to have shapes resembling sinusoidal functions, even when the signal is as anharmonic as a sawtooth wave, or a square wave. For a highly anharmonic signal such as the stretched square wave described by (10.44), the SSA approach is clearly better than the Fourier approach (Section 10.6). Furthermore, in the real world, except for the seasonal cycle and tidal cycles, signals tend not to have a precise frequency like the sawtooth wave. Let us take the El Niño-Southern Oscillation (ENSO) as our next example.

The *Southern Oscillation Index* (SOI) is defined as the normalized air pressure difference between Tahiti and Darwin. There are several slight variations in the SOI values calculated at various centres; here we use the SOI calculated by the Climate Research Unit at the University of East Anglia, based on the method of Ropelewski and Jones (1987). The SOI measures the seesaw oscillations of the sea level air pressure between the eastern and western equatorial Pacific. When the

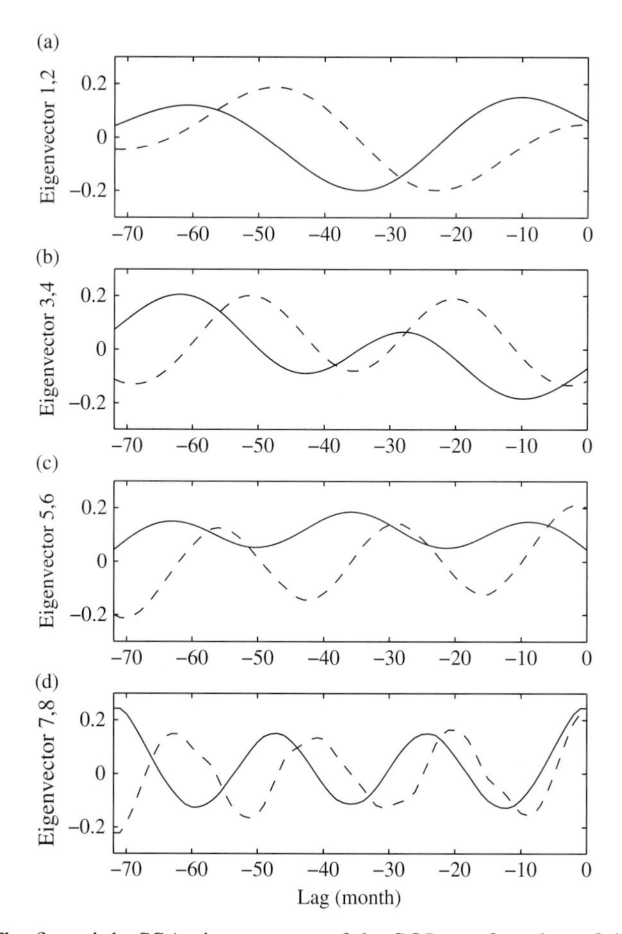

Fig. 3.7 The first eight SSA eigenvectors of the SOI as a function of time lag. (a) Mode 1 (solid curve) and mode 2 (dashed curve). (b) Mode 3 (solid) and mode 4 (dashed). (c) Mode 5 (solid) and mode 6 (dashed). (d) Mode 7 (solid) and mode 8 (dashed).

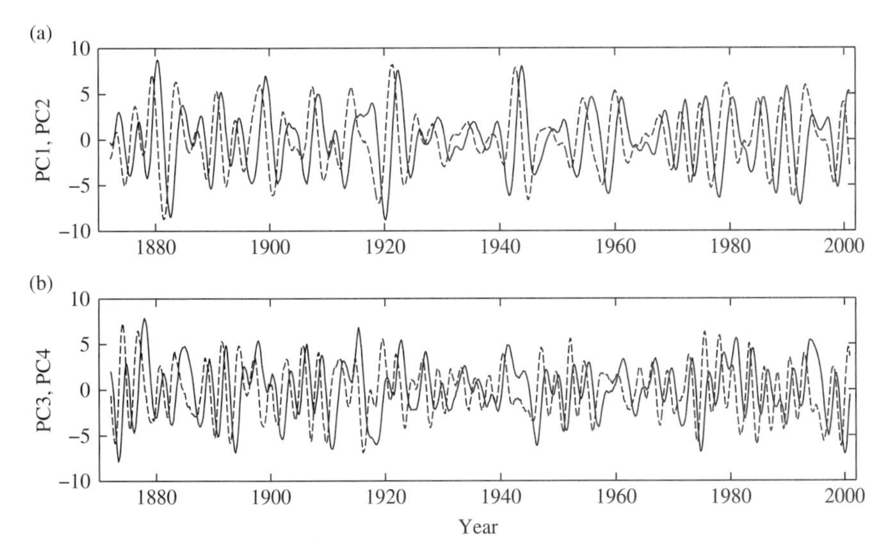

Fig. 3.8 The PC time series of (a) SSA mode 1 (solid curve) and mode 2 (dashed curve) for the SOI. (b) Mode 3 (solid) and mode 4 (dashed).

SOI is strongly negative, the eastern equatorial Pacific sea surface also becomes warm (i.e. an El Niño episode occurs); when SOI is strongly positive, the central equatorial Pacific becomes cool (i.e. a La Niña episode occurs). The SOI is known to have the main spectral peak at a period of about 4–5 years (Troup, 1965). For SSA, the window L needs to be long enough to accommodate this main spectral period, hence $L = 72$ months was chosen for the SSA.

The monthly SOI time series, from January 1866 to December 2000, was analyzed by the SSA (Hsieh and Wu, 2002), with the first eight eigenvectors shown in Fig. 3.7. These first eight modes account for 11.0, 10.5, 9.0, 8.1, 6.4, 5.6, 3.3 and 2.0%, respectively, of the variance of the SOI. In contrast, because the SO phenomenon does not have a precise frequency, Fourier analysis led to energy being spread between many frequency bands, with the strongest band accounting for only 4.0% of the variance, far less than the 11.0% of the first SSA mode. The first pair of modes displays oscillations with a period slightly above 50 months, while the next pair of modes manifests a period slightly above 30 months (Fig. 3.7). The oscillations also display anharmonic features. The four leading PCs are plotted in pairs in Fig. 3.8, revealing the longer oscillatory time scale in the first pair and the shorter time scale in the second pair.

Another advantage of SSA over the Fourier approach lies in the multivariate situation – the Fourier approach does not generalize naturally to large multivariate datasets, whereas the SSA, based on the PCA method, does.

3.5 Multichannel singular spectrum analysis

In the multivariate case, there are m variables $y_k(t_j) \equiv y_{kj}$, $(k = 1, \dots, m;\ j = 1, \dots, n)$. The data matrix time lagged by l $(l = 0, 1, 2, \dots, L-1)$ is defined by

$$\mathbf{Y}_{(l)} = \begin{bmatrix} y_{1,1+l} & \cdots & y_{1,n-L+l+1} \\ \vdots & \ddots & \vdots \\ y_{m,1+l} & \cdots & y_{m,n-L+l+1} \end{bmatrix}. \tag{3.57}$$

Again, treat the time lagged data as extra variables, so the augmented data matrix is

$$\mathbf{Y} = \begin{bmatrix} \mathbf{Y}_{(0)} \\ \mathbf{Y}_{(1)} \\ \mathbf{Y}_{(2)} \\ \vdots \\ \mathbf{Y}_{(L-1)} \end{bmatrix}, \tag{3.58}$$

i.e.

$$\mathbf{Y} = \begin{bmatrix} y_{11} & y_{12} & \cdots & y_{1,n-L+1} \\ \vdots & \vdots & \vdots & \vdots \\ y_{m1} & y_{m2} & \cdots & y_{m,n-L+1} \\ \vdots & \vdots & \vdots & \vdots \\ y_{1L} & y_{1,L+1} & \cdots & y_{1n} \\ \vdots & \vdots & \vdots & \vdots \\ y_{mL} & y_{m,L+1} & \cdots & y_{mn} \end{bmatrix}. \tag{3.59}$$

Note that PCA can again be applied to the augmented data matrix $\mathbf{Y}$ to get the SSA modes.

If there is more than one variable, the method is called the *space–time PCA* (ST–PCA), or *multichannel singular spectrum analysis* (MSSA). In this book, we will, for brevity, use the term SSA to denote both the univariate and the multivariate cases. The term *extended empirical orthogonal function* (EEOF) analysis is also used in the literature, especially when the number of lags (L) is small. So far, we have assumed the time series was lagged one time step at a time. To save computational time, larger lag intervals can be used, i.e. lags can be taken over several time steps at a time. Monahan *et al.* (1999) pointed out that the ST–PCA can have degenerate modes for some choices of lag interval.

For an example of the application of SSA to a larger multivariate dataset, consider the analysis of the tropical Pacific monthly sea surface temperature anomalies (SSTA) data from 1950–2000 by Hsieh and Wu (2002), where the climatological seasonal cycle and the linear trend have been removed from the SST data to give the SSTA. As we want to resolve the ENSO variability, a window of 73 months was chosen. With a lag interval of 3 months, the original plus 24 lagged copies of the SSTA data formed the augmented SSTA dataset. (Note that if a lag interval of 1 month were used instead, then to cover the window of 73 months, the original plus 72 copies of the SSTA data would have produced a much bigger augmented data matrix.) The first eight SSA modes respectively explain 12.4%, 11.7%, 7.1%, 6.7%, 5.4%, 4.4%, 3.5% and 2.8% of the total variance of the augmented dataset (with the first six modes shown in Fig. 3.9). The first two modes have space–time eigenvectors (i.e. loading patterns) showing an oscillatory time scale of about 48 months, comparable to the ENSO time scale, with the mode 1 anomaly pattern occurring about 12 months before a very similar mode 2 pattern, i.e. the two patterns are in quadrature. The PC time series also show similar time scales for modes 1 and 2. Modes 3 and 5 show longer time scale fluctuations, while modes 4 and 6 show shorter time scale fluctuations – around the 30 month time scale.

Similarly, the tropical Pacific monthly sea level pressure anomalies (SLPA) have been analyzed. The first eight SSA modes of the SLPA accounted for 7.9%, 7.1%, 5.0%, 4.9%, 4.0%, 3.1%, 2.5% and 1.9% respectively, of the total variance of the augmented data. The first two modes displayed the Southern Oscillation (SO), the east–west seesaw of SLPA at around the 50-month period, while the higher modes displayed fluctuations at around the average period of 28 months (Fig. 3.10).

3.6 Principal oscillation patterns

For some datasets containing multiple time series, one would like to find a low order linear dynamical system to account for the behaviour of the data. The *principal oscillation pattern* (POP) method, proposed by Hasselmann (1988), is one such technique.

Consider a simple system with two variables y_1 and y_2, obeying the linear dynamical equations

$$\frac{\mathrm{d}y_1}{\mathrm{d}t} = L_1(y_1, y_2), \qquad \frac{\mathrm{d}y_2}{\mathrm{d}t} = L_2(y_1, y_2), \tag{3.60}$$

where L_1 and L_2 are linear functions. The discretized version of the dynamical equations is of the form

$$y_1(t+1) = a_{11}y_1(t) + a_{12}y_2(t), \tag{3.61}$$

$$y_2(t+1) = a_{21}y_1(t) + a_{22}y_2(t), \tag{3.62}$$

where the a_{ij} are parameters. For an m-variable first order linear dynamical system, the discretized governing equations can be expressed as

$$\mathbf{y}(t+1) = \mathbf{A}\mathbf{y}(t), \tag{3.63}$$

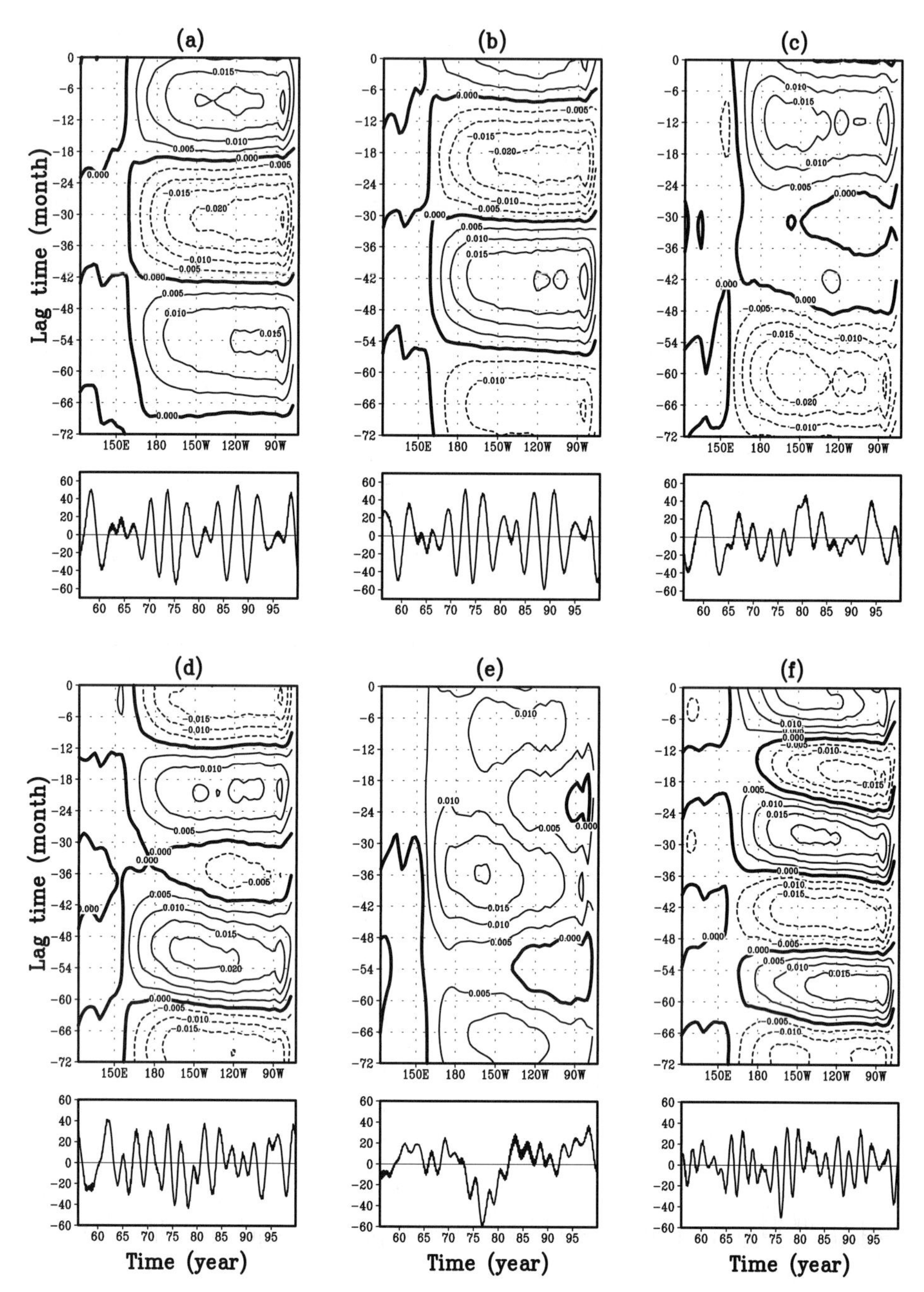

Fig. 3.9 The SSA modes 1–6 for the tropical Pacific SSTA are shown in (a)–(f), respectively. The contour plots display the space–time eigenvectors (loading patterns), showing the SSTA along the equator as a function of the lag. Solid contours indicate positive anomalies and dashed contours, negative anomalies, with the zero contour indicated by the thick solid curve. In a separate panel beneath each contour plot, the principal component (PC) of each SSA mode is also plotted as a time series. The time of the PC is synchronized to the lag time of 0 month in the space–time eigenvector. (Reproduced from Hsieh and Wu (2002) with permission of the American Geophysical Union.)

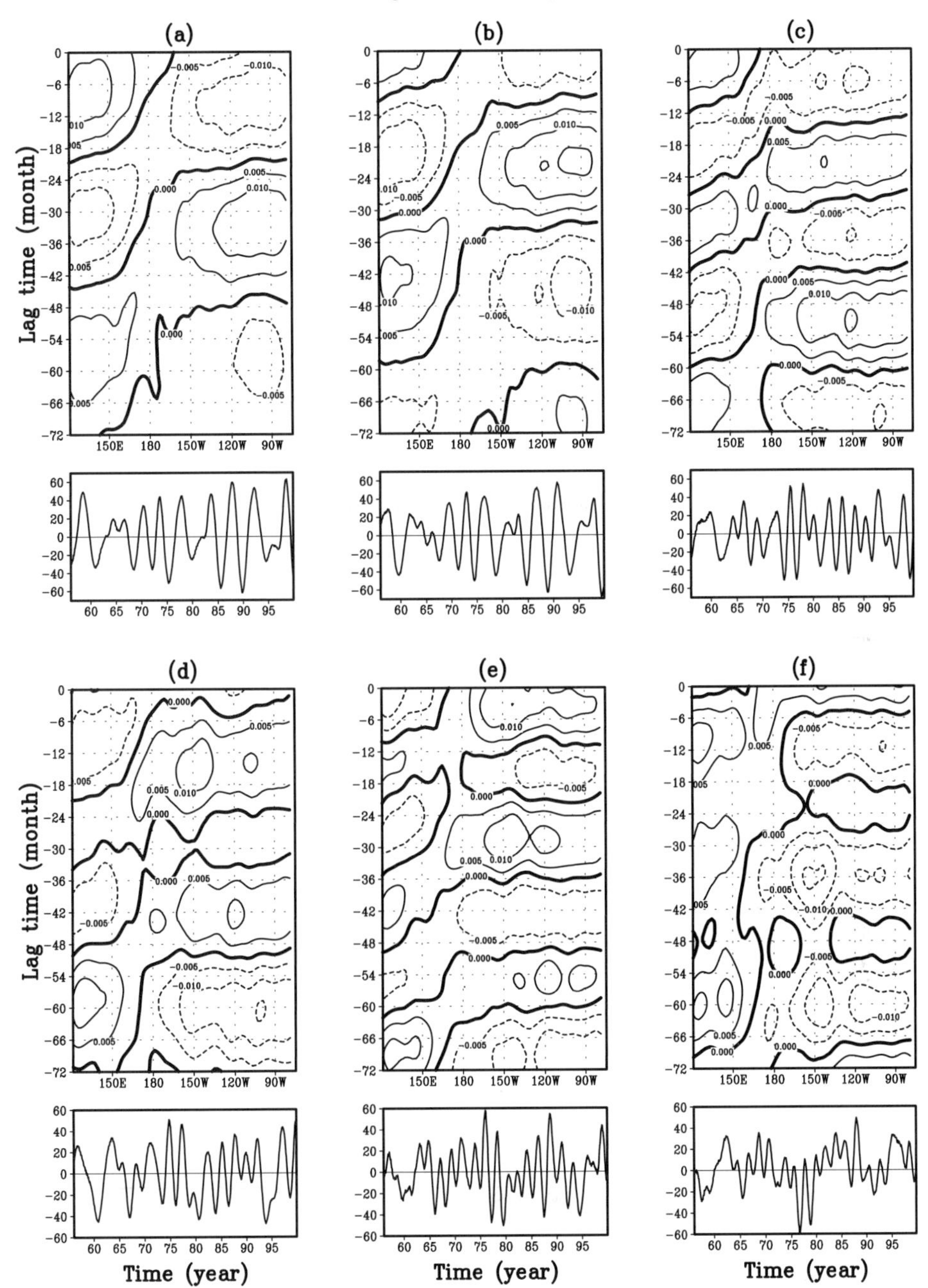

Fig. 3.10 The SSA modes 1–6 for the tropical Pacific SLPA are shown in (a)–(f), respectively. The contour plots display the space–time eigenvectors, showing the SLPA along the equator as a function of the lag. The PC of each SSA mode is plotted as a time series beneath each contour plot. (Reproduced from Hsieh (2004) with permission of the American Meteorological Society.)

where $\mathbf{y}$ is an m-element column vector, and $\mathbf{A}$ is an $m \times m$ matrix. Note that $\mathbf{A}$ is a real matrix, but generally not symmetric; hence, its eigenvalues λ and eigenvectors $\mathbf{p}$ are in general complex. Taking the complex conjugate of the eigenvector equation

$$\mathbf{A}\mathbf{p} = \lambda\mathbf{p}, \tag{3.64}$$

yields $\mathbf{A}\mathbf{p}^* = \lambda^*\mathbf{p}^*$, (as $\mathbf{A}$ is real), which means λ^* and $\mathbf{p}^*$ are also eigenvalues and eigenvectors of $\mathbf{A}$. Assuming that the eigenvalues are distinct and non-zero, the eigenvectors form a linear basis.

We introduce a matrix $\mathbf{P}$, where its jth column is simply the jth eigenvector $\mathbf{p}_j$, i.e.

$$\mathbf{P} = [\mathbf{p}_1|\mathbf{p}_2|\cdots|\mathbf{p}_m], \tag{3.65}$$

then

$$\mathbf{P}^{-1}\mathbf{A}\mathbf{P} = \mathbf{\Lambda} = \begin{bmatrix} \lambda_1 & 0 & \cdots & 0 \\ 0 & \lambda_2 & 0 & \cdots \\ \vdots & \vdots & \ddots & \vdots \\ 0 & \cdots & 0 & \lambda_m \end{bmatrix}, \tag{3.66}$$

where the matrix $\mathbf{\Lambda}$ has non-zero elements only along the diagonal, made up of the eigenvalues $\lambda_1, \ldots, \lambda_m$.

Applying $\mathbf{P}^{-1}$ to (3.63) yields

$$\mathbf{P}^{-1}\mathbf{y}(t+1) = \mathbf{P}^{-1}\mathbf{A}\mathbf{y}(t) = \mathbf{P}^{-1}\mathbf{A}\mathbf{P}\,\mathbf{P}^{-1}\mathbf{y}(t), \tag{3.67}$$

which can be expressed as

$$\mathbf{z}(t+1) = \mathbf{\Lambda}\,\mathbf{z}(t), \tag{3.68}$$

where

$$\mathbf{z} = \mathbf{P}^{-1}\mathbf{y}, \tag{3.69}$$

and (3.66) has been invoked.

By applying $\mathbf{P}$ to (3.69), the inverse transform is obtained:

$$\mathbf{y}(t) = \mathbf{P}\,\mathbf{z}(t), \quad \text{i.e.} \tag{3.70}$$

$$\mathbf{y}(t) = \sum_{j=1}^{m} \mathbf{p}_j\, z_j(t). \tag{3.71}$$

The eigenvector

$$\mathbf{p}_j = \mathbf{p}_j^{\mathrm{r}} + \mathrm{i}\mathbf{p}_j^{\mathrm{i}}, \tag{3.72}$$

is called a Principal Oscillation Pattern (POP) of $\mathbf{y}(t)$.

The corresponding POP coefficient

$$z_j(t) = z_j^{\mathrm{r}}(t) + \mathrm{i}z_j^{\mathrm{i}}(t), \tag{3.73}$$

obeys (3.68), i.e.

$$z_j(t+1) = \lambda_j\, z_j(t), \tag{3.74}$$

where

$$\lambda_j = |\lambda_j|\mathrm{e}^{\mathrm{i}\theta_j} \equiv \mathrm{e}^{-1/\tau_j}\,\mathrm{e}^{\mathrm{i}2\pi/T_j}, \tag{3.75}$$

as $|\lambda_j| < 1$ in the real world (von Storch and Zwiers, 1999, pp. 336–337). Hence

$$z_j(t+1) = \mathrm{e}^{-1/\tau_j}\,\mathrm{e}^{\mathrm{i}2\pi/T_j}\,z_j(t), \tag{3.76}$$

where τ_j is an e-folding decay time scale, and T_j is the oscillatory period, i.e. z_j will evolve in time displaying exponential decay and oscillatory behaviour, governed by the parameters τ_j and T_j, respectively. As $\mathbf{y}(t)$ is real, (3.70) gives

$$\mathbf{y}(t) = \mathrm{Re}[\mathbf{Pz}(t)] = \sum_j [\mathbf{p}_j^{\mathrm{r}} z_j^{\mathrm{r}}(t) - \mathbf{p}_j^{\mathrm{i}} z_j^{\mathrm{i}}(t)]. \tag{3.77}$$

As t progresses, the signs of z_j^{r} and z_j^{i} oscillate, resulting in an evolving $\mathbf{p}_j^{\mathrm{r}} z_j^{\mathrm{r}} - \mathbf{p}_j^{\mathrm{i}} z_j^{\mathrm{i}}$ pattern:

$$z_j^{\mathrm{r}} : \cdots \to + \to 0 \to - \to 0 \to + \to \cdots, \tag{3.78}$$

$$z_j^{\mathrm{i}} : \cdots \to 0 \to + \to 0 \to - \to 0 \to \cdots, \tag{3.79}$$

$$\mathbf{p}_j^{\mathrm{r}} z_j^{\mathrm{r}} - \mathbf{p}_j^{\mathrm{i}} z_j^{\mathrm{i}} : \cdots \to \mathbf{p}_j^{\mathrm{r}} \to -\mathbf{p}_j^{\mathrm{i}} \to -\mathbf{p}_j^{\mathrm{r}} \to \mathbf{p}_j^{\mathrm{i}} \to \mathbf{p}_j^{\mathrm{r}} \to \cdots. \tag{3.80}$$

Unlike PCA, POP allows representation of propagating waves using only one POP mode. For instance, Fig. 3.11(a) shows a POP pattern representing a wave propagating from right to left, while Fig. 3.11(b) represents a pair of eddies rotating around the centre of the figure. An arbitrary phase α_j exists in the POP representation,

$$\mathbf{y} = \sum_j \mathbf{p}_j \mathrm{e}^{\mathrm{i}\alpha_j}\,\mathrm{e}^{-\mathrm{i}\alpha_j}\,z_j = \sum_j \mathbf{p}'_j z'_j\,; \tag{3.81}$$

this phase can be adjusted to make the interpretation of the jth mode easier (as shown in an example below).

For real data, noise ϵ must be added to the dynamical system,

$$\mathbf{y}(t+1) = \mathbf{A}\mathbf{y}(t) + \epsilon. \tag{3.82}$$

From E[(3.82) $\mathbf{y}^{\mathrm{T}}(t)$] and the fact that E[$\epsilon$]= 0, one can estimate $\mathbf{A}$ from

$$\mathbf{A} = \mathrm{E}[\mathbf{y}(t+1)\,\mathbf{y}^{\mathrm{T}}(t)]\,\{\mathrm{E}[\mathbf{y}(t)\,\mathbf{y}^{\mathrm{T}}(t)]\}^{-1}. \tag{3.83}$$

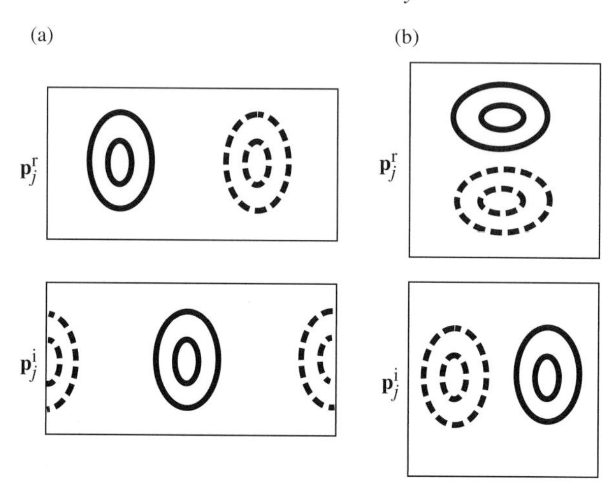

Fig. 3.11 Illustrating how the POP mode evolves with time, manifesting the patterns: $\cdots \to \mathbf{p}_j^r \to -\mathbf{p}_j^i \to -\mathbf{p}_j^r \to \mathbf{p}_j^i \to \mathbf{p}_j^r \to \cdots$. In (a), a wave will propagate to the left, while in (b), the pair of anomalies will rotate counterclockwise around each other, where positive and negative anomalies are denoted by solid and dashed contours.

From $\mathbf{A}$, one can compute the eigenvectors $\mathbf{p}_j$, eigenvalues λ_j, and z_j from (3.69).

When the dataset is large, one often prefilters with the PCA, i.e. the $\mathbf{y}(t)$ are the first few leading PCs. Then POP is performed on these PC time series.

The POP technique has been used to study the tropical atmospheric 30–60 day oscillation, a.k.a. the Madden-Julian Oscillation (MJO) (von Storch *et al.*, 1988), the equatorial stratospheric Quasi-Biennial Oscillation (QBO) (Xu, 1992), the El Niño-Southern Oscillation (ENSO) (Xu and von Storch, 1990; Tang, 1995), and Arctic variability (Tang *et al.*, 1994). Penland and Magorian (1993) combined many POP modes to forecast the tropical Pacific sea surface temperatures, and called the approach *linear inverse modelling*. Note that POP has been reviewed by von Storch *et al.* (1995) and von Storch and Zwiers (1999).

The analysis of the tropical Pacific wind stress by Tang (1995) provided a good illustration of the POP technique. From a combined PCA (Section 2.3) of the wind stress, he took the first nine PC time series for his POP analysis, as using more or fewer PCs reduced the forecast skill of the POP model. Choosing λ_1 to be the λ_j with the largest $|\lambda_j|$, which corresponded to the longest decay time, he obtained $\tau_1 = 12$ months, and $T_1 = 41$ months. These decay and oscillatory time scales are consistent with the observed behaviour of the ENSO oscillation, where El Niño warm episodes appear every few years, decaying away in about a year. The arbitrary phase α_1 was chosen so that z_1^r was positive and $z_1^i = 0$ in December, 1982, during the peak of a major El Niño (Fig. 3.12). As z_1^r is correlated at 0.76 with the SST anomalies in the Niño 3 region (Fig. 2.3) in the equatorial eastern Pacific, it is

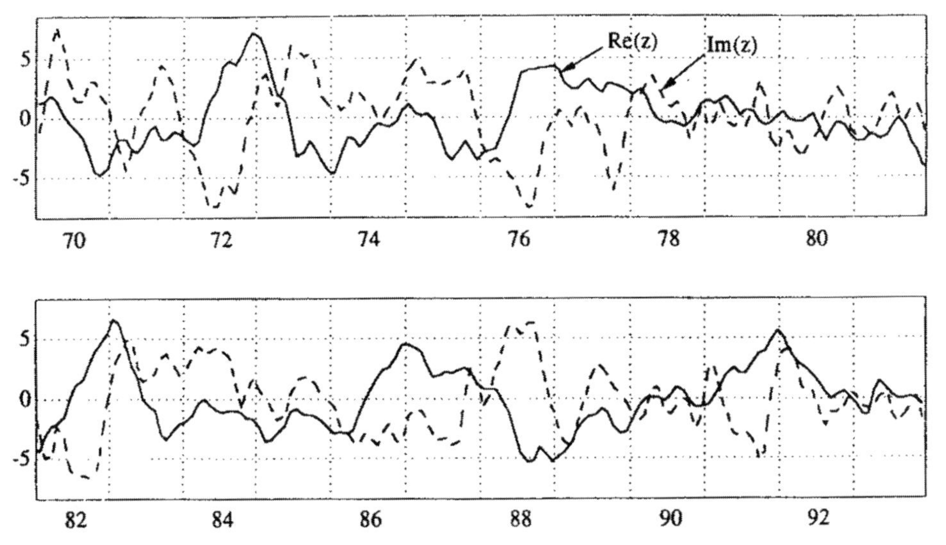

Fig. 3.12 The real (solid) and imaginary (dashed) components of z_1, the POP coefficient for the tropical Pacific wind stress anomaly field from 1970 to 1993. The real part of z_1 is high during the warm episodes of the El Niño-Southern Oscillation (ENSO). (Reproduced from Tang (1995, Fig. 1) with permission of the American Meteorological Society.)

an ENSO POP. The real part of the first eigenvector shows the wind stress anomalies during the peak of a warm episode, while the imaginary part, displaying the anomalies two to three seasons before the peak of a warm episode, can be regarded as the precursor pattern (Fig. 3.13). The evolution of the POP coefficient z_1 with time on the complex plane is shown for the years 1972, 1982, 1985 and 1988 in Fig. 3.14. Major El Niño warming developed in 1972 and 1982, and a major La Niña cooling in 1988, while 1985 is a neutral year. It is evident that the El Niño episodes evolved into the lower-right corner of the complex plane, while the La Niña episode, the upper-left corner. In a neutral year such as 1985, the amplitude of z_1 is small. Tang (1995) defined precursor areas for warm and cold episodes in the complex plane – once the system enters one of the precursor regions, it evolves to a warm or cold episode over the next few months. Hence, by monitoring whether the POP coefficient enters the precursor regions, one can forecast a warm episode or a cold episode.

From applying (3.74) k times, one can forecast the POP coefficient at time t_0 into the future by k time steps:

$$z_1(t_0 + k) = (\lambda_1)^k z_1(t_0) = \mathrm{e}^{-k/\tau_1}\, \mathrm{e}^{\mathrm{i}2\pi k/T_1} z_1(t_0), \tag{3.84}$$

although Tang (1995) found somewhat better ENSO forecast skills with the exponential decay term omitted from the right hand side of (3.84).

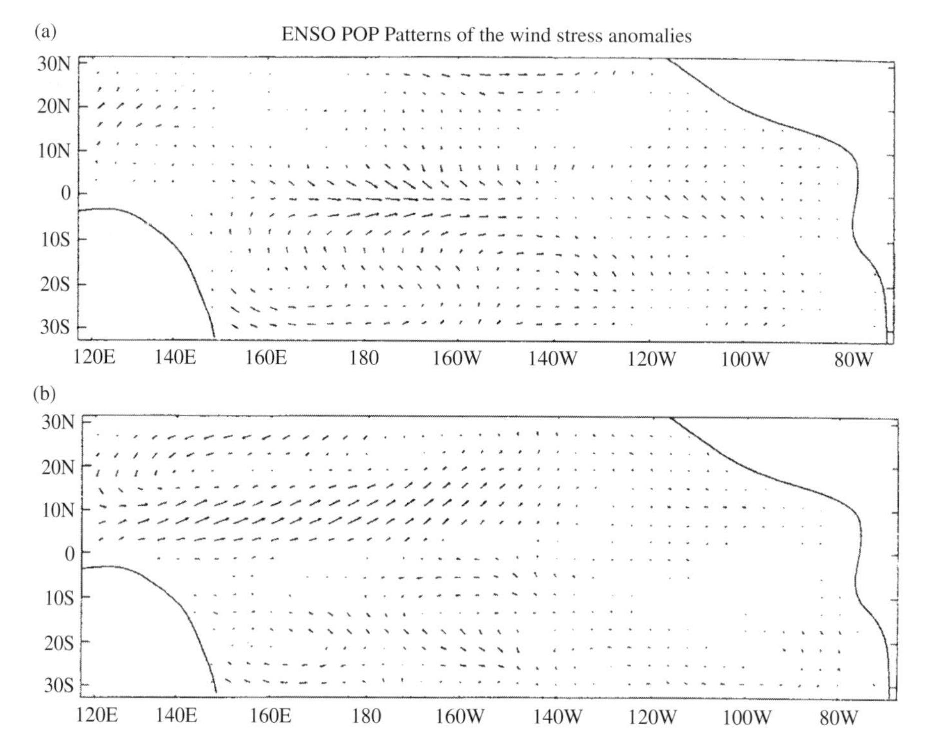

Fig. 3.13 The first eigenvector $\mathbf{p}_1$ from the POP analysis of the tropical Pacific wind stress anomalies. (a) The real part of $\mathbf{p}_1$, which shows up during the peak of a warm episode, thus giving the pattern of wind stress anomalies (as indicated by the small vectors at the grid points) during the peak of a warm episode. (b) The imaginary part of $\mathbf{p}_1$, which appears two to three seasons before the peak of a warm episode. (Reproduced from Tang (1995, Fig. 4) with permission of the American Meteorological Society.)

3.7 Spectral principal component analysis

The theory for generalizing the PCA technique to the frequency domain was developed by Wallace and Dickinson (1972). The technique has been used to analyze tropical wave disturbances (Wallace, 1972) and tropical wind systems (Barnett, 1983), and has been reviewed by Horel (1984) and von Storch and Zwiers (1999). A scalar field $y(t)$ (with mean removed) has the Fourier representation

$$y(t) = \sum_l [a_l \cos(\omega_l t) + b_l \sin(\omega_l t)]. \tag{3.85}$$

Let y^{h} be the Hilbert transform of y, defined as

$$y^{\mathrm{h}}(t) = \sum_l \left[a_l \cos\left(\omega_l t + \frac{\pi}{2}\right) + b_l \sin\left(\omega_l t + \frac{\pi}{2}\right)\right], \tag{3.86}$$

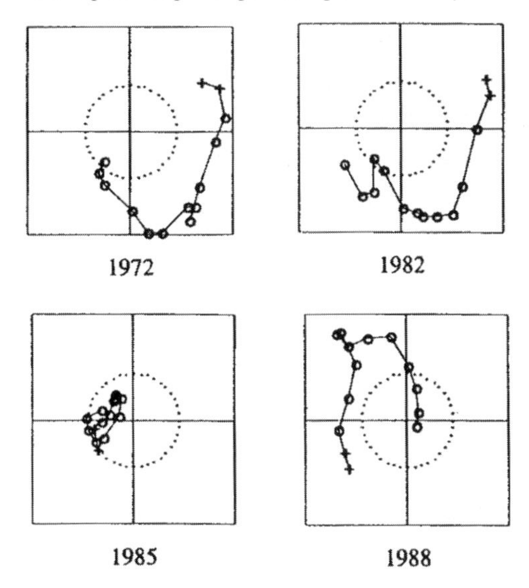

Fig. 3.14 The POP coefficient z_1 shown in the complex plane for the months in the years 1972, 1982, 1985 and 1988. The horizontal and vertical axes represent the real and imaginary components of z_1. Each o represents 1 month, and the two + in each plane are for the January and February of the following year. The dotted circle is $|z| = 3.42$. (Reproduced from Tang (1995, Fig. 2) with permission of the American Meteorological Society.)

which supplies the 90° out of phase information. This can be rewritten as

$$y^{\mathrm{h}}(t) = \sum_{m} [\, b_l \cos(\omega_l t) - a_l \sin(\omega_l t)]. \tag{3.87}$$

As

$$\frac{\mathrm{d}}{\mathrm{d}t}[a_l \cos(\omega_l t) + b_l \sin(\omega_l t)] = \omega_l [b_l \cos(\omega_l t) - a_l \sin(\omega_l t)], \tag{3.88}$$

this means $y^{\mathrm{h}}(t)$ supplies information on the time derivative of y. Let

$$y^{\mathrm{H}}(t) = y(t) + \mathrm{i}\, y^{\mathrm{h}}(t). \tag{3.89}$$

This complex y^{H} supplies information on y and its time derivative, reminiscent of Hamiltonian mechanics, where the canonical variables are position and momentum.

With m stations, we have $y_j(t),\ j = 1, \ldots, m$. Using (3.85) and (3.87) to represent y_j and $y_j^{\mathrm{h}}(t)$ respectively, we have

$$y_j^{\mathrm{H}}(t) = \sum_{l} \left[(a_{jl} + \mathrm{i} b_{jl}) \cos(\omega_l t) + (b_{jl} - \mathrm{i} a_{jl}) \sin(\omega_l t) \right]. \tag{3.90}$$

To perform PCA in the frequency domain, the procedure is as follows: as with spectral methods, it is advisable first to taper the ends of the time series before performing Fourier transform, as discussed in Section 3.2 on windows. After Fourier transforming $y_j^{\mathrm{H}}(t)$, select a given frequency band, and integrate (or sum) the spectral components $\hat{y}_j^{\mathrm{H}}(\omega_m)$ over the frequency band. Generate the cross-spectral matrix $\mathbf{C}$ of the particular frequency band,

$$C_{ij} = \frac{N\Delta t}{4\pi} \hat{y}_i^{\mathrm{H}*} \hat{y}_j^{\mathrm{H}}, \tag{3.91}$$

where the cross-spectrum can be regarded as basically the (complex) covariance matrix for this frequency band. This method is the *spectral PCA*. The method has also been used with all frequencies retained – then the integration over all frequency bands yields the covariance matrix.

Apply PCA to extract the (complex) eigenvectors $\mathbf{e}_j$ of the cross-spectral matrix. This method is called the *Hilbert PCA*, or sometimes *complex PCA*, though this latter name is ambiguous as PCA is also applied to complex variables (e.g. the horizontal wind $w = u + \mathrm{i}v$ in Section 2.3), whereas with Hilbert PCA, we are interested in analyzing real variables, though the real variables are first complexified by the Hilbert transform (3.89), before being analyzed by PCA.

The complex time series $y_j^{\mathrm{H}}(t)$ can be expanded in terms of the eigenvectors

$$\mathbf{y}^{\mathrm{H}}(t) = \sum_j \mathbf{e}_j \mathbf{a}_j^{*\mathrm{T}}, \tag{3.92}$$

where the time coefficients $\mathbf{a}_j$ for the jth mode are generally complex, and the superscript ${}^{*\mathrm{T}}$ denotes the complex conjugate transpose. Let us focus on the contributions by the jth mode only. Taking the real part, we have

$$\mathbf{y}(t) = \mathbf{e}_j^{\mathrm{R}} a_j^{\mathrm{R}}(t) - \mathbf{e}_j^{\mathrm{I}} a_j^{\mathrm{I}}(t), \tag{3.93}$$

where the superscripts R and I denote the real and imaginary parts, respectively. Since an arbitrary phase factor $\mathrm{e}^{\mathrm{i}\phi_j}$ can be multiplied to $\mathbf{e}_j$ and $\mathrm{e}^{-\mathrm{i}\phi_j}$ to $\mathbf{a}_j$ without changing (3.92), one may choose the arbitrary phase to ensure zero correlation between $a_j^{\mathrm{R}}(t)$ and $a_j^{\mathrm{I}}(t)$. Then as $a_j^{\mathrm{R}}(t)$ and $a_j^{\mathrm{I}}(t)$ take turns manifesting themselves, the spatial patterns $\mathbf{e}_j^{\mathrm{R}}$ and $\mathbf{e}_j^{\mathrm{I}}$ also take turns manifesting themselves. Suppose, as t progresses, we have at t_1, $a_j^{\mathrm{R}}(t) > 0$, $a_j^{\mathrm{I}}(t) = 0$; at t_2, $a_j^{\mathrm{R}}(t) = 0$, $a_j^{\mathrm{I}}(t) < 0$; at t_3, $a_j^{\mathrm{R}}(t) < 0$, $a_j^{\mathrm{I}}(t) = 0$; and at t_4, $a_j^{\mathrm{R}}(t) = 0$, $a_j^{\mathrm{I}}(t) > 0$, then the evolution of the spatial patterns of $\mathbf{y}$ at these four times will be $\mathbf{e}_j^{\mathrm{R}}(t) \to -\mathbf{e}_j^{\mathrm{I}}(t) \to -\mathbf{e}_j^{\mathrm{R}}(t) \to \mathbf{e}_j^{\mathrm{I}}(t)$. This sequence of spatial patterns allows the representation of propagating waves, somewhat analogous to the POP technique in Section 3.6.

For vector fields, e.g. $[u_j(t), v_j(t)]$ at $j = 1, \ldots, m$ stations, let

$$u_j^{\mathrm{H}}(t) = u_j(t) + \mathrm{i}\, u_j^{\mathrm{h}}(t), \qquad v_j^{\mathrm{H}}(t) = v_j(t) + \mathrm{i}\, v_j^{\mathrm{h}}(t). \tag{3.94}$$

Treat u_j^{H} and v_j^{H} as separate variables, i.e. treat the dataset as though there are $2m$ stations each contributing a scalar variable. This approach obviously generalizes to higher dimensional vector fields.

Exercises

(3.1) Fourier spectral analysis is performed on hourly sea level height data. The main tidal period is around 12 hours, but there are actually two components, M_2 from the moon at 12.42 hours, and S_2 from the sun at 12.00 hours. What is the minimum length of the data record required in order to see two distinct peaks around 12 hours in your spectrum due to the M_2 and S_2 tidal components?

(3.2) If the data are dominated by grid-scale noise, i.e. the data in 1-dimension have adjacent grid points simply flipping signs like $-1, +1, -1, +1, -1, +1, \ldots$, show that one application of the triangular filter (3.49) eliminates the grid-scale noise completely, whereas one application of the 3-point moving-average filter (3.42) does not.

(3.3) The POP (Principal Oscillation Pattern) method was used to analyze the evolution of monthly sea surface temperature anomalies. If the time step in the POP analysis is 1 month, and the two most important POP modes have eigenvalues $\lambda_1 = 0.9355 \exp(\mathrm{i}\, 0.1047)$ and $\lambda_2 = 0.8825 \exp(\mathrm{i}\, 1.571)$, what can we learn from the eigenvalues?

(3.4) Analyze the data file provided in the book website using Fourier spectral analysis. (a) Compute the autospectrum for the time series x_1 (with the time series t giving the time of the observation in days). (b) Compute the autospectrum for time series x_2, and compare with that in (a).

(3.5) Using the same time series x_1 and x_2 from Exercise 3.4, perform singular spectrum analysis (SSA) on (a) x_1 and (b) x_2, with the time series lagged by an extra 50 days in SSA (so that there is a total of 51 time series). Briefly discuss the difference between the SSA result and the Fourier spectral result in Exercise 3.4 – in particular, which method is more efficient in capturing the variance?

4

Feed-forward neural network models

The human brain is an organ of marvel – with a massive network of about 10^{11} interconnecting neural cells called neurons, it performs highly parallel computing. The brain is exceedingly robust and fault tolerant. After all, we still recognize a friend whom we have not seen in a decade, though many of our brain cells have since died. A neuron is only a very simple processor, and its 'clockspeed' is actually quite slow, of the order of a millisecond, about a million times slower than that of a computer, yet the human brain beats the computer on many tasks involving vision, motor control, common sense, etc. Hence, the power of the brain lies not in its clockspeed, nor in the computing power of a neuron, but in its massive network structure. What computational capability is offered by a massive network of interconnected neurons is a question which has greatly intrigued scientists, leading to the development of the field of *neural networks* (NN) .

Of course there are medical researchers who are interested in how real neural networks function. However, there are far more scientists from all disciplines who are interested in *artificial* neural networks, i.e. how to borrow ideas from neural network structures to develop better techniques in computing, artificial intelligence, data analysis, modelling and prediction. In fact, neural network methods have spread beyond medicine and science into engineering, economics and commerce.

In NN literature, there are two main types of learning problem, *supervised* and *unsupervised* learning (Section 1.7). In the supervised learning case, one is provided with the predictor data, $\mathbf{x}_1, \ldots, \mathbf{x}_n$, and the response data, $\mathbf{y}_1, \ldots, \mathbf{y}_n$. Given the predictor data as input, the model produces outputs, $\mathbf{y}'_1, \ldots, \mathbf{y}'_n$. The model learning is 'supervised' in the sense that the model output $(\mathbf{y}'_1, \ldots, \mathbf{y}'_n)$ is guided towards the given response data $(\mathbf{y}_1, \ldots, \mathbf{y}_n)$, usually by minimizing an *objective function* (also called a *cost function* or *error function*). Regression and classification problems involve supervised learning. In contrast, for unsupervised learning, only input data are provided, and the model discovers the natural patterns or structure in the input data. Clustering and principal component analysis involve unsupervised learning.

There are many types of neural network, some mainly of interest to the biomedical researcher, others with broad applications. The most widely used type of NN is the *feed-forward* neural network, where the signal in the model only proceeds forward from the inputs through any intermediate layers to the outputs without any feedback. The most common feed-forward NN is represented by the multi-layer perceptron model. The next two sections present some of the historical developments leading to the multi-layer perceptron model.

4.1 McCulloch and Pitts model

The earliest NN model of significance is the McCulloch and Pitts (1943) model. From neurobiology, it is known that a neuron receives stimulus (signals) from its neighbours, and if the total stimulus exceeds some threshold, the neuron becomes activated and fires off (outputs) a signal. Their model neuron is a binary threshold unit, i.e. it receives a weighed sum of its inputs from other units, and outputs either 1 or 0 depending on whether the sum exceeds a threshold. This simulates a neuron receiving signals from its neighbouring neurons, and depending on whether the strength of the overall stimulus exceeds the threshold, the neuron either becomes activated (outputs the value 1) or remains at rest (value 0). For a neuron, if x_i denotes the input signal from the ith neighbour, which is weighed by a *weight parameter* w_i, the output of the neuron y is given by

$$y = H\left(\sum_i w_i x_i + b\right), \tag{4.1}$$

where b is called an *offset* or *bias* parameter, and H is the Heaviside step function,

$$H(z) = \begin{cases} 1 & \text{if } z \geq 0 \\ 0 & \text{if } z < 0. \end{cases} \tag{4.2}$$

By adjusting b, the threshold level for the firing of the neuron can be changed. McCulloch and Pitts proved that networks made up of such neurons are capable of performing any computation a digital computer can, though there is no provision that such a NN computer is necessarily faster or easier. In the McCulloch and Pitts model, the neurons are very similar to conventional logical gates, and there is no algorithm for finding the appropriate weight and offset parameters for a particular problem.

4.2 Perceptrons

The next major advance is the *perceptron* model of Rosenblatt (1958, 1962) (and similar work by Widrow and Hoff (1960)). The perceptron model consists of an

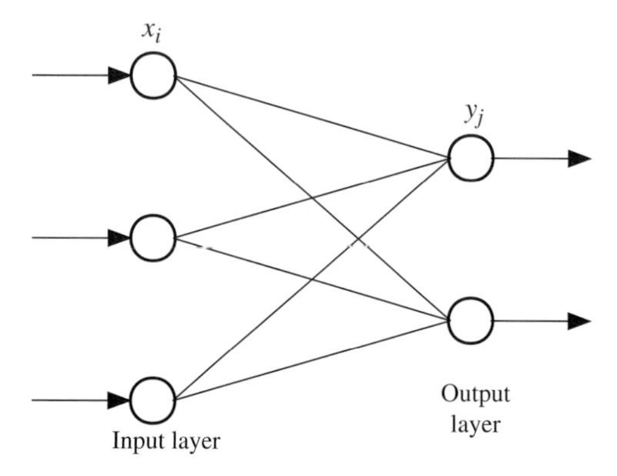

Fig. 4.1 The perceptron model consists of a layer of input neurons connected directly to a layer of output neurons.

input layer of neurons connected to an output layer of neurons (Fig. 4.1). Neurons are also referred to as *nodes* or *units* in the NN literature. The key advance is the introduction of a *learning algorithm*, which finds the weight and offset parameters of the network for a particular problem. An output neuron

$$y_j = f\left(\sum_i w_{ji} x_i + b_j\right), \tag{4.3}$$

where x_i denotes an input, f a specified transfer function known as an *activation function*, w_{ji} the weight parameter connecting the ith input neuron to the jth output neuron, and b_j the offset parameter of the jth output neuron. The offset or bias parameter is completely unrelated to the statistical bias of a model, and $-b_j$ is also called the threshold parameter by some authors.

A more compact notation eliminates the distinction between weight and offset parameters by expressing $\sum_i w_i x_i + b$ as $\sum_i w_i x_i + w_0 = \sum_i w_i x_i + w_0 1$, i.e. b can be regarded as simply the weight w_0 of an extra constant input $x_0 = 1$, and (4.3) can be written as

$$y_j = f\left(\sum_i w_{ji} x_i\right), \tag{4.4}$$

with the summation of i starting from $i = 0$. This convenient notation incorporating the offsets into the weights will often be used in this book to simplify equations.

The step function was originally used as the activation function, but other continuous functions can also be used. Given data of the inputs and outputs, one can train the network so that the model output values y_j derived from the inputs using (4.3) are as close as possible to the data y_{dj} (also called the *target*), by finding the appropriate weight and offset parameters in (4.3) – i.e. try to fit the model to the data by adjusting the parameters. With the parameters known, (4.3) gives the output variable y_j as a function of the input variables. Detailed explanation of the training process will be given later.

In situations where the outputs are not binary variables, a commonly used activation function is the *logistic sigmoidal function*, or simply the logistic function, where 'sigmoidal' means an S-shaped function:

$$f(x) = \frac{1}{1 + \mathrm{e}^{-x}}. \tag{4.5}$$

This function has an asymptotic value of 0 when $x \to -\infty$, and rises smoothly as x increases, approaching the asymptotic value of 1 as $x \to +\infty$. The logistic function is used because it is nonlinear and differentiable. From a biological perspective, the shape of the logistic function resembles the activity of a neuron. Suppose the 'activity' of a neuron is 0 if it is at rest, or 1 if it is activated. A neuron is subjected to stimulus from its neighbouring neurons, each with activity x_i, transmitting a stimulus of $w_i x_i$ to this neuron. With the summed stimulus from all its neighbours being $\sum_i w_i x_i$, whether this neuron is activated depends on whether the summed stimulus exceeds a threshold c or not, i.e. the activity of this neuron is 1 if $\sum_i w_i x_i \geq c$, but is 0 otherwise. This behaviour is captured by the Heaviside step function $H\left(\sum_i w_i x_i - c\right)$. To have a differentiable function with a smooth transition between 0 and 1 instead of an abrupt step, one can use the logistic function f in (4.5), i.e. $f\left(\sum_i w_i x_i + b\right)$, where $b = -c$ is the offset parameter. The role of the weight and offset parameters can be readily seen in the univariate form $f(wx + b)$, where a large w gives a steeper transition from 0 to 1 ($w \to \infty$ approaches the Heaviside function), while increasing b slides the logistic curve to the left along the x-axis.

We will next show that the logistic activation function also arises naturally from Bayes decision (Section 1.6) with two classes C_1 and C_2. For such classification problems, the perceptron network can have two output neurons y_1 and y_2 (Fig. 4.1). If for a given input $\mathbf{x}$, the network yields $y_j > y_i$, then class j is chosen. We will show that the outputs y_i $(i = 1, 2)$ can be made to match the Bayes discriminant function $P(C_j|\mathbf{x})$ of Section 1.6. Recall Bayes theorem:

$$P(C_1|\mathbf{x}) = \frac{p(\mathbf{x}|C_1)P(C_1)}{p(\mathbf{x}|C_1)P(C_1) + p(\mathbf{x}|C_2)P(C_2)} \tag{4.6}$$

$$= \frac{1}{1 + \frac{p(\mathbf{x}|C_2)P(C_2)}{p(\mathbf{x}|C_1)P(C_1)}} \tag{4.7}$$

$$= \frac{1}{1 + \mathrm{e}^{-u}}, \tag{4.8}$$

which has the form of a logistic sigmoidal function, with

$$u = \ln\left[\frac{p(\mathbf{x}|C_1)P(C_1)}{p(\mathbf{x}|C_2)P(C_2)}\right]. \tag{4.9}$$

If we further assume that the two classes both have Gaussian distributions, with means $\mathbf{m}_1$ and $\mathbf{m}_2$, and equal covariance matrices $\mathbf{C}$, then for $i = 1,\ 2$,

$$p(\mathbf{x}|C_i) = \frac{1}{(2\pi)^{l/2}|\mathbf{C}|^{1/2}} \exp\left\{-\frac{1}{2}(\mathbf{x}-\mathbf{m}_i)^{\mathrm{T}}\mathbf{C}^{-1}(\mathbf{x}-\mathbf{m}_i)\right\}, \tag{4.10}$$

with l the dimension of the $\mathbf{x}$ vector and $|\mathbf{C}|$ the determinant of $\mathbf{C}$. Substituting this into (4.9), we have

$$u = \mathbf{w}^{\mathrm{T}}\mathbf{x} + \mathbf{b}, \tag{4.11}$$

where

$$\mathbf{w} = \mathbf{C}^{-1}(\mathbf{m}_1 - \mathbf{m}_2), \tag{4.12}$$

$$b = -\frac{1}{2}\mathbf{m}_1^{\mathrm{T}}\mathbf{C}^{-1}\mathbf{m}_1 + \frac{1}{2}\mathbf{m}_2^{\mathrm{T}}\mathbf{C}^{-1}\mathbf{m}_2 + \ln\frac{P(C_1)}{P(C_2)}. \tag{4.13}$$

Hence, with a logistic function as the activation function, (4.8) renders the output

$$y_1 = \frac{1}{1+\exp\{-(\mathbf{w}^{\mathrm{T}}\mathbf{x}+\mathbf{b})\}} \tag{4.14}$$

$$= P(C_1|\mathbf{x}), \tag{4.15}$$

and similarly $y_2 = P(C_2|\mathbf{x})$. By using the logistic activation function, the outputs of this perceptron network, serving as discriminant functions, can be regarded as posterior probabilities. In fact, when there are only two classes, it is enough to use only a single output neuron y_1. Since we want $P(C_1|\mathbf{x}) + P(C_2|\mathbf{x}) = 1$, we can obtain $P(C_2|\mathbf{x})$ from $P(C_2|\mathbf{x}) = 1 - y_1$. The model (4.14) is also called *logistic regression*. How to use NN models for classification, especially when there are more than two classes, is discussed in detail in Chapter 8.

The advent of the perceptron model led to great excitement; however, the serious limitations of the perceptron model were soon recognized (Minsky and Papert, 1969). Simple examples are provided by the use of perceptrons to model the Boolean logical operators AND and XOR (the exclusive OR). For $z = x$.AND.y, z is TRUE only when both x and y are TRUE. For $z = x$.XOR.y, z is TRUE only when exactly one of x or y is TRUE. Let 0 denote FALSE and 1 denote TRUE, then Fig. 4.2 shows the simple perceptron model which represents $z = x$.AND.y, mapping from (x, y) to z in the following manner:

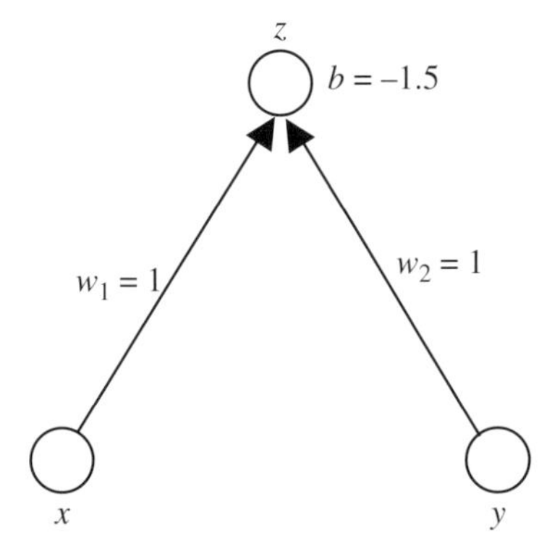

Fig. 4.2 The perceptron model for computing $z = x$.AND.y. The activation function used is the step function.

$$(0, 0) \rightarrow 0$$
$$(0, 1) \rightarrow 0$$
$$(1, 0) \rightarrow 0$$
$$(1, 1) \rightarrow 1.$$

However, a perceptron model for $z = x$.XOR.y does not exist! One cannot find a perceptron model which will map:

$$(0, 0) \rightarrow 0$$
$$(0, 1) \rightarrow 1$$
$$(1, 0) \rightarrow 1$$
$$(1, 1) \rightarrow 0.$$

The difference between the two problems is shown in Fig. 4.3. The AND problem is *linearly separable*, i.e. the input data can be classified correctly with a linear (i.e. hyperplanar) decision boundary, whereas the XOR problem is not linearly separable. It is easy to see why the perceptron model is limited to a linearly separable problem. If the activation function $f(z)$ has a decision boundary at $z = c$, (4.3) implies that the decision boundary for the jth output neuron is given by $\sum_i w_{ji} x_i + b_j = c$, which is the equation of a straight line in the input $\mathbf{x}$-space.

For an input $\mathbf{x}$-space with dimension $n = 2$, there are 16 possible Boolean functions (among them AND and XOR), and 14 of the 16 are linearly separable. When $n = 3$, 104 out of 256 Boolean functions are linearly separable. When $n = 4$, the fraction of Boolean functions which are linearly separable drops further – only 1882 out of 65 536 are linearly separable (Rojas, 1996). As n gets large, the set of linearly separable functions forms a very tiny subset of the total set (Bishop, 1995).

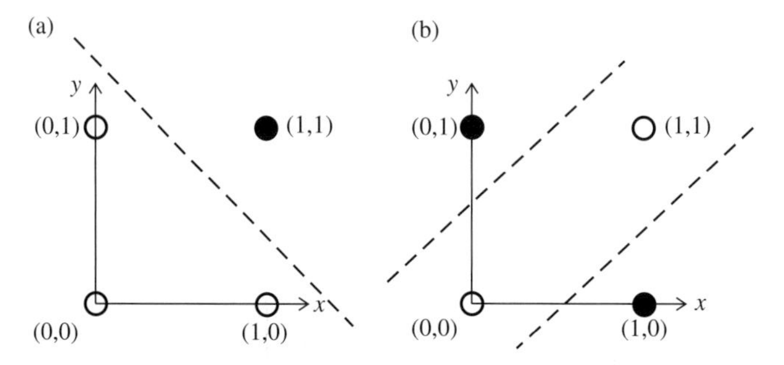

Fig. 4.3 The classification of the input data (x, y) by the Boolean logical operator (a) AND, and (b) XOR (exclusive OR). In (a), the decision boundary separating the TRUE domain (black circle) from the FALSE domain (white circles) can be represented by a straight (dashed) line, hence the problem is linearly separable; whereas in (b), two lines are needed, rendering the problem not linearly separable.

Interest in NN research waned following the realization that the perceptron model is restricted to linearly separable problems.

4.3 Multi-layer perceptrons (MLP)

When the limitations of the perceptron model were realized, it was felt that the NN might have greater power if additional '*hidden*' layers of neurons were placed between the input layer and the output layer. Unfortunately, there was then no algorithm which would solve for the parameters of the *multi-layer perceptrons* (MLP). Revival of interest in NN did not occur until the mid 1980s – largely through the highly influential paper, Rumelhart *et al.* (1986), which rediscovered the *back-propagation algorithm* to solve the MLP problem, as the algorithm had actually been derived earlier by Werbos (1974).

Figure 4.4 illustrates a MLP with one hidden layer. The input signals x_i are mapped to the hidden layer of neurons h_j by

$$h_j = f\left(\sum_i w_{ji}x_i + b_j\right), \tag{4.16}$$

and then on to the output y_k,

$$y_k = g\left(\sum_j \tilde{w}_{kj}h_j + \tilde{b}_k\right), \tag{4.17}$$

where f and g are activation functions, w_{ji} and $\tilde{w}_{kj}$ weight parameter matrices, and b_j and $\tilde{b}_k$ are offset parameters.

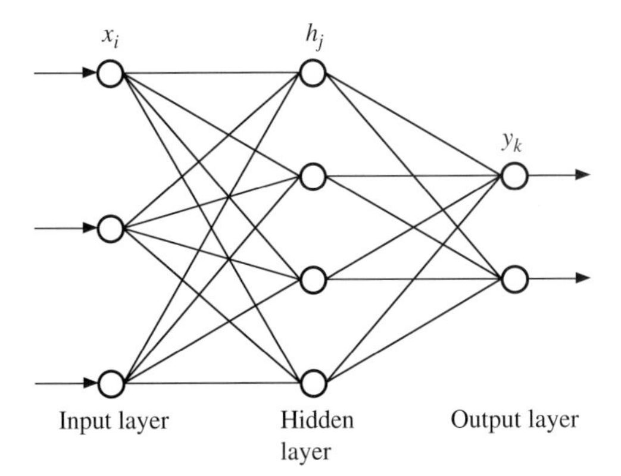

Fig. 4.4 The multi-layer perceptron (MLP) model with one 'hidden layer' of neurons sandwiched between the input layer and the output layer.

To train the NN to learn from a dataset (the target), we need to minimize the *objective function* J (also referred to as the *cost function*, *error function*, or *loss function*), defined here to be one half the *mean squared error* (MSE) between the model output and the target,

$$J = \frac{1}{N} \sum_{n=1}^{N} \left\{ \frac{1}{2} \sum_{k} \left[y_k^{(n)} - y_{dk}^{(n)} \right]^2 \right\}, \tag{4.18}$$

where y_{dk} is the target data, and there are $n = 1, \ldots, N$ observations or measurements. In back-propagation literature, a conventional scale factor of $\frac{1}{2}$ is usually included, though it and the factor $\frac{1}{N}$ can be replaced by any positive constant. (In other parts of this book, the constant factors $\frac{1}{2}$ and $\frac{1}{N}$ may be dropped from the objective function.) An optimization algorithm is needed to find the weight and offset parameter values which minimize the objective function, hence the MSE between the model output and the target. The MSE is the most common form used for the objective function in nonlinear regression problems, as minimizing the MSE is equivalent to maximizing the likelihood function assuming Gaussian error distribution (see Section 6.1). Details on how to perform the nonlinear optimization will be presented later in this chapter and in the next one, as there are many choices of optimization scheme. In general, nonlinear optimization is difficult, and convergence can be drastically slowed or numerically inaccurate if the input variables are poorly scaled. For instance if an input variable has a mean exceeding its standard deviation, it is strongly recommended that the mean be subtracted from the variable to remove the systematic bias. Hence it is common to *standardize* the data before applying the NN model, i.e. each variable has its mean value subtracted, and then is divided by its standard deviation, so the standardized variable will have zero mean and unit standard deviation.

Besides the logistic function, another commonly used sigmoidal activation function is the *hyperbolic tangent function*

$$f(x) = \tanh x = \frac{e^x - e^{-x}}{e^x + e^{-x}}. \tag{4.19}$$

This function has an asymptotic value of -1 when $x \to -\infty$, and rises smoothly as x increases, approaching the asymptotic value of 1 as $x \to +\infty$. The tanh function can be viewed as a scaled version of the logistic function (4.5), as the two are related by:

$$\tanh(x) = 2\ \text{logistic}(2x) - 1. \tag{4.20}$$

While the range of the logistic function is (0, 1), the range of the tanh function is $(-1, 1)$. Hence the output of a tanh function does not have the positive systematic bias found in the output of a logistic function, which if input to the next layer of neurons, is somewhat difficult for the network to handle, resulting in slower convergence during NN training. It has been found empirically (LeCun *et al.*, 1991) that a network using logistic activation functions in the hidden layer(s) tends to converge slower during NN training than a corresponding network using tanh activation functions. Hence the tanh activation function is to be preferred over the logistic activation function in the hidden layer(s).

Since the range of the tanh function is $(-1, 1)$, the range of the network output will be similarly bounded if one of these activation functions is used for g in (4.17). This is useful if the NN tries to classify the output into one of two classes, but may cause a problem if the output variables are unbounded. Even if the output range is bounded within $[-1, 1]$, values such as -1 and 1 can only be represented by the tanh function at its asymptotic limit. One possible solution is to scale the output variables, so they all lie within the range $(-0.9, 0.9)$ – the range $(-0.9, 0.9)$ is preferred to $(-1, 1)$ as it avoids using the asymptotic range of the tanh function, resulting in faster convergence during NN training.

Another possibility when the output variables are within $[-1, 1]$, is to use a scaled tanh function, $f(x) = 1.7159 \tanh(\frac{2}{3}x)$ as the activation function. Bounded within $(-1.7159, 1.7159)$, this $f(x)$ has an almost linear slope for $-1 < x < 1$, with $f(1) = 1$, and $f(-1) = -1$.

When the output is not restricted to a bounded interval, the identity activation function is commonly used for g, i.e. the output is simply a linear combination of the hidden neurons in the layer before,

$$y_k = \sum_j \tilde{w}_{kj} h_j + \tilde{b}_k. \tag{4.21}$$

For the 1-hidden-layer NN, this means the output is just a linear combination of sigmoidal shaped functions.

There is some confusion in the literature on how to count the number of layers. The most common convention is to count the number of layers of mapping functions, which is equivalent to the number of layers of neurons excluding the input layer. The 1-hidden-layer NN in Fig. 4.4 will be referred to as a 2-layer NN. Some researchers count the total number of layers of neurons (which is a little misleading since the complexity of the NN is determined by the layers of mappings), and refer to the 1-hidden-layer NN as a 3-layer NN. In this book, we will refer to the NN in Fig. 4.4 as either a 1-hidden-layer NN or a 2-layer NN. A useful shorthand notation for describing the number of inputs, hidden and output neurons is the m_1-m_2-m_3 notation, where a 3-4-2 network denotes a 1-hidden-layer NN with 3 input, 4 hidden and 2 output neurons.

The total number of (weight and offset) parameters in an m_1-m_2-m_3 network is $N_p = (m_1 + 1)m_2 + (m_2 + 1)m_3$, of which $m_1m_2 + m_2m_3 = m_2(m_1 + m_3)$ are weight parameters, and m_2+m_3 are offset parameters. In multiple linear regression problems with m predictors and one response variable, i.e. $y = a_0 + a_1x_1 + \ldots + a_mx_m$, there are $m + 1$ parameters. For corresponding nonlinear regression with an m-m_2-1 MLP network, there will be $N_p = (m+1)m_2+(m_2+1)$ parameters, usually greatly exceeding the number of parameters in the multiple linear regression model. Furthermore, the parameters of an NN model are in general extremely difficult to interpret, unlike the parameters in a multiple linear regression model, which have straightfoward interpretations.

Incidentally, in a 1-hidden-layer MLP, if the activation functions at both the hidden and output layers are linear, then it is easily shown that the outputs are simply linear combinations of the inputs. The hidden layer is therefore redundant and can be deleted altogether. In this case, the MLP model simply reduces to multiple linear regression. This demonstrates that the presence of a nonlinear activation function at the hidden layer is essential for the MLP model to have nonlinear modelling capability.

While there can be exceptions (Weigend and Gershenfeld, 1994), MLP are usually employed with $N_p < N$, N being the number of observations in the dataset. Ideally, one would like to have $N_p \ll N$, but in many environmental problems, this is unattainable. In most climate problems, decent climate data have been available only after World War II. Another problem is that the number of predictors can be very large, although there can be strong correlations between predictors. In this case, principal component analysis (PCA) (Section 2.1) is commonly applied to the predictor variables, and the leading few principal components (PCs) are extracted, and serve as inputs to the MLP network, to reduce greatly the number of input neurons and therefore N_p.

If there are multiple outputs, one has two choices: build a single NN with multiple outputs, or build multiple NN models each with a single output. If the output variables are correlated between themselves, then the single NN approach often leads to higher skills, since training separate networks for individual outputs does not take into account the relations between the output variables (see the example in Section 12.1.3). On the other hand, if the output variables are uncorrelated (e.g. the outputs are principal components), then training separate networks often leads to slightly higher skills, as this approach focuses the single-output NN (with fewer parameters than the multiple-output NN) on one output variable without the distraction from the other uncorrelated output variables.

For modelling a nonlinear regression relation such as $y = f(\mathbf{x})$, why can one not use the familiar Taylor series expansion instead of MLP? The Taylor expansion is of the form

$$y = a_0 + \sum_{i_1=1}^{m} a_{i_1} x_{i_1} + \sum_{i_1=1}^{m}\sum_{i_2=1}^{m} a_{i_1 i_2} x_{i_1} x_{i_2} + \sum_{i_1=1}^{m}\sum_{i_2=1}^{m}\sum_{i_3=1}^{m} a_{i_1 i_2 i_3} x_{i_1} x_{i_2} x_{i_3} + \cdots . \tag{4.22}$$

In practice, the series is truncated, and only terms up to order k are kept, i.e. y is approximated by a kth order polynomial, and can be written more compactly as

$$y = \sum_{i=1}^{L} c_i \phi_i + c_0, \tag{4.23}$$

where ϕ_i represent the terms x_{i_1}, $x_{i_1} x_{i_2}$, ..., and c_i the corresponding parameters. There are L terms in the sum, with $L \sim m^k$ (Bishop, 1995), hence there are about m^k parameters. In practice, m^k means the number of model parameters rises at an unacceptable rate as m increases. Barron (1993) showed that for the m-m_2-1 MLP model, the summed squared error is of order $O(1/m_2)$ (independent of m), while for the polynomial approximation, the summed squared error is at least $O(1/L^{2/m}) = O(1/m^{2k/m})$, yielding an unacceptably slow rate of convergence for large m.

Let us examine how the summed squared error reduces with increasing number of parameters in the MLP and in the polynomial representation. For MLP, since the error is $O(1/m_2)$, if $1/m_2$ is to be halved, we need a new $m_2' = 2\,m_2$, and the total number of model parameters $\sim m\, m_2' = 2\, m\, m_2$, i.e. a doubling of the original number of parameters. In the polynomial approximation, for $1/m^{2k/m}$ to be halved, we need a new k', satisfying

$$m^{2k'/m} = 2\, m^{2k/m}, \quad \text{i.e. } k' = k + \frac{m}{2}\frac{\ln 2}{\ln m}. \tag{4.24}$$

The number of parameters in the new model compared with that in the old model is

$$\frac{m^{k'}}{m^k} = m^{(m \ln 2)/(2 \ln m)}. \tag{4.25}$$

For $m = 10$, this ratio is 32, and for $m = 100$, this ratio is 10^{15} – for comparison, the ratio of parameters in the MLP model is 2, independent of m. This is known as the *curse of dimensionality*, i.e. the polynomial approximation requires an astronomical number of parameters even for a moderate input dimension m, since the number of model parameters is $\sim m^k$, and the error converges very slowly at $O(1/m^{2k/m})$. The form (4.23) does have one advantage over MLP, namely that y depends on the parameters linearly, and so only a linear optimization is needed to solve for the parameters. In contrast, for the MLP model, because of the nonlinear activation function, y does not depend on the parameters linearly, hence a nonlinear optimization process, which is much harder than linear optimization, is needed to solve for the parameters. Nonetheless, this advantage of the polynomial approximation is not enough to compensate for the dreaded curse of dimensionality. The advent of kernel methods (Section 7.5) in the mid 1990s revived many methods previously discarded due to the curse of dimensionality.

4.4 Back-propagation

We now turn to the crucial problem of how to find the optimal weight and offset parameters which would minimize the objective function J. To minimize J, one needs to know the gradient of J with respect to the parameters. The *back-propagation* algorithm gives the gradient of J through the backward propagation of the model errors – the reader familiar with data assimilation will notice the similarity with the backward error propagation in the adjoint data assimilation method. In fact, the MLP problem could not be solved until the introduction of the back-propagation algorithm by Rumelhart *et al.* (1986), though the algorithm had actually been discovered in the Ph.D. thesis of Werbos (1974).

The back-propagation algorithm is composed of two parts: the first part computes the gradient of J by backward propagation of the model errors, while the second part descends along the gradient towards the minimum of J. This descent method is called *gradient descent* or *steepest descent*, and is notoriously inefficient, thus rendering the original back-propagation algorithm a very slow method. Nowadays, the term 'back-propagation' is used somewhat ambiguously – it could mean the original back-propagation algorithm, or it could mean using only the first part involving the backward error propagation to compute the gradient of J, to be followed by a much more efficient descent algorithm, such as the *conjugate gradient* algorithm, resulting in much faster convergence.

While the original back-propagation algorithm is not recommended for actual computation because of its slowness, it is presented here because of its historical significance, and more importantly, because it is the easiest algorithm to understand.

In (4.18), the objective function J is evaluated over all observations. It is convenient to define the objective function $J^{(n)}$ associated with the nth observation or training pattern, i.e.

$$J^{(n)} = \frac{1}{2} \sum_k [y_k^{(n)} - y_{dk}^{(n)}]^2. \tag{4.26}$$

The objective function J in (4.18) is simply the mean of $J^{(n)}$ over all N observations. In the following derivation, we will be dealing with $J^{(n)}$, though for brevity, we will simply use the symbol J. In the 2-layer network described by (4.16) and (4.17), letting $w_{j0} \equiv b_j$, $\tilde{w}_{k0} \equiv \tilde{b}_k$, and $x_0 = 1 = h_0$ allows the equations to be expressed more compactly as

$$h_j = f\left(\sum_i w_{ji} x_i\right), \tag{4.27}$$

$$y_k = g\left(\sum_j \tilde{w}_{kj} h_j\right), \tag{4.28}$$

where the summation indices start from 0 to include the offset terms. We can introduce an even more compact notation by putting all the w_{ji} and $\tilde{w}_{kj}$ parameters into a single parameter vector or weight vector $\mathbf{w}$. In back-propagation, the parameters are first assigned random initial values, and then adjusted according to the gradient of the objective function, i.e.

$$\Delta \mathbf{w} = -\eta \frac{\partial J}{\partial \mathbf{w}}. \tag{4.29}$$

The parameters are adjusted by $\Delta \mathbf{w}$, proportional to the objective function gradient vector by the scale factor η, which is called the *learning rate*. In component form, we have

$$\Delta w_{ji} = -\eta \frac{\partial J}{\partial w_{ji}}, \quad \text{and} \quad \Delta \tilde{w}_{kj} = -\eta \frac{\partial J}{\partial \tilde{w}_{kj}}. \tag{4.30}$$

Note that η determines the size of the step taken by the optimization algorithm as it descends along the direction of the objective function gradient, with the parameters adjusted for the next step $n + 1$ by

$$\mathbf{w}(n+1) = \mathbf{w}(n) + \Delta \mathbf{w}(n). \tag{4.31}$$

Let us introduce the symbols

$$s_j = \sum_i w_{ji} x_i, \quad \text{and} \quad \tilde{s}_k = \sum_j \tilde{w}_{kj} h_j, \tag{4.32}$$

hence (4.27) and (4.28) become

$$h_j = f(s_j), \quad \text{and} \quad y_k = g(\tilde{s}_k). \tag{4.33}$$

The objective function gradients in (4.30) can be solved by applying the chain rule in differentiation, e.g.

$$\frac{\partial J}{\partial \tilde{w}_{kj}} = \frac{\partial J}{\partial \tilde{s}_k}\frac{\partial \tilde{s}_k}{\partial \tilde{w}_{kj}} \equiv -\tilde{\delta}_k \frac{\partial \tilde{s}_k}{\partial \tilde{w}_{kj}}, \tag{4.34}$$

where $\tilde{\delta}_k$ is called the *sensitivity* of the kth output neuron. Further application of the chain rule yields

$$\tilde{\delta}_k \equiv -\frac{\partial J}{\partial \tilde{s}_k} = -\frac{\partial J}{\partial y_k}\frac{\partial y_k}{\partial \tilde{s}_k} = (y_{\mathrm{d}k} - y_k) g'(\tilde{s}_k), \tag{4.35}$$

where (4.26) has been differentiated, and g' is the derivative of g from (4.33).

From (4.32), we have

$$\frac{\partial \tilde{s}_k}{\partial \tilde{w}_{kj}} = h_j. \tag{4.36}$$

Substituting this and (4.35) into (4.34) and (4.30), we obtain the weight update or learning rule for the weights connecting the hidden and output layers:

$$\Delta \tilde{w}_{kj} = \eta \tilde{\delta}_k h_j = \eta (y_{\mathrm{d}k} - y_k) g'(\tilde{s}_k) h_j. \tag{4.37}$$

If linear activation functions are used at the output neurons, i.e. g is the identity map, then $g' = 1$.

Similarly, to obtain the learning rule for the weights connecting the input layer to the hidden layer, we use the chain rule repeatedly:

$$\frac{\partial J}{\partial w_{ji}} = \sum_k \frac{\partial J}{\partial y_k}\frac{\partial y_k}{\partial \tilde{s}_k}\frac{\partial \tilde{s}_k}{\partial h_j}\frac{\partial h_j}{\partial s_j}\frac{\partial s_j}{\partial w_{ji}} = -\sum_k \tilde{\delta}_k \tilde{w}_{kj} f'(s_j) x_i. \tag{4.38}$$

Hence, the learning rule for the weights connecting the input layer to the hidden layer is

$$\Delta w_{ji} = \eta \sum_k \tilde{\delta}_k \tilde{w}_{kj} f'(s_j) x_i. \tag{4.39}$$

This can be expressed in a similar form as (4.37), i.e.

$$\Delta w_{ji} = \eta \delta_j x_i, \quad \text{with} \quad \delta_j = \left(\sum_k \tilde{\delta}_k \tilde{w}_{kj} \right) f'(s_j). \tag{4.40}$$

Equations (4.37) and (4.40) give the back-propagation algorithm. The model output error $(y_k - y_{\mathrm{d}k})$ is propagated backwards by (4.37) to update the weights connecting the hidden to output layers, and then further back by (4.40) to update the weights connecting the input to hidden layers.

In general, the weights are randomly initialized at the start of the optimization process. The inputs are mapped forward by the network, and the output model error is obtained. The error is then back-propagated to update the weights. This process of mapping the inputs forward and then back-propagating the error is iterated until the objective function satisfies some convergence criterion. The derivation of the back-propagation algorithm here is for a network with only one hidden layer. The algorithm can be extended in a straightforward manner to a network with two (or more) hidden layers.

There are two main training protocols, *sequential training* and *batch training*. In sequential training, each pattern or observation is presented to the network and the weights are updated. The next pattern is then presented, until all patterns in the training dataset have been presented once to the network – called an *epoch*. The same patterns are then presented repeatedly for many epochs until the objective function convergence criterion is satisfied. A variant of sequential training is *stochastic training*, where a pattern is randomly selected from the training dataset, presented to the network for weight updating, and the process is repeated. For very large patterns presenting storage problems, *on-line training* can be used, where a pattern is presented to the network, and the weights updated repeatedly, before moving on to the next pattern. As each pattern is presented only once, there is no need to store all the patterns on the computer.

In batch training, all patterns in the training dataset are presented to the network before the weights are updated. This process is repeated for many epochs. The objective function J is the mean of $J^{(n)}$ over all N observations, as defined in (4.18). Some second-order optimization algorithms (see Chapter 5) are more readily applied to batch training. However, if the training data are redundant, i.e. a pattern may be presented several times in the dataset, then stochastic training can be more efficient than batch training.

The objective function convergence criterion is a subtle issue. Many MLP applications do not train until convergence to the global minimum – this can be a great shock to researchers trained in classical optimization. The reason is that data contain both signal and noise. Given enough hidden neurons, an MLP can have enough parameters to fit the training data to arbitrary accuracy, which means it is also fitting to the noise in the data, an undesirable condition known as *overfitting*. When one obtains an overfitted solution, it will not fit new data well (Fig. 4.5). In other words, one is interested not in using NNs to fit a given dataset to arbitrary accuracy, but in using NN to learn the underlying relationship in the given data, i.e. be able to generalize from a given dataset, so that the extracted relationship even fits new data not used in training the NN model.

To prevent overfitting, the dataset is often divided into two parts, one for training, the other for *validation*. As the number of training epochs increases, the objective function evaluated over the training data decreases. However, the objective function evaluated over the validation data will drop but eventually increase as training epochs increase (Fig. 4.6), indicating that the training dataset is already overfitted. When the objective function evaluated over the validation data reaches a minimum,

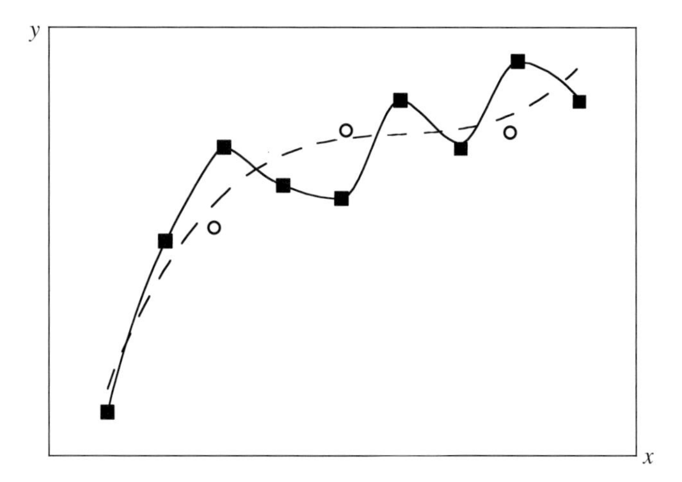

Fig. 4.5 Illustrating the problem of overfitting: the dashed curve illustrates a good fit to noisy data (indicated by the squares), while the solid curve illustrates overfitting – where the fit is perfect on the training data (squares), but is poor on the validation data (circles). Often the NN model begins by fitting the training data as the dashed curve, but with further iterations, ends up overfitting as the solid curve. (Follows Hsieh and Tang (1998).)

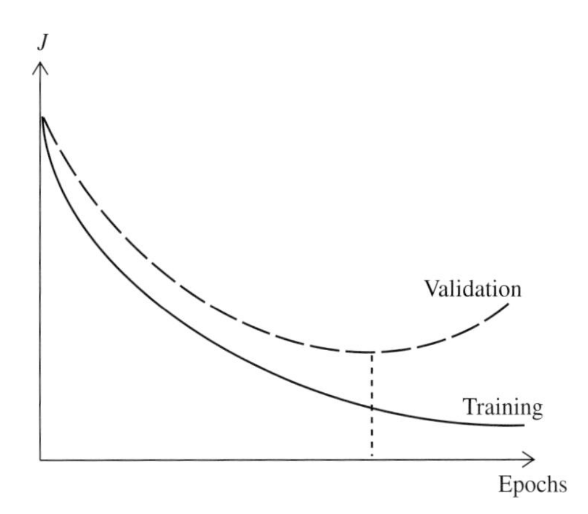

Fig. 4.6 Illustrating the behaviour of the objective function J as the number of training epochs increases. Evaluated over the training data, the objective function (solid curve) decreases with increasing number of epochs; however, evaluated over an independent set of validation data, the objective function (dashed curve) initially drops but eventually rises with increasing number of epochs, indicating that overfitting has occurred when a large number of training epochs is used. The minimum in the objective function evaluated over the validation data indicates when training should be stopped to avoid overfitting (as marked by the vertical dotted line).

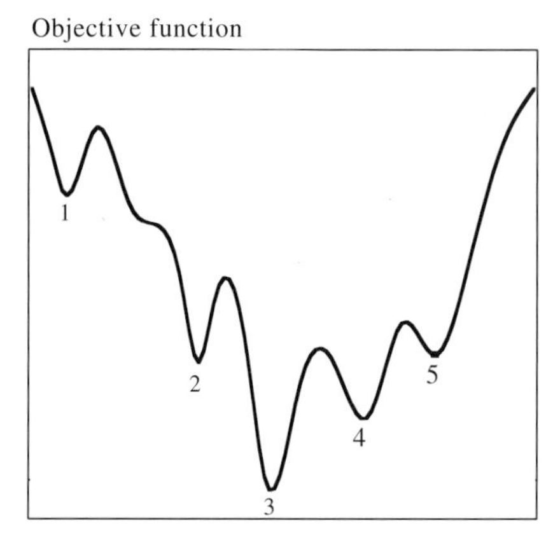

Fig. 4.7 Illustrating the objective function surface where, depending on the starting condition, the search algorithm often gets trapped in one of the numerous deep local minima. The local minima labelled 2, 4 and 5 are likely to be reasonable local minima, while the minimum labelled 1 is likely to be a bad one (in that the data were not at all well fitted). The minimum labelled 3 is the global minimum, which could correspond to an overfitted solution (i.e. fitted closely to the noise in the data), and may in fact be a poorer solution than the minima labelled 2, 4 and 5. (Reproduced from Hsieh and Tang (1998) with permission of the American Meteorological Society.)

it gives a useful signal that this is the appropriate time to stop the training, as additional training epochs only contribute to overfitting. This very common approach is called the *early stopping* (a.k.a. stopped training) method. What fraction of the data one should reserve for validation is examined in Section 6.4.

Similarly, the objective function evaluated over the training data generally drops as the number of hidden neurons increases. Again, the objective function evaluated over a validation set will drop initially but eventually increase due to overfitting from excessive number of hidden neurons. Hence the minimum of the objective function over the validation data may give an indication on how many hidden neurons to use.

A further complication is the common presence of multiple local minima in the objective function (Fig. 4.7) – the algorithm may converge to a shallow local minimum, yielding a poor fit to the training data. The local minima problem will be discussed further in the next two chapters.

4.5 Hidden neurons

So far, we have described only an NN with one hidden layer. It is not uncommon to use NN with more than one hidden layer, and the back-propagation algorithm can be readily generalized to more than one hidden layer. Usually each hidden layer uses a sigmoidal-shaped activation function and receives input from the preceding layer of neurons, similar to (4.16). How many layers of hidden neurons does one need? And how many neurons in each hidden layer?

Studies such as Cybenko (1989), Hornik *et al.* (1989) and Hornik (1991) have shown that given enough hidden neurons in an MLP with one hidden layer, the network can approximate arbitrarily well any continuous function $y = f(\mathbf{x})$. Thus even though the original perceptron is of limited power, by adding one hidden layer, the MLP has become a successful universal function approximator. There is, however, no guidance as to how many hidden neurons are needed. Experience tells us that a complicated continuous function f with many bends needs many hidden neurons, while an f with few bends needs fewer hidden neurons.

Intuitively, it is not hard to understand why a 1-hidden-layer MLP with enough hidden neurons can approximate any continuous function. For instance, the function can be approximated by a sum of sinusoidal functions under Fourier decomposition. One can in turn approximate a sinusoidal curve by a series of small steps, i.e. a sum of Heaviside step functions (or sigmoidal shaped functions) (for details, see Bishop (1995), or Duda *et al.* (2001)). Hence the function can be approximated by a linear combination of step functions or sigmoidal shaped functions, which is exactly the architecture of the 1-hidden-layer MLP. Alternatively, one can think of the continuous function as being composed of localized bumps, each of which can be approximated by a sum of sigmoidal functions. Of course, the sinusoidal and bump constructions are only conceptual aids, the actual NN does not try deliberately to sum sigmoidal functions to yield sinusoidals or bumps.

In real world applications, one may encounter some very complicated nonlinear relations where a very large number of hidden neurons are needed if a single hidden layer is used, whereas if two hidden layers are used, a more modest number of hidden neurons suffices, and gives greater accuracy. In Chapter 12, there are several examples of 2-hidden-layer MLPs used in satellite remote sensing, e.g. for synthetic aperture radar (SAR) data, MLPs with two and even three hidden layers are used (Section 12.1.4). Later, 3-hidden-layer MLPs are encountered in nonlinear principal component analysis (Chapter 10).

Hidden neurons have been somewhat of a mystery, especially to new users of NN models. They are intermediate variables needed to carry out the computation from the inputs to the outputs, and are generally not easy to interpret. If the hidden neurons are few, then they might be viewed as a low-dimensional phase space describing the state of the system. For example, Hsieh and Tang (1998) considered a simple MLP network for forecasting the tropical Pacific wind stress field. The input consists of the first 11 principal components (PCs) from a singular spectrum analysis of the wind stress field, plus a sine and cosine function of annual period to

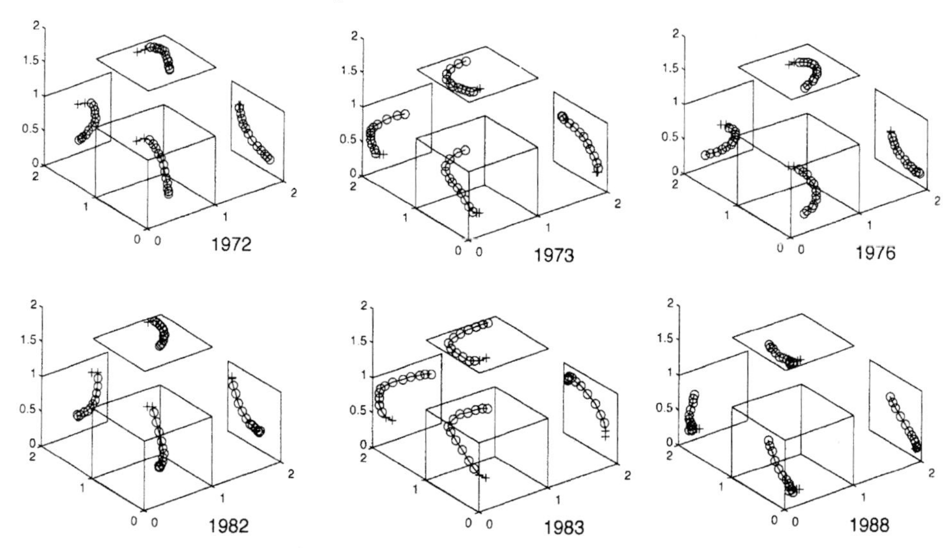

Fig. 4.8 The values of the three hidden neurons plotted in 3-D space for the years 1972, 1973, 1976, 1982, 1983 and 1988. Projections onto 2-D planes are also shown. The small circles are for the months from January to December, and the two '+' signs for January and February of the following year. El Niño warm episodes occurred during 1972, 1976 and 1982, while a cool episode occurred in 1988. In 1973 and 1983, the tropics returned to cooler conditions from an El Niño. Notice the similarity between the trajectories during 1972, 1976 and 1982, and during 1973, 1983 and 1988. In years with neither warm nor cold episodes, the trajectories oscillate randomly near the centre of the cube. From these trajectories, one can identify the precursor phase regions for warm episodes and cold episodes; and when the system enters one of these precursor phase regions, a forecast for either a warm or a cool episode can be issued. (Reproduced from Hsieh and Tang (1998) with permission of the American Meteorological Society.)

indicate the phase of the annual cycle, as the El Niño-Southern Oscillation (ENSO) fluctuations are often phase locked to the annual cycle. The single hidden layer has three neurons, and the output layer the same 11 PC time series one month later. As the values of the three hidden neurons can be plotted in 3-D space, Fig. 4.8 shows their trajectory for selected years. From Fig. 4.8 and the trajectories of other years (not shown), we can identify regions in the 3-D phase space as the El Niño warm episode phase and its precursor phase, and the cold episode and its precursor. One can issue warm episode or cold episode forecasts whenever the system enters the warm episode precursor region or the cold episode precursor region, respectively. Thus the hidden layer spans a 3-D phase space for the ENSO system, and is thus a higher dimension generalization of the 2-D phase space based on the principal oscillation pattern (POP) method (Section 3.6).

Hence in this example, we can interpret the NN as a projection from the input space onto a phase space, as spanned by the neurons in the hidden layer. The state of the system in the phase space then allows a projection onto the output space, which can be the same input variables some time in the future, or other variables in the future. This interpretation provides a guide for choosing the appropriate number of neurons in the hidden layer, namely, the number of hidden neurons should be the same as the embedding manifold for the system. Since the ENSO system is thought to have only a few degrees of freedom (Grieger and Latif, 1994), the number of hidden neurons needed should be about 3–7 in NN models for forecasting ENSO – using more could easily lead to overfitting.

For most NN applications, it is not worth spending time to find interpretations for the hidden neurons, especially when there are many hidden neurons in the network. One is also tempted to interpret the weights in an NN as nonlinear regression coefficients; however, in general there are no easy ways to find meaningful interpretations of the large number of weights in a typical NN.

4.6 Radial basis functions (RBF)

While sigmoidal-shaped activation functions (e.g. the hyperbolic tangent function) are widely used in feed-forward NN, *radial basis functions* (RBF) involving Gaussian-shaped functions are also commonly used.

Radial basis function methods originated in the problem of *exact interpolation*, where every input vector is required to be mapped exactly to the corresponding target vector (Powell, 1987). First, consider a 1-D target space. The output of the mapping f is a linear combination of basis functions g

$$f(\mathbf{x}) = \sum_{j=1}^{k} w_j \, g(\|\mathbf{x} - \mathbf{c}_j\|, \sigma), \tag{4.41}$$

where each basis function is specified by its centre $\mathbf{c}_j$ and a width parameter σ. In the case of exact interpolation, if there are n observations, then there are $k = n$ basis functions to allow for the exact interpolation, and each $\mathbf{c}_j$ corresponds to one of the input data vectors. There is a number of choices for the basis functions, the most common being the Gaussian form

$$g(r, \sigma) = \exp\left(-\frac{r^2}{2\sigma^2}\right). \tag{4.42}$$

In NN applications, exact interpolation is undesirable, as it would mean an exact fit to noisy data. The remedy is simply to choose $k < n$, i.e. use fewer (often far fewer) basis functions than the number of observations. This prevents an exact fit, but allows a smooth interpolation of the noisy data. By adjusting the number of basis functions used, one can obtain the desired level of closeness of fit. The mapping is now

$$f(\mathbf{x}) = \sum_{j=1}^{k} w_j \, g(\|\mathbf{x} - \mathbf{c}_j\|, \sigma_j) + w_0, \tag{4.43}$$

where (i) the centres $\mathbf{c}_j$ are no longer given by the input data vectors but are determined during training, (ii) instead of using a uniform σ for all the basis functions, each basis function has its own width σ_j, determined from training, and (iii) an offset parameter w_0 has been added.

If the output is multivariate, the mapping simply generalizes to

$$f_i(\mathbf{x}) = \sum_{j=1}^{k} w_{ji} \, g(\|\mathbf{x} - \mathbf{c}_j\|, \sigma_j) + w_{0i}, \tag{4.44}$$

for the ith output variable.

It is also common to use *renormalized* (or simply *normalized*) radial basis functions (Moody and Darken, 1989), where the RBF $g(\|\mathbf{x} - \mathbf{c}_j\|, \sigma_j)$ in (4.43) is replaced by the renormalized version

$$\frac{g(\|\mathbf{x} - \mathbf{c}_j\|, \sigma_j)}{\sum_{m=1}^{k} g(\|\mathbf{x} - \mathbf{c}_m\|, \sigma_m)}. \tag{4.45}$$

Using RBF can lead to holes, i.e. regions where the basis functions all give little support (Fig. 4.9). This problem is avoided by using renormalized basis functions.

While RBF neural networks can be trained like MLP networks by back-propagation (termed adaptive RBFs), RBFs are most commonly used in a non-adaptive manner, i.e. the training is performed in two separate stages, where the first stage uses unsupervised learning to find the centres and widths of the RBFs, followed by supervised learning via linear least squares to minimize the MSE between the network output and the target data.

Let us describe the procedure of the non-adaptive RBF: first choose k, the number of RBFs to be used. To find the centres of the RBFs, one commonly uses *K-means clustering* (Section 1.7), or self-organizing maps (SOMs) (Section 10.3).

Next, estimate the width parameters σ_j. For the jth centre $\mathbf{c}_j$, find the distance r_j to the closest neighbouring centre, then set $\sigma_j = \alpha r_j$, where the factor α is typically chosen in the range $1 \leq \alpha \leq 3$.

With the basis functions $g(\|\mathbf{x} - \mathbf{c}_j\|, \sigma_j)$ now determined, the only task left is to find the weights w_{ji} in the equation

$$f_i(\mathbf{x}) = \sum_{j=0}^{k} w_{ji} \, g_j(\mathbf{x}), \tag{4.46}$$

which is the same as (4.44), with $g_j(\mathbf{x}) = g(\|\mathbf{x} - \mathbf{c}_j\|, \sigma_j)$ $(j = 1, \ldots, k)$, $g_0(\mathbf{x}) = 1$, and the summation starting from $j = 0$ to incorporate the offset parameter w_{0i}.

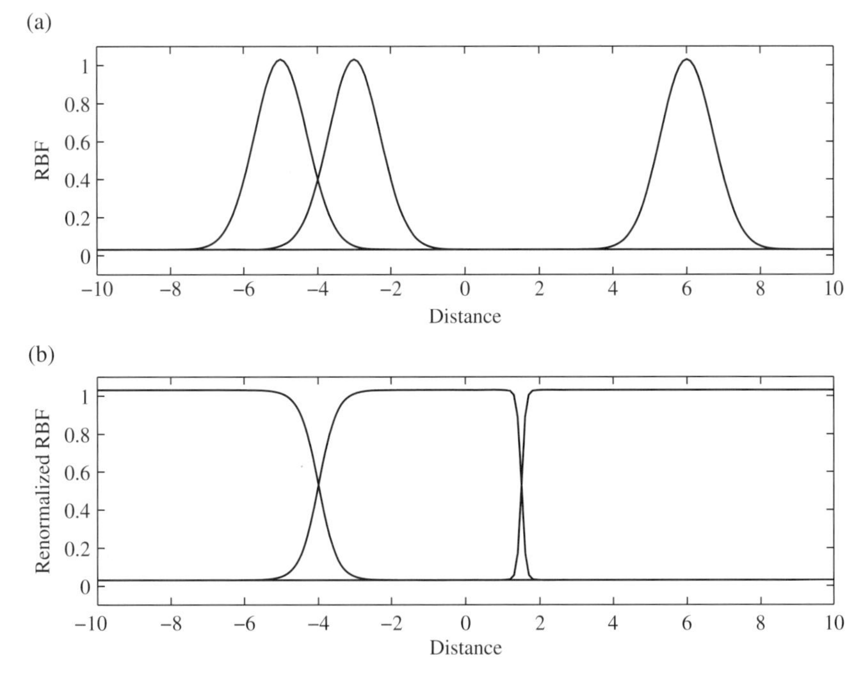

Fig. 4.9 (a) Radial basis functions (RBFs) and (b) renormalized RBFs. Holes are present in (a), where RBFs with fixed width σ are used. This problem is avoided in (b) with the renormalized RBFs.

The network output $f_i(\mathbf{x})$ is to approximate the target data $y_i(\mathbf{x})$ by minimizing the MSE, which is simply a linear least squares problem. In matrix notation, this can be written as

$$\mathbf{Y} = \mathbf{GW} + \mathbf{E}, \tag{4.47}$$

where $(\mathbf{Y})_{li} = y_i(\mathbf{x}^{(l)})$, (with $l = 1, \ldots, n$), $(\mathbf{G})_{lj} = g_j(\mathbf{x}^{(l)})$, $(\mathbf{W})_{ji} = w_{ji}$, and $\mathbf{E}$ is the error or residual in the least squares fit. The linear least squares solution (analogous to (1.46)) is given by

$$\mathbf{W} = (\mathbf{G}^{\mathrm{T}}\mathbf{G})^{-1}\mathbf{G}^{\mathrm{T}}\mathbf{Y}. \tag{4.48}$$

In summary, the RBF NN is most commonly trained in two distinct stages. The first stage involves finding the centres and widths of radial basis functions by unsupervised learning of the input data (with no consideration of the output target data). The second stage involves finding the best linear least squares fit to the output target data (supervised learning). In contrast, in the multi-layer perceptron (MLP) NN approach, all weights are trained together under supervised learning. The supervised learning in MLP involves nonlinear optimizations which usually have multiple minima in the objective function, whereas the supervised learning in

RBF involves only optimization of *linear* least squares, hence no multiple minima in the objective function – a main advantage of the RBF over the MLP approach. However, that the basis functions are computed in the RBF NN approach without considering the output target data can be a major drawback, especially when the input dimension is large. The reason is that many of the input variables may have significant variance but have no influence on the output target, yet these irrelevant inputs introduce a large number of basis functions. The second stage training may then involve solving a very large, poorly conditioned matrix problem, which can be computationally very expensive or even intractable. The advent of kernel methods has alleviated this problem (see Section 8.4 for classification and Section 9.1 for regression).

4.7 Conditional probability distributions

So far, the NN models have been used for nonlinear regression, i.e. the output y is related to the inputs $\mathbf{x}$ by a nonlinear function, $y = f(\mathbf{x})$. In many applications, one is less interested in a single predicted value for y given by $f(\mathbf{x})$ than in $p(y|\mathbf{x})$, a conditional probability distribution of y given $\mathbf{x}$. With $p(y|\mathbf{x})$, one can easily obtain a single predicted value for y by taking the mean, the median or the mode (i.e. the location of the peak) of the distribution $p(y|\mathbf{x})$. In addition, the distribution provides an estimate of the uncertainty in the predicted value for y. For managers of utility companies, the forecast that tomorrow's air temperature will be 25°C is far less useful than the same forecast accompanied by the additional information that there is a 10% chance that the temperature will be higher than 30°C and 10% chance lower than 22°C.

Many types of non-Gaussian distribution are encountered in the environment. For instance, variables such as precipitation and wind speed have distributions which are skewed to the right, since one cannot have negative values for precipitation and wind speed. The gamma distribution and the Weibull distributions have been commonly used to model precipitation and wind speed respectively (Wilks, 1995). Variables such as relative humidity and cloud amount (measured as covering a fraction of the sky) are limited to lie within the interval [0, 1], and are commonly modelled by the beta distribution (Wilks, 1995). The Johnson system of distributions is a system of flexible functions used to fit a wide variety of empirical data (Niermann, 2006).

Knowledge of the distribution of *extreme values* is also vital to the design of safe buildings, dams, floodways, roads and bridges. For instance, levees need to be built to handle say the strongest hurricane expected in a century to prevent a city from drowning. Insurance companies also need to know the risks involved in order to set insurance premiums at a profitable level. In global warming studies, one is also interested in the change in the extremes – e.g. the extremes of the daily maximum temperature. Long term changes in the distribution of extreme events from global climate change have also been investigated, e.g. for storms (Lambert and Fyfe, 2006) and heavy precipitation events (Zhang *et al.*, 2001). The Gumbel distribution, which is skewed to the right, is commonly used in extreme value analysis (von Storch and Zwiers, 1999).

Suppose we have selected an appropriate distribution function, which is governed by some parameters $\boldsymbol{\theta}$. For instance, the gamma distribution is governed by two parameters ($\boldsymbol{\theta} = [c, s]^{\mathrm{T}}$):

$$g(y|c, s) = \frac{1}{Z}\left(\frac{y}{s}\right)^{c-1} \exp\left(-\frac{y}{s}\right), \quad 0 \le y < \infty, \tag{4.49}$$

where $c > 0, s > 0$ and $Z = \Gamma(c)s$, with Γ denoting the gamma function, an extension of the factorial function to a real or complex variable. We allow the parameters to be functions of the inputs $\mathbf{x}$, i.e. $\boldsymbol{\theta} = \boldsymbol{\theta}(\mathbf{x})$. The conditional distribution $p(y|\mathbf{x})$ is now replaced by $p(y|\boldsymbol{\theta}(\mathbf{x}))$. The functions $\boldsymbol{\theta}(\mathbf{x})$ can be approximated by an NN (e.g. an MLP or an RBF) model, i.e. inputs of the NN model are $\mathbf{x}$ while the outputs are $\boldsymbol{\theta}$. Using NN to model the parameters of a conditional probability density distribution is sometimes called a *conditional density network* (CDN) model. To train the NN model, we need an objective function.

To obtain an objective function, we turn to the principle of *maximum likelihood*. If we have a probability distribution $p(\mathbf{y}|\boldsymbol{\theta})$, and we have observed values $\mathbf{y}_{\mathrm{d}}$ given by the dataset D, then the parameters $\boldsymbol{\theta}$ can be found by maximizing the likelihood function $p(D|\boldsymbol{\theta})$, i.e. the parameters $\boldsymbol{\theta}$ should be chosen so that the likelihood of observing D is maximized. Note that $p(D|\boldsymbol{\theta})$ is a function of $\boldsymbol{\theta}$ as D is known, and the output $\mathbf{y}$ can be multivariate.

Since the observed data have $n = 1, \ldots, N$ observations, and if we assume independent observations so we can multiply their probabilities together, the likelihood function is then

$$L = p(D|\boldsymbol{\theta}) = \prod_{n=1}^{N} p(\mathbf{y}^{(n)}|\boldsymbol{\theta}^{(n)}) = \prod_{n=1}^{N} p(\mathbf{y}^{(n)}|\boldsymbol{\theta}(\mathbf{x}^{(n)})), \tag{4.50}$$

where the observed data $\mathbf{y}_{\mathrm{d}}^{(n)}$ are simply written as $\mathbf{y}^{(n)}$, and $\boldsymbol{\theta}(\mathbf{x}^{(n)})$ are determined by the weights $\mathbf{w}$ (including all weight and offset parameters) of the NN model. Hence

$$L = \prod_{n=1}^{N} p(\mathbf{y}^{(n)}|\mathbf{w}, \mathbf{x}^{(n)}). \tag{4.51}$$

Mathematically, maximizing the likelihood function is equivalent to minimizing the negative logarithm of the likelihood function, since the logarithm is a monotonically increasing function. Hence we choose the objective function to be

$$J = -\ln L = -\sum_{n=1}^{N} \ln p(\mathbf{y}^{(n)}|\mathbf{w}, \mathbf{x}^{(n)}), \tag{4.52}$$

where we have converted the (natural) logarithm of a product of N terms to a sum of N logarithmic terms. Since $\mathbf{x}^{(n)}$ and $\mathbf{y}^{(n)}$ are known from the given data, the unknowns $\mathbf{w}$ are optimized to attain the minimum J. Once the weights of the NN model are solved, then for any input $\mathbf{x}$, the NN model outputs $\boldsymbol{\theta}(\mathbf{x})$, which gives the conditional distribution $p(\mathbf{y}|\mathbf{x})$ via $p(\mathbf{y}|\boldsymbol{\theta}(\mathbf{x}))$.

In our example, where $p(y|\boldsymbol{\theta})$ is the gamma distribution (4.49), we have to ensure that the outputs of the NN model satisfy the restriction that both parameters (c and s) of the gamma distribution have to be positive. This can be accommodated easily by letting

$$c = \exp(z_1), \quad s = \exp(z_2), \tag{4.53}$$

where z_1 and z_2 are the NN model outputs.

4.7.1 Mixture models

The disadvantage of specifying a parametric form for the conditional distribution is that even adjusting the parameters may not lead to a good fit to the observed data. One way to produce an extremely flexible distribution function is to use a mixture (i.e. a weighted sum) of simple distribution functions to produce a *mixture model*:

$$p(\mathbf{y}|\mathbf{x}) = \sum_{k=1}^{K} a_k(\mathbf{x})\phi_k(\mathbf{y}|\mathbf{x}), \tag{4.54}$$

where K is the number of components (also called *kernels*) in the mixture, $a_k(\mathbf{x})$ is the (non-negative) *mixing coefficient* and $\phi_k(\mathbf{y}|\mathbf{x})$ the conditional distribution from the kth kernel. There are many choices for the kernel distribution function ϕ, the most popular choice being the Gaussian function

$$\phi_k(\mathbf{y}|\mathbf{x}) = \frac{1}{(2\pi)^{M/2}\,\sigma_k^M(\mathbf{x})} \exp\left(-\frac{\|\mathbf{y} - \boldsymbol{\mu}_k(\mathbf{x})\|^2}{2\sigma_k^2(\mathbf{x})}\right), \tag{4.55}$$

where the Gaussian kernel function is centred at $\boldsymbol{\mu}_k(\mathbf{x})$ with variance $\sigma_k^2(\mathbf{x})$, and M is the dimension of the output vector $\mathbf{y}$. With large enough K, and with properly chosen $\boldsymbol{\mu}_k(\mathbf{x})$ and $\sigma_k(\mathbf{x})$, $p(\mathbf{y}|\mathbf{x})$ of any form can be approximated to arbitrary accuracy by the Gaussian mixture model.

An NN model approximates the parameters of the Gaussian mixture model, namely $\boldsymbol{\mu}_k(\mathbf{x})$, $\sigma_k(\mathbf{x})$ and $a_k(\mathbf{x})$. There are a total of $M \times K$ parameters in $\boldsymbol{\mu}_k(\mathbf{x})$, and

K parameters in each of $\sigma_k(\mathbf{x})$ and $a_k(\mathbf{x})$, hence a total of $(M+2)K$ parameters. Let $\mathbf{z}$ represent the $(M+2)K$ outputs of the NN model. Since there are constraints on $\sigma_k(\mathbf{x})$ and $a_k(\mathbf{x})$, they cannot simply be the direct outputs from the NN model. As $\sigma_k(\mathbf{x}) > 0$, we need to represent them as

$$\sigma_k = \exp\left(z_k^{(\sigma)}\right), \tag{4.56}$$

where $z_k^{(\sigma)}$ are the NN model outputs related to the σ parameters.

From the normalization condition

$$\int p(\mathbf{y}|\mathbf{x})\mathrm{d}\mathbf{y} = 1, \tag{4.57}$$

we obtain, via (4.54) and (4.55), the constraints

$$\sum_{k=1}^{K} a_k(\mathbf{x}) = 1, \quad 0 \le a_k(\mathbf{x}) \le 1. \tag{4.58}$$

To satisfy these constraints, $a_k(\mathbf{x})$ is related to the NN output $z_k^{(a)}$ by a *softmax function*, i.e.

$$a_k = \frac{\exp\left(z_k^{(a)}\right)}{\sum_{k'=1}^{K} \exp\left(z_{k'}^{(a)}\right)}. \tag{4.59}$$

As the softmax function is widely used in NN models involved in classification problems, it is discussed in detail in Chapter 8. Since there are no constraints on $\boldsymbol{\mu}_k(\mathbf{x})$, they can simply be the NN model outputs directly, i.e.

$$\mu_{jk} = z_{jk}^{(\mu)}. \tag{4.60}$$

The objective function for the NN model is again obtained via the likelihood as in (4.52), with

$$J = -\sum_n \ln\left(\sum_{k=1}^{K} a_k(\mathbf{x}^{(n)})\phi_k(\mathbf{y}^{(n)}|\mathbf{x}^{(n)})\right). \tag{4.61}$$

Once the NN model weights $\mathbf{w}$ are solved from minimizing J, we obtain the mixture model parameters $\boldsymbol{\mu}_k(\mathbf{x})$, $\sigma_k(\mathbf{x})$ and $a_k(\mathbf{x})$ from the NN model outputs, hence the conditional distribution $p(\mathbf{y}|\mathbf{x})$ via (4.54).

To get a specific $\mathbf{y}$ value given $\mathbf{x}$, we calculate the mean of the conditional distribution using (4.54) and (4.55), yielding

$$\mathrm{E}[\mathbf{y}|\mathbf{x}] = \int \mathbf{y}\, p(\mathbf{y}|\mathbf{x})\mathrm{d}\mathbf{y} = \sum_k a_k(\mathbf{x}) \int \mathbf{y}\, \phi_k(\mathbf{y}|\mathbf{x})\mathrm{d}\mathbf{y} = \sum_k a_k(\mathbf{x})\boldsymbol{\mu}_k(\mathbf{x}). \tag{4.62}$$

We can also calculate the variance of the conditional distribution about the conditional mean using (4.54), (4.55) and (4.62):

$$s^2(\mathbf{x}) = \mathrm{E}\left[\, \|\mathbf{y} - \mathrm{E}[\mathbf{y}|\mathbf{x}]\,\|^2 \,|\, \mathbf{x}\,\right]$$
$$= \sum_k a_k(\mathbf{x}) \left[\sigma_k(\mathbf{x})^2 + \left\| \boldsymbol{\mu}_k(\mathbf{x}) - \sum_{k=1}^{K} a_k(\mathbf{x})\boldsymbol{\mu}_k(\mathbf{x}) \right\|^2 \right]. \qquad (4.63)$$

Hence the Gaussian mixture model not only gives the conditional distribution $p(\mathbf{y}|\mathbf{x})$, but also conveniently provides, for a given $\mathbf{x}$, a specific estimate for $\mathbf{y}$ and a measure of its uncertainty via the conditional mean (4.62) and variance (4.63).

Exercises

(4.1) It is given that x, y and z are binary variables with the value of 0 or 1, and $z = f(x, y)$. We have encountered in Section 4.2 two special cases of f, namely the AND logical function, and the XOR (i.e. the exclusive OR) function. There are a total of 16 such possible $f(x, y)$ logical functions. Which of these 16 functions cannot be represented by a perceptron model with two input neurons connected directly to an output neuron by a step function?

(4.2) Show that for a multi-layer perceptron (MLP) NN with one hidden layer, if the activation function for the hidden layer is linear, then the NN reduces to one without any hidden layer.

(4.3) Consider an MLP NN model containing m_1 inputs, one hidden layer with m_2 neurons and m_3 outputs. (a) If the activation function in the hidden layer is the tanh function, what happens if we flip the signs of all the weights (including offset parameters) feeding into a particular hidden neuron, and also flip the signs of the weights leading out of that neuron? For a given set of weights for the NN, how many equivalent sets of weights are there due to the 'sign-flip' symmetries? (b) Furthermore, the weights associated with one hidden neuron can be interchanged with those of another hidden neuron without affecting the NN outputs. Hence show that there are a total of $m_2!\,2^{m_2}$ equivalent sets of weights from the 'sign-flip' and the 'interchange' symmetries.

(4.4) For the logistic sigmoidal activation function $f(x)$ in (4.5), show that its derivative $f'(x)$ can be expressed in terms of $f(x)$. Also show that for the activation function $\tanh(x)$, its derivative can also be expressed in terms of the tanh function.

5

Nonlinear optimization

As mentioned in the previous chapter, the polynomial fit, which suffers from the curse of dimensionality, requires only linear optimization, while the MLP NN model needs nonlinear optimization. To appreciate the vast difference between linear optimization and nonlinear optimization, consider the relation

$$y = w_0 + \sum_{l=1}^{L} w_l f_l, \tag{5.1}$$

where $f_l = f_l(x_1, \ldots, x_m)$, and the polynomial fit is a special case. Although the response variable y is nonlinearly related to the predictor variables $x_1, \ldots, x_m$ (as f_l is in general a nonlinear function), y is a linear function of the parameters $\{w_l\}$. It follows that the objective function

$$J = \sum (y - y_{\mathrm{d}})^2, \tag{5.2}$$

(with y_{d} the target data and the summation over all observations) is a quadratic function of the $\{w_l\}$, which means that the objective function $J(w_0, \ldots, w_L)$ is a parabolic surface, which has a single minimum, the global minimum.

In contrast, when y is a nonlinear function of $\{w_l\}$, the objective function surface is in general filled with numerous hills and valleys, i.e. there are usually many local minima besides the global minimum. (If there are symmetries among the parameters, there can even be multiple global minima.) Thus nonlinear optimization involves finding a global minimum among many local minima. The difficulty faced by the optimization algorithm is similar to that encountered by a robot rover sent to explore the rugged surface of a planet. The rover can easily fall into a hole or a valley and be unable to escape from it, thereby never reaching its final objective, the global minimum. Thus nonlinear optimization is vastly more tricky than linear optimization, with no guarantee that the algorithm actually finds the global minimum, as it may become trapped by a local minimum.

In essence, with NN models, one needs to minimize the objective function J with respect to $\mathbf{w}$ (which includes all the weight and offset/bias parameters), i.e. find the optimal parameters which will minimize J. It is common to solve the minimization problem using an iterative procedure. Suppose the current approximation of the solution is $\mathbf{w}_0$. A Taylor series expansion of $J(\mathbf{w})$ around $\mathbf{w}_0$ yields

$$J(\mathbf{w}) = J(\mathbf{w}_0) + (\mathbf{w} - \mathbf{w}_0)^{\mathrm{T}} \nabla J(\mathbf{w}_0) + \frac{1}{2}(\mathbf{w} - \mathbf{w}_0)^{\mathrm{T}} \mathbf{H} (\mathbf{w} - \mathbf{w}_0) + \ldots, \tag{5.3}$$

where ∇J has components $\partial J/\partial w_i$, and $\mathbf{H}$ is the *Hessian matrix*, with elements

$$(\mathbf{H})_{ij} \equiv \left. \frac{\partial^2 J}{\partial w_i \partial w_j} \right|_{\mathbf{w}_0}. \tag{5.4}$$

Applying the gradient operator to (5.3), we obtain

$$\nabla J(\mathbf{w}) = \nabla J(\mathbf{w}_0) + \mathbf{H} (\mathbf{w} - \mathbf{w}_0) + \ldots \tag{5.5}$$

If $\mathbf{w}_0$ is a minimum, then $\nabla J(\mathbf{w}_0) = 0$, and (5.3) (with the higher order terms ignored) reduces to an equation describing a parabolic surface. Hence, near a minimum, assuming the Hessian matrix is non-zero, the objective function has an approximately parabolic surface.

Next, let us derive an iterative scheme for finding the optimal $\mathbf{w}$. At the optimal $\mathbf{w}$, $\nabla J(\mathbf{w}) = 0$, and (5.5), with higher order terms ignored, yields

$$\mathbf{H} (\mathbf{w} - \mathbf{w}_0) = -\nabla J(\mathbf{w}_0), \quad \text{i.e. } \mathbf{w} = \mathbf{w}_0 - \mathbf{H}^{-1} \nabla J(\mathbf{w}_0). \tag{5.6}$$

This suggests the following iterative scheme for proceding from step k to step $k+1$:

$$\mathbf{w}_{k+1} = \mathbf{w}_k - \mathbf{H}_k^{-1} \nabla J(\mathbf{w}_k). \tag{5.7}$$

This is known as *Newton's method*. In the 1-dimensional case, (5.7) reduces to the familiar form

$$w_{k+1} = w_k - \frac{J'(w_k)}{J''(w_k)}, \tag{5.8}$$

for finding a root of $J'(w) = 0$, where the prime and double prime denote respectively the first and second derivatives.

In the multi-dimensional case, if $\mathbf{w}$ is of dimension L, then the Hessian matrix $\mathbf{H}_k$ is of dimension $L \times L$. Computing $\mathbf{H}_k^{-1}$, the inverse of an $L \times L$ matrix, may be computationally too costly. Simplification is needed, resulting in quasi-Newton methods. Unlike quasi-Newton and other deterministic optimization methods, stochastic optimization methods (e.g. simulated annealing and genetic algorithms) introduce some stochastic element into the search for the global minimum.

5.1 Gradient descent method

We have already encountered the *gradient descent* or *steepest descent* method in the back-propagation algorithm, where the parameters are updated at the kth step by

$$\mathbf{w}_{k+1} = \mathbf{w}_k - \eta \nabla J(\mathbf{w}_k), \tag{5.9}$$

with η the learning rate. Clearly (5.9) is a major simplification of Newton's method (5.7), with the learning rate η replacing $\mathbf{H}^{-1}$, the inverse of the Hessian matrix. One also tries to reach the optimal $\mathbf{w}$ by descending along the negative gradient of J in (5.9), hence the name gradient descent or steepest descent, as the negative gradient gives the direction of steepest descent. An analogy is a hiker trying to descend in thick fog from a mountain to the bottom of a valley by taking the steepest descending path. One might be tempted to think that following the direction of steepest descent should allow the hiker to reach the bottom most efficiently; alas, this approach is surprisingly inefficient, as we shall see.

The learning rate η can be either a fixed constant, or calculated by a line minimization algorithm. In the former case, one simply takes a step of fixed size along the direction of the negative gradient of J. In the latter, one proceeds along the negative gradient of J until one reaches the minimum of J along that direction (Fig. 5.1). More precisely, suppose at step k, we have estimated parameters $\mathbf{w}_k$. We then descend along the negative gradient of the objective function, i.e. travel along the direction

$$\mathbf{d}_k = -\nabla J(\mathbf{w}_k). \tag{5.10}$$

We then travel along $\mathbf{d}_k$, with our path described by $\mathbf{w}_k + \eta\mathbf{d}_k$, until we reach the minimum of J along this direction. Going further along this direction would mean

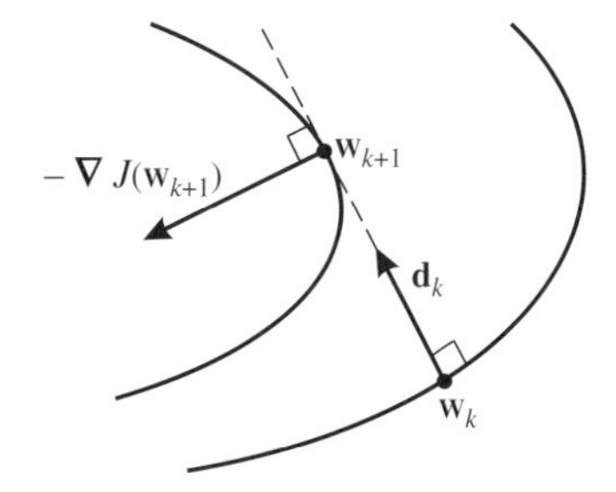

Fig. 5.1 The gradient descent approach starts from the parameters $\mathbf{w}_k$ estimated at step k of an iterative optimization process. The descent path $\mathbf{d}_k$ is chosen along the negative gradient of the objective function J, which is the steepest descent direction. Note that $\mathbf{d}_k$ is perpendicular to the J contour where $\mathbf{w}_k$ lies. The descent along $\mathbf{d}_k$ proceeds until it is tangential to a second contour at $\mathbf{w}_{k+1}$, where the direction of steepest descent is given by $-\nabla J(\mathbf{w}_{k+1})$. The process is iterated.

we would actually be ascending rather than descending, so we should stop at this minimum of J along $\mathbf{d}_k$, which occurs at

$$\frac{\partial}{\partial \eta} J(\mathbf{w}_k + \eta \mathbf{d}_k) = 0, \tag{5.11}$$

thereby yielding the optimal step size η. The differentiation by η gives

$$\mathbf{d}_k^{\mathrm{T}} \nabla J(\mathbf{w}_k + \eta \mathbf{d}_k) = 0. \tag{5.12}$$

With

$$\mathbf{w}_{k+1} = \mathbf{w}_k + \eta \mathbf{d}_k, \tag{5.13}$$

we can rewrite the above equation as

$$\mathbf{d}_k^{\mathrm{T}} \nabla J(\mathbf{w}_{k+1}) = 0, \quad \text{i.e. } \mathbf{d}_k \perp \nabla J(\mathbf{w}_{k+1}). \tag{5.14}$$

But since $\mathbf{d}_{k+1} = -\nabla J(\mathbf{w}_{k+1})$, we have

$$\mathbf{d}_k^{\mathrm{T}} \mathbf{d}_{k+1} = 0, \quad \text{i.e. } \mathbf{d}_k \perp \mathbf{d}_{k+1}. \tag{5.15}$$

As the new direction $\mathbf{d}_{k+1}$ is orthogonal to the previous direction $\mathbf{d}_k$, this results in an inefficient zigzag path of descent (Fig. 5.2(a)).

The other alternative of using fixed step size is also inefficient, as a small step size results in taking too many steps (Fig. 5.2(b)), while a large step size results in an even more severely zigzagged path of descent (Fig. 5.2(c)).

One way to reduce the zigzag in the gradient descent scheme is to add '*momentum*' to the descent direction, so

$$\mathbf{d}_k = -\nabla J(\mathbf{w}_k) + \mu \mathbf{d}_{k-1}, \tag{5.16}$$

with μ the momentum parameter. Here the momentum or memory of $\mathbf{d}_{k-1}$ prevents the new direction $\mathbf{d}_k$ from being orthogonal to $\mathbf{d}_{k-1}$, thereby reducing the zigzag (Fig. 5.2(d)). The next estimate for the parameters in the momentum method is also given by (5.13). The following scheme, the conjugate gradient method, automatically chooses the momentum parameter μ.

5.2 Conjugate gradient method

The linear conjugate gradient method was developed by Hestenes and Stiefel (1952), and extended to nonlinear problems by Fletcher and Reeves (1964). Assume that at step k, (5.12) is satisfied. Now we want to find the next direction $\mathbf{d}_{k+1}$, which preserves what was achieved in (5.12) – in (5.12), the gradient of J in the direction of $\mathbf{d}_k$ has been made 0; now starting from $\mathbf{w}_{k+1}$, we want to find the

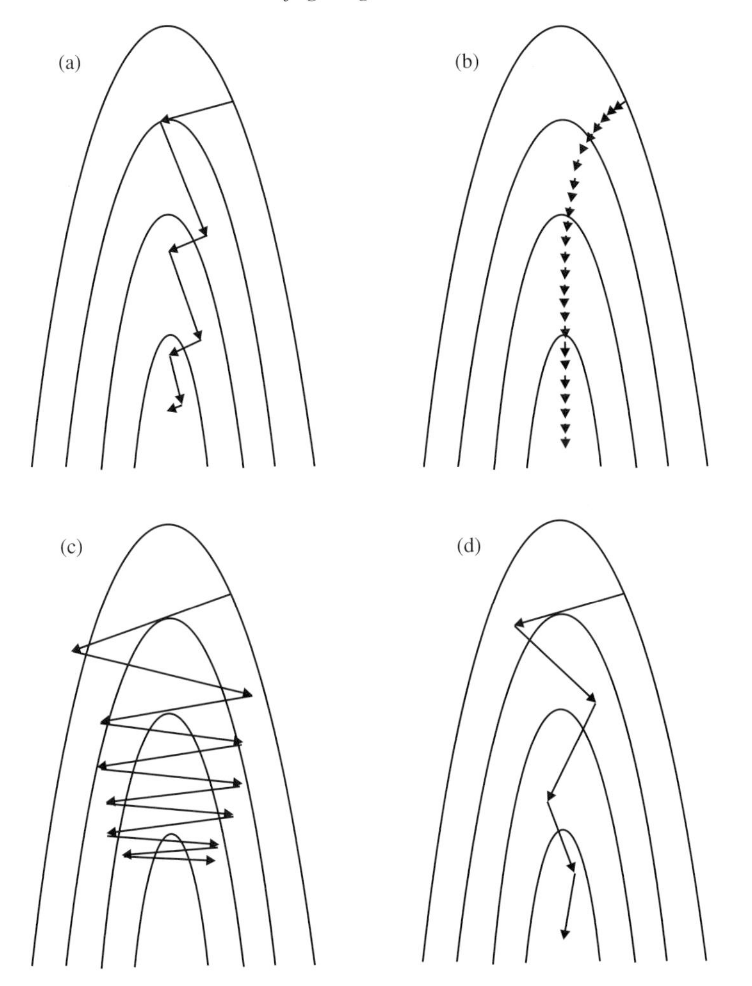

Fig. 5.2 The gradient descent method with (a) line minimization (i.e. optimal step size η), (b) fixed step size which is too small, (c) fixed step size which is too large, and (d) momentum, which reduces the zigzag behaviour during descent. (Adapted from Masters (1995).)

new direction $\mathbf{d}_{k+1}$, such that the gradient of J in the direction of $\mathbf{d}_k$ remains 0 (to lowest order), as we travel along $\mathbf{d}_{k+1}$, i.e.

$$\mathbf{d}_k^{\mathrm{T}} \nabla J(\mathbf{w}_{k+1} + \eta \mathbf{d}_{k+1}) = 0. \tag{5.17}$$

Using (5.5), we can write

$$\nabla J(\mathbf{w}_{k+1} + \eta\, \mathbf{d}_{k+1}) \approx \nabla J(\mathbf{w}_{k+1}) + \mathbf{H}\, \eta\, \mathbf{d}_{k+1}. \tag{5.18}$$

Hence (5.17) becomes

$$0 = \mathbf{d}_k^{\mathrm{T}} \nabla J(\mathbf{w}_{k+1} + \eta\, \mathbf{d}_{k+1}) \approx \mathbf{d}_k^{\mathrm{T}} \nabla J(\mathbf{w}_{k+1}) + \eta\, \mathbf{d}_k^{\mathrm{T}}\, \mathbf{H}\, \mathbf{d}_{k+1}. \tag{5.19}$$

Invoking (5.14), we obtain (to lowest order) the conjugate direction property

$$\mathbf{d}_k^{\mathrm{T}}\,\mathbf{H}\,\mathbf{d}_{k+1} = 0, \tag{5.20}$$

where $\mathbf{d}_{k+1}$ is said to be *conjugate* to $\mathbf{d}_k$.

Next, we try to obtain an estimate for the momentum parameter μ. Let

$$\mathbf{g}_k = \nabla J(\mathbf{w}_k). \tag{5.21}$$

Substituting (5.16) for $\mathbf{d}_{k+1}$ in (5.20) yields

$$\mathbf{d}_k^{\mathrm{T}}\,\mathbf{H}\,(-\mathbf{g}_{k+1} + \mu\mathbf{d}_k) = 0. \tag{5.22}$$

Hence

$$\mu\mathbf{d}_k^{\mathrm{T}}\,\mathbf{H}\,\mathbf{d}_k = \mathbf{d}_k^{\mathrm{T}}\,\mathbf{H}\,\mathbf{g}_{k+1} = \mathbf{g}_{k+1}^{\mathrm{T}}\,\mathbf{H}\,\mathbf{d}_k, \tag{5.23}$$

as $\mathbf{H}^{\mathrm{T}} = \mathbf{H}$. While μ can be calculated from this equation, it involves the Hessian matrix, which is in general not known and is computationally costly to estimate.

Ignoring higher order terms and following (5.5), we have

$$\mathbf{g}_{k+1} - \mathbf{g}_k = \mathbf{H}(\mathbf{w}_{k+1} - \mathbf{w}_k) = \eta\mathbf{H}\mathbf{d}_k, \tag{5.24}$$

where (5.13) has been invoked. Substituting this equation into (5.23) gives an estimate for the momentum parameter μ

$$\mu = \frac{\mathbf{g}_{k+1}^{\mathrm{T}}(\mathbf{g}_{k+1} - \mathbf{g}_k)}{\mathbf{d}_k^{\mathrm{T}}(\mathbf{g}_{k+1} - \mathbf{g}_k)}. \tag{5.25}$$

This way to estimate μ is called the *Hestenes–Stiefel* method.

A far more commonly used conjugate gradient algorithm is the *Polak–Ribiere* method: (5.12) can be written as

$$\mathbf{d}_k^{\mathrm{T}}\,\mathbf{g}_{k+1} = 0. \tag{5.26}$$

In accordance with (5.16),

$$\mathbf{d}_k^{\mathrm{T}}\,\mathbf{g}_k = -\mathbf{g}_k^{\mathrm{T}}\,\mathbf{g}_k + \mu\,\mathbf{d}_{k-1}^{\mathrm{T}}\,\mathbf{g}_k = -\mathbf{g}_k^{\mathrm{T}}\,\mathbf{g}_k, \tag{5.27}$$

where (5.26) has been invoked. Equation (5.25) becomes

$$\mu = \frac{\mathbf{g}_{k+1}^{\mathrm{T}}(\mathbf{g}_{k+1} - \mathbf{g}_k)}{\mathbf{g}_k^{\mathrm{T}}\mathbf{g}_k}, \tag{5.28}$$

which is the Polak–Ribiere method (Polak and Ribiere, 1969; Polak, 1971).

Another commonly used algorithm is the *Fletcher–Reeves* method (Fletcher and Reeves, 1964), with

$$\mu = \frac{\mathbf{g}_{k+1}^{\mathrm{T}}\mathbf{g}_{k+1}}{\mathbf{g}_k^{\mathrm{T}}\mathbf{g}_k}, \tag{5.29}$$

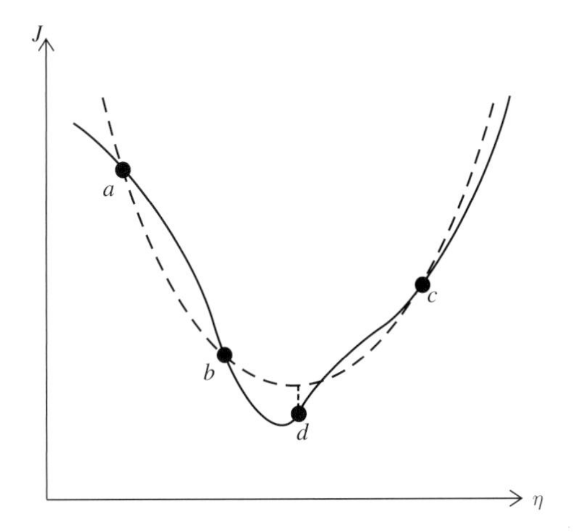

Fig. 5.3 Using line search to find the minimum of the function $J(\eta)$. First, three points a, b and c are found with $J(a) > J(b)$ and $J(b) < J(c)$, so that the minimum is bracketed within the interval (a, c). Next a parabola is fitted to pass through the three points (dashed curve). The minimum of the parabola is at $\eta = d$. Next the three points among a, b, c and d with the three lowest values of J are selected, and a new parabola is fitted to the three selected points, with the process iterated till convergence to the minimum of J.

which follows from the Polak–Ribiere version, as it can be shown that $\mathbf{g}_{k+1}^{\mathrm{T}}\mathbf{g}_k = 0$ to lowest order (Bishop, 1995). Since the objective function is not exactly a quadratic, the higher order terms cause the two versions to differ from each other. Which version is better depends on the particular problem, though the Polak–Ribiere version is generally considered to have the better performance (Luenberger, 1984; Haykin, 1999).

We still have to find the optimal step size η along the search direction $\mathbf{d}_k$, i.e. the minimum of the objective function J along $\mathbf{d}_k$ has to be located by finding the η which minimizes $J(\mathbf{w}_k + \eta\mathbf{d}_k)$. For notational simplicity, we will write $J(\mathbf{w}_k + \eta\mathbf{d}_k)$ as $J(\eta)$. To avoid dealing with costly Hessian matrices, a *line search* algorithm is commonly used. The basic line search procedure is as follows:

(1) Find three points a, b and c along the search direction such that $J(a) > J(b)$ and $J(b) < J(c)$. As the objective function is continuous, this then guarantees a minimum has been *bracketed* within the interval (a, c).
(2) Fit a parabolic curve to pass through $J(a)$, $J(b)$ and $J(c)$ (Fig. 5.3). The minimum of the parabola is at $\eta = d$.
(3) Next choose the three points among a, b, c and d with the three lowest values of J. Repeat (2) till convergence to the minimum. More sophisticated algorithms include the widely used Brent's method (Brent, 1973; Press *et al.*, 1986).

5.3 Quasi-Newton methods

Earlier in this chapter, we have encountered Newton's method as (5.7), which can be expressed as

$$\mathbf{w}_{k+1} = \mathbf{w}_k - \mathbf{G}_k \mathbf{g}_k, \tag{5.30}$$

with $\mathbf{g}_k (\equiv \nabla J(\mathbf{w}_k))$, the gradient of the objective function, and $\mathbf{G}_k (\equiv \mathbf{H}_k^{-1})$, the inverse of the Hessian matrix. As this form was derived by ignoring terms above the quadratic in the objective function, this form is highly effective near a minimum, where the objective function has generally a parabolic surface, but may not be effective further away from the minimum because of the higher order terms in the objective function. Hence, a simple modification of Newton's method is to have

$$\mathbf{w}_{k+1} = \mathbf{w}_k - \eta_k \mathbf{G}_k \mathbf{g}_k, \tag{5.31}$$

for some scalar step size η_k.

Newton's method is extremely expensive for higher dimension problems, since at each iteration, the inverse of the Hessian matrix has to be computed. *Quasi-Newton methods* try to reduce computational costs by making simpler estimates of $\mathbf{G}_k$. Quasi-Newton methods are also related to gradient and conjugate gradient methods. In fact, the simplest approximation, i.e. replacing $\mathbf{G}_k$ by the identity matrix $\mathbf{I}$ in (5.31), yields the gradient descent method. The commonly used quasi-Newton methods also preserve the conjugate direction property (5.20) as in conjugate gradient methods.

The first successful quasi-Newton method is the *Davidon–Fletcher–Powell* (DFP) method (Davidon, 1959; Fletcher and Powell, 1963). The procedure is to start at step $k = 0$, with any $\mathbf{w}_0$ and any symmetric positive definite matrix $\mathbf{G}_0$. Then iterate the following steps.

(1) Set $\mathbf{d}_k = -\mathbf{G}_k \mathbf{g}_k$.
(2) Minimize $J(\mathbf{w}_k + \eta_k \mathbf{d}_k)$ with respect to $\eta_k \geq 0$. One then computes $\mathbf{w}_{k+1}$, $\mathbf{p}_k \equiv \eta_k \mathbf{d}_k$, and $\mathbf{g}_{k+1}$.
(3) Set

$$\mathbf{q}_k = \mathbf{g_{k+1}} - \mathbf{g_k}, \text{ and}$$
$$\mathbf{G}_{k+1} = \mathbf{G}_k + \frac{\mathbf{p}_k \mathbf{p}_k^{\mathrm{T}}}{\mathbf{p}_k^{\mathrm{T}} \mathbf{q}_k} - \frac{\mathbf{G}_k \mathbf{q}_k \mathbf{q}_k^{\mathrm{T}} \mathbf{G}_k}{\mathbf{q}_k^{\mathrm{T}} \mathbf{G}_k \mathbf{q}_k}. \tag{5.32}$$

(4) Update k, then return to (1) if needed.

The conjugate direction property (5.20) is preserved in the DFP method (Luenberger, 1984). If one chooses the initial approximation $\mathbf{G}_0 = \mathbf{I}$, the DFP method becomes the conjugate gradient method.

The most popular quasi-Newton method is the *Broyden–Fletcher–Goldfarb–Shanno* (BFGS) method (Broyden, 1970; Fletcher, 1970; Goldfarb, 1970; Shanno, 1970). Instead of (5.32), the update for the BFGS method (Luenberger, 1984) is

$$\mathbf{G}_{k+1}^{\mathrm{BFGS}} = \mathbf{G}_{k+1}^{\mathrm{DFP}} + \mathbf{v}_k \mathbf{v}_k^{\mathrm{T}}, \tag{5.33}$$

where $\mathbf{G}_{k+1}^{\mathrm{DFP}}$ is given by (5.32), and

$$\mathbf{v}_k = (\mathbf{q}_k^{\mathrm{T}} \mathbf{G}_k \mathbf{q}_k)^{1/2} \left(\frac{\mathbf{p}_k}{\mathbf{p}_k^{\mathrm{T}} \mathbf{q}_k} - \frac{\mathbf{G}_k \mathbf{q}_k}{\mathbf{q}_k^{\mathrm{T}} \mathbf{G}_k \mathbf{q}_k} \right). \tag{5.34}$$

The conjugate direction property (5.20) is also preserved in the BFGS method.

In summary, both conjugate gradient and quasi-Newton methods avoid using the inverse of the Hessian matrix. However, quasi-Newton methods do try to approximate the inverse Hessian, while preserving the conjugate direction property of the conjugate gradient methods. Thus quasi-Newton methods can be regarded as a further extension of the conjugate gradient methods by incorporating an approximation of the inverse Hessian to simulate Newton's method, which leads to faster convergence than the conjugate gradient methods. Another advantage is that the line search, which is of critical importance in the conjugate gradient methods, need not be performed to high accuracy in the quasi-Newton methods, as its role is not as critical. The major disadvantage of the quasi-Newton methods is the large storage associated with carrying the $L \times L$ matrix $\mathbf{G}_k$ (as $\mathbf{w}$ has L elements), i.e. the memory requirement for the quasi-Newton methods is of order $O(L^2)$ versus $O(L)$ for the conjugate gradient methods. For L of a thousand or more weights, the memory requirement of $O(L^2)$ may become prohibitive, and the conjugate gradient method has the advantage.

To reduce the large memory requirement, Shanno (1978) proposed *limited memory quasi-Newton methods*. In the limited memory version of the BFGS method, the matrix $\mathbf{G}_k$ in (5.32) and (5.34) is replaced by the identity matrix $\mathbf{I}$ (Luenberger, 1984), reducing the memory requirement to $O(L)$. It turns out that when an exact line search is used, this new method is equivalent to the Polak–Ribiere form of the conjugate gradient method – of course, the line search is not required to be highly accurate in this limited memory quasi-Newton method.

5.4 Nonlinear least squares methods

So far, the optimization methods presented have not been limited to an objective function of a specific form. In many situations, the objective function or error function $\mathcal{E}$ involves a sum of squares, i.e.

$$\mathcal{E} = \frac{1}{2} \sum_n (\epsilon^{(n)})^2 = \frac{1}{2} \|\boldsymbol{\epsilon}\|^2, \tag{5.35}$$

where we have used $\mathcal{E}$ instead of J to avoid confusion with the Jacobian matrix $\mathbf{J}$ later, $\epsilon^{(n)}$ is the error associated with the nth observation or pattern, $\boldsymbol{\epsilon}$ is a vector with elements $\epsilon^{(n)}$, (and the conventional scale factor $\frac{1}{2}$ has been added to avoid the appearance of a factor of 2 in the derivatives). We now derive optimization methods specially designed to deal with such nonlinear least squares problems.

Suppose at the kth step, we are at the point $\mathbf{w}_k$ in parameter space, and we take the next step to $\mathbf{w}_{k+1}$. To first order, the Taylor expansion of $\boldsymbol{\epsilon}$ is given by

$$\boldsymbol{\epsilon}(\mathbf{w}_{k+1}) = \boldsymbol{\epsilon}(\mathbf{w}_k) + \mathbf{J}_k(\mathbf{w}_{k+1} - \mathbf{w}_k), \tag{5.36}$$

where $\mathbf{J}_k$ is the Jacobian matrix $\mathbf{J}$ at step k, with the elements of $\mathbf{J}$ given by

$$(\mathbf{J})_{ni} = \frac{\partial \epsilon^{(n)}}{\partial w_i}. \tag{5.37}$$

Substituting (5.36) into (5.35), we have

$$\begin{aligned}
\frac{1}{2}\|\boldsymbol{\epsilon}(\mathbf{w}_{k+1})\|^2 &= \frac{1}{2}\|\boldsymbol{\epsilon}(\mathbf{w}_k) + \mathbf{J}_k(\mathbf{w}_{k+1} - \mathbf{w}_k)\|^2 \\
&= \frac{1}{2}\|\boldsymbol{\epsilon}(\mathbf{w}_k)\|^2 + \boldsymbol{\epsilon}^{\mathrm{T}}(\mathbf{w}_k)\mathbf{J}_k(\mathbf{w}_{k+1} - \mathbf{w}_k) \\
&\quad + \frac{1}{2}(\mathbf{w}_{k+1} - \mathbf{w}_k)^{\mathrm{T}}\mathbf{J}_k^{\mathrm{T}}\mathbf{J}_k(\mathbf{w}_{k+1} - \mathbf{w}_k).
\end{aligned} \tag{5.38}$$

To find the optimal $\mathbf{w}_{k+1}$, differentiate the right hand side of the above equation by $\mathbf{w}_{k+1}$ and set the result to zero, yielding

$$\mathbf{J}_k^{\mathrm{T}}\boldsymbol{\epsilon}(\mathbf{w}_k) + \mathbf{J}_k^{\mathrm{T}}\mathbf{J}_k(\mathbf{w}_{k+1} - \mathbf{w}_k) = 0. \tag{5.39}$$

Solving this equation for $\mathbf{w}_{k+1}$, we get

$$\mathbf{w}_{k+1} = \mathbf{w}_k - (\mathbf{J}_k^{\mathrm{T}}\mathbf{J}_k)^{-1}\mathbf{J}_k^{\mathrm{T}}\boldsymbol{\epsilon}(\mathbf{w}_k), \tag{5.40}$$

which is known as the *Gauss–Newton method*. Equation (5.40) resembles the normal equations (1.46) encountered previously in the multiple linear regression problem. In fact, one can regard the Gauss–Newton method as replacing a nonlinear least squares problem with a sequence of linear least squares problems (5.40).

For the sum-of-squares error function (5.35), the gradient can be written as

$$\nabla \mathcal{E} = \mathbf{J}^{\mathrm{T}}\boldsymbol{\epsilon}(\mathbf{w}), \tag{5.41}$$

while the Hessian matrix has elements

$$(\mathbf{H})_{ij} = \frac{\partial^2 \mathcal{E}}{\partial w_i \partial w_j} = \sum_n \left\{ \frac{\partial \epsilon^{(n)}}{\partial w_i}\frac{\partial \epsilon^{(n)}}{\partial w_j} + \epsilon^{(n)}\frac{\partial^2 \epsilon^{(n)}}{\partial w_i \partial w_j} \right\}. \tag{5.42}$$

If the error function depends on the weights linearly, then the second derivatives in (5.42) vanish. Even when the error function is not a linear function of the weights, we will ignore the second order terms in (5.42), and approximate the Hessian by

$$\mathbf{H} = \mathbf{J}^{\mathrm{T}}\mathbf{J}. \tag{5.43}$$

By (5.41) and (5.43), we can regard (5.40) as an approximation of Newton's method (5.7), with the inverse of the Hessian matrix being approximated by $(\mathbf{J}^{\mathrm{T}}\mathbf{J})^{-1}$.

While (5.40) can be used repeatedly to reach the minimum of the error function, the pitfall is that the step size may become too large, so the first order Taylor approximation (5.36) becomes inaccurate, and the Gauss–Newton method may not converge at all. To correct this problem, the *Levenberg–Marquardt* method (Levenberg, 1944; Marquardt, 1963) adds a penalty term to the error function (5.38), i.e.

$$\frac{1}{2}\|\boldsymbol{\epsilon}(\mathbf{w}_{k+1})\|^2 = \frac{1}{2}\|\boldsymbol{\epsilon}(\mathbf{w}_k) + \mathbf{J}_k(\mathbf{w}_{k+1} - \mathbf{w}_k)\|^2 + \lambda\|\mathbf{w}_{k+1} - \mathbf{w}_k\|^2, \tag{5.44}$$

where large step size is penalized by the last term, with a larger parameter λ tending to give smaller step size. Again minimizing this penalized error function with respect to $\mathbf{w}_{k+1}$ yields

$$\mathbf{w}_{k+1} = \mathbf{w}_k - (\mathbf{J}_k^{\mathrm{T}}\mathbf{J}_k + \lambda\mathbf{I})^{-1}\mathbf{J}_k^{\mathrm{T}}\boldsymbol{\epsilon}(\mathbf{w}_k), \tag{5.45}$$

where $\mathbf{I}$ is the identity matrix. For small λ, this Levenberg–Marquardt formula reduces to the Gauss–Newton method, while for large λ, this reduces to the gradient descent method. While the Gauss–Newton method converges very quickly near a minimum, the gradient descent method is robust even when far from a minimum. A common practice is to change λ during the optimization procedure: start with some arbitrary value of λ, say $\lambda = 0.1$. If the error function decreases after taking a step by (5.45), reduce λ by a factor of 10, and the process is repeated. If the error function increases after taking a step, discard the new $\mathbf{w}_{k+1}$, increase λ by a factor of 10, and repeat the step (5.45). The whole process is repeated till convergence. In essence, the Levenberg–Marquardt method improves on the robustness of the Gauss–Newton method by switching to gradient descent when far from a minimum, and then switching back to the Gauss–Newton method for fast convergence when close to a minimum.

Note that if $\mathbf{w}$ is of dimension L, the Hessian $\mathbf{H}$ is of dimension $L \times L$, while the Jacobian $\mathbf{J}$ is of dimension $N \times L$ where N is the number of observations or patterns. Since N is likely to be considerably larger than L, the Jacobian matrix may require even more memory storage than the Hessian matrix. Hence the storage of the Jacobian in Gauss–Newton and Levenberg–Marquardt methods renders them most demanding on memory, surpassing even the quasi-Newton methods, which require the storage of an approximate Hessian matrix.

5.5 Evolutionary computation and genetic algorithms

All the optimization methods presented so far belong to the class of methods known as *deterministic optimization*, in that each step of the optimization process is determined by explicit formulas. While such methods tend to converge to a minimum efficiently, they often converge to a nearby local minimum. To find a global minimum, one usually has to introduce some stochastic element into the search. A simple way is to repeat the optimization process many times, each starting from different random initial weights. These multiple runs will find multiple minima, and one hopes that the lowest minimum among them is the desired global minimum. Of course, there is no guarantee that the global minimum has been found. Nevertheless by using a large enough number of runs and broadly distributed random initial weights, the global minimum can usually be found with such an approach.

Unlike deterministic optimization, *stochastic optimization* methods repeatedly introduce randomness during the search process to avoid getting trapped in a local minimum. Such methods include *simulated annealing* and *evolutionary computation*.

Simulated annealing was inspired by the metallurgical process of annealing of steel (iron–carbon alloy), which is used to produce less brittle metal by gradually cooling very hot metal. When the metal is hot, rapid cooling locks atoms into whatever position they were in when the cooling was applied, thereby producing brittle metal. When the metal is cooled slowly, the atoms tend to align properly, resulting in less brittle metal. If convergence to a minimum is analogous to a metal cooling down, simulated annealing slows the 'cooling' by allowing the search to make random jumps in the parameter space during the convergence process. This slows the convergence process but allows the search to jump out of local minima and to (hopefully) settle down eventually in a global minimum (van Laarhoven and Aarts, 1987; Masters, 1995).

Intelligence has emerged in Nature via biological evolution, so it is not surprising that *evolutionary computation* (EC) (Fogel, 2005) has become a significant branch of Computational Intelligence. Among EC methods, *genetic algorithms* (GA) (Haupt and Haupt, 2004) were inspired by biological evolution where crossover of genes from parents and genetic mutations result in a stochastic process which can lead to superior descendants after many generations. The weight vector $\mathbf{w}$ of a model can be treated as a long strand of DNA, and an ensemble of solutions is treated like a population of organisms. A part of the weight vector of one solution can be exchanged with a part from another solution to form a new solution, analogous to the *cross-over* of DNA material from two parents. For instance, two parents have weight vectors $\mathbf{w}$ and $\mathbf{w}'$. A random position is chosen (in this example just before the third weight parameter) for an incision, and the second part of $\mathbf{w}'$ is connected to the first part of $\mathbf{w}$ and vice versa in the offspring, i.e.

$$\begin{array}{ccc} [w_1, w_2, w_3, w_4, w_5, w_6] & & [w_1', w_2', w_3, w_4, w_5, w_6] \\ & -\text{ cross-over} \rightarrow & \\ [w_1', w_2', w_3', w_4', w_5', w_6'] & & [w_1, w_2, w_3', w_4', w_5', w_6']. \end{array} \tag{5.46}$$

Genetic *mutation* can be simulated by randomly perturbing one of the weights w_j in the weight vector $\mathbf{w}$, i.e. randomly choose a j and replace w_j by $w_j + \epsilon$ for some random ϵ (usually a small random number). These two processes introduce many new offspring, but only the relatively fit offspring have a high probability of surviving to reproduce. With the 'survival of the fittest' principle pruning the offspring, successive generations eventually converge towards the global optimum.

One must specify a *fitness function* f to evaluate the fitness of the individuals in a population. If for the ith individual in the population, its fitness is $f(i)$, then a fitness probability $P(i)$ can be defined as

$$P(i) = \frac{f(i)}{\sum_{i=1}^{N} f(i)}, \tag{5.47}$$

where N is the total number of individuals in a population. Individuals with high P will be given greater chances to reproduce, while those with low P will be given greater chances to die off.

Thus, the basic GA is structured as set out below.

(1) Choose the population size (N) and the number of generations (N_g). Initialize the weight vectors of the population. Repeat the following steps N_g times.
(2) Calculate the fitness function f and the fitness probability P for each individual in the population.
(3) Select a given number of individuals from the population, where the chance of an individual getting selected is given by its fitness probability P.
(4) Duplicate the weight vectors of these individuals, then apply either the cross-over or the mutation operation on the various duplicated weight vectors to produce new offspring.
(5) To keep the population size constant, individuals with poor fitness are selected (based on the probability $1 - P$) to die off, and are replaced by the new offspring. (The fittest individual is never chosen to die.)

Finally, after N_g generations, the individual with the greatest fitness is chosen as the solution. To monitor the evolutionary progress over successive generations, one can check the average fitness of the population, simply by averaging the fitness f over all individuals in the population.

GA can be used to perform the nonlinear optimization in NN problems, where the fitness function can for instance be the negative of the mean squared error. In general, deterministic optimization using gradient descent methods would converge much quicker than stochastic optimization methods such as GA. However, there are three advantages with GA.

(1) In problems where the fitness function cannot be expressed in closed analytic form, gradient descent methods cannot be used effectively, whereas GA works well.
(2) When there are many local optima in the fitness function, gradient descent methods can be trapped too easily.
(3) GA can utilize parallel processors much more readily than gradient descent algorithms, since in GA different individuals in a population can be computed simultaneously on different processors.

In some NN applications, the individuals in a population do not all have the same network topology, e.g. they can have different numbers of hidden neurons. In such cases, GA can be used to find not only the optimal weights but also the optimal network topology. Examples include development of the NERO video games, where the NERO agents (soldiers) evolve their NN topology and weights using GA (Miikkulainen *et al.*, 2006).

In GA applications, the weights $\mathbf{w}$ need not be restricted to continuous variables. However, for $\mathbf{w}$ restricted to continuous variables, much faster EC optimization algorithms are available, e.g. the *differential evolution* algorithm (Storn and Price, 1997; Price *et al.*, 2005), which, like GA, also uses mutations and cross-overs.

Exercises

(5.1) Find the minimum of the function $f(x, y) = x - x^2 + \frac{1}{4}x^4 + 5\sin(xy) - y + y^2$, using (a) stochastic optimization (e.g. genetic algorithm or differential evolution), and (b) deterministic optimization. Compare the cpu times required by the various methods.
(5.2) Train an MLP NN model using either genetic algorithm or differential evolution to perform the nonlinear optimization.

6

Learning and generalization

In Chapter 4, we have learned that NN models are capable of approximating any nonlinear relation $\mathbf{y} = \mathbf{f}(\mathbf{x})$ to arbitrary accuracy by using enough model parameters. However, data generally contain both signal and noise. In the process of fitting the highly flexible NN models to the data to find the underlying relation, one can easily fit to the noise in the data. Like Ulysses who had to balance the twin evils of Scylla and Charybdis, the NN modeller must also steer a careful course between using a model with too little flexibility to model the underlying nonlinear relation adequately (*underfitting*), and using a model with too much flexibility, which readily fits to the noise (*overfitting*). Finding the closest fit to the data – an objective adopted when using linear models – often leads to overfitting when using nonlinear NN models. It needs to be replaced by a wiser objective, that of learning the underlying relation accurately with the NN model. When the NN has found an overfitted solution, it will not fit new data well (Fig. 4.5), but if the NN has learned the underlying relationship well, it will be able to generalize from the original dataset, so that the extracted relationship even fits new data not used in training the NN model. This chapter surveys the various approaches which lead to proper learning and generalization. A comparison of different approaches to estimating the predictive uncertainty of a model is discussed in Section 6.9. Finally, in Section 6.10, we examine why nonlinear machine learning methods often have difficulty outperforming linear methods in climate applications. It turns out that computing climate data by averaging over daily data effectively linearizes the relations in the dataset due to the central limit theorem in statistics.

6.1 Mean squared error and maximum likelihood

In Section 4.3, we have discussed multi-layer perceptron NN models, where minimizing the objective function J involves minimizing the *mean squared error* (MSE) between the model outputs $\mathbf{y}$ and the target data $\mathbf{y}_{\mathrm{d}}$, i.e.

$$J = \frac{1}{N}\sum_{n=1}^{N}\left\{\frac{1}{2}\sum_{k}[y_k^{(n)} - y_{dk}^{(n)}]^2\right\}, \tag{6.1}$$

where there are $k = 1, \ldots, M$ output variables y_k, and there are $n = 1, \ldots, N$ observations. While minimizing the MSE is quite intuitive and is used in many types of model besides NN (e.g. linear regression), it can be derived from the broader principle of *maximum likelihood* under the assumption of Gaussian noise distribution.

If we assume that the multivariate target data y_{dk} are independent random variables, then the conditional probability distribution of $\mathbf{y}_d$ given predictors $\mathbf{x}$ can be written as

$$p(\mathbf{y}_d|\mathbf{x}) = \prod_{k=1}^{M} p(y_{dk}|\mathbf{x}). \tag{6.2}$$

The target data are made up of noise ϵ_k plus an underlying signal (which we are trying to simulate by a model with parameters $\mathbf{w}$ and outputs y_k), i.e.

$$y_{dk} = y_k(\mathbf{x}; \mathbf{w}) + \epsilon_k. \tag{6.3}$$

We assume that the noise ϵ_k obeys a Gaussian distribution with zero mean and standard deviation σ, with σ independent of k and $\mathbf{x}$, i.e.

$$p(\boldsymbol{\epsilon}) = \prod_{k=1}^{M} p(\epsilon_k) = \frac{1}{(2\pi)^{M/2}\sigma^M}\exp\left(-\frac{\sum_k \epsilon_k^2}{2\sigma^2}\right). \tag{6.4}$$

From (6.3) and (6.4), the conditional probability distribution

$$p(\mathbf{y}_d|\mathbf{x}; \mathbf{w}) = \frac{1}{(2\pi)^{M/2}\sigma^M}\exp\left[-\frac{\sum_k (y_k(\mathbf{x}; \mathbf{w}) - y_{dk})^2}{2\sigma^2}\right]. \tag{6.5}$$

The principle of *maximum likelihood* says: if we have a conditional probability distribution $p(\mathbf{y}_d|\mathbf{x}; \mathbf{w})$, and we have observed values $\mathbf{y}_d$ given by the dataset D and $\mathbf{x}$ by the dataset X, then the parameters $\mathbf{w}$ can be found by maximizing the likelihood function $p(D|X; \mathbf{w})$, i.e. the parameters $\mathbf{w}$ should be chosen so that the likelihood of observing D given X is maximized. Note that $p(D|X; \mathbf{w})$ is a function of $\mathbf{w}$ only as D and X are known.

The datasets X and D contain the observations $\mathbf{x}^{(n)}$ and $\mathbf{y}_d^{(n)}$, with $n = 1, \ldots, N$. The likelihood function L is then

$$L = p(D|X; \mathbf{w}) = \prod_{n=1}^{N} p\left(\mathbf{y}_d{}^{(n)}|\mathbf{x}^{(n)}; \mathbf{w}\right). \tag{6.6}$$

Instead of maximizing the likelihood function, it is more convenient to minimize the negative log of the likelihood, as the logarithm function is a monotonic function. From (6.5) and (6.6), we end up minimizing the following objective function with respect to **w**:

$$\tilde{J} = -\ln L = \frac{1}{2\sigma^2}\sum_{n=1}^{N}\sum_{k=1}^{M}\left[y_k(\mathbf{x}^{(n)};\mathbf{w}) - y_{\mathrm{d}k}^{(n)}\right]^2 + NM\ln\sigma + \frac{NM}{2}\ln(2\pi). \tag{6.7}$$

Since the last two terms are independent of **w**, they are irrelevant to the minimization process and can be omitted. Other than a constant multiplicative factor, the remaining term in $\tilde{J}$ is the same as the MSE objective function J in (6.1). Hence minimizing MSE is equivalent to maximizing likelihood assuming Gaussian noise distribution.

6.2 Objective functions and robustness

In this section, we examine where the model outputs converge to, under the MSE objective function (6.1) in the limit of infinite sample size N with a flexible enough model (Bishop, 1995). However, the MSE is not the only way to incorporate information about the error between the model output y_k and the target data $y_{\mathrm{d}k}$ into J. We could minimize the *mean absolute error* (MAE) instead of the MSE, i.e. define

$$J = \frac{1}{N}\sum_{n=1}^{N}\sum_{k}\left|y_k^{(n)} - y_{\mathrm{d}k}^{(n)}\right|. \tag{6.8}$$

Any data point $y_{\mathrm{d}k}$ lying far from the mean of the distribution of $y_{\mathrm{d}k}$ would exert far more influence in determining the solution under the MSE objective function than in the MAE objective function. We will show that unlike the MAE, the MSE objective function is not robust to *outliers* (i.e. data points lying far away from the mean, which might have resulted from defective measurements or from exceptional events).

Let us first study the MSE objective function (6.1). With infinite N, the sum over the N observations in the objective function can be replaced by integrals, i.e.

$$J = \frac{1}{2}\sum_{k}\int\int[y_k(\mathbf{x};\mathbf{w}) - y_{\mathrm{d}k}]^2\, p(y_{\mathrm{d}k},\mathbf{x})\,\mathrm{d}y_{\mathrm{d}k}\,\mathrm{d}\mathbf{x}, \tag{6.9}$$

where $\mathbf{x}$ and $\mathbf{w}$ are the model inputs and model parameters respectively, and $p(y_{\mathrm{d}k},\mathbf{x})$, a joint probability distribution. Since

$$p(y_{\mathrm{d}k},\mathbf{x}) = p(y_{\mathrm{d}k}|\mathbf{x})\,p(\mathbf{x}), \tag{6.10}$$

where $p(\mathbf{x})$ is the probability density of the input data, and $p(y_{\mathrm{d}k}|\mathbf{x})$ is the probability density of the target data conditional on the inputs, we have

$$J = \frac{1}{2}\sum_k \int\int [y_k(\mathbf{x};\mathbf{w}) - y_{dk}]^2\, p(y_{dk}|\mathbf{x})\, p(\mathbf{x})\, dy_{dk}\, d\mathbf{x}. \tag{6.11}$$

Next we introduce the following conditional averages of the target data:

$$\langle y_{dk}|\mathbf{x}\rangle = \int y_{dk}\; p(y_{dk}|\mathbf{x})\, dy_{dk}, \tag{6.12}$$

$$\langle y_{dk}^2|\mathbf{x}\rangle = \int y_{dk}^2\; p(y_{dk}|\mathbf{x})\, dy_{dk}, \tag{6.13}$$

so we can write

$$[y_k - y_{dk}]^2 = [y_k - \langle y_{dk}|\mathbf{x}\rangle + \langle y_{dk}|\mathbf{x}\rangle - y_{dk}]^2 \tag{6.14}$$

$$= [y_k - \langle y_{dk}|\mathbf{x}\rangle]^2 + 2[y_k - \langle y_{dk}|\mathbf{x}\rangle][\langle y_{dk}|\mathbf{x}\rangle - y_{dk}] + [\langle y_{dk}|\mathbf{x}\rangle - y_{dk}]^2. \tag{6.15}$$

Upon substituting (6.15) into (6.11), we note that the second term of (6.15) vanishes from the integration over y_{dk} and from (6.12). Invoking (6.13), the objective function can be written as

$$J = \frac{1}{2}\sum_k \left\{ \int [y_k(\mathbf{x};\mathbf{w}) - \langle y_{dk}|\mathbf{x}\rangle]^2 p(\mathbf{x}) d\mathbf{x} + [\langle y_{dk}^2|\mathbf{x}\rangle - \langle y_{dk}|\mathbf{x}\rangle^2]\, p(\mathbf{x}) d\mathbf{x} \right\}. \tag{6.16}$$

The second term does not depend on the model output y_k, hence it is independent of the model weights $\mathbf{w}$. Thus during the search for the optimal weights to minimize J, the second term in (6.16) can be ignored. In the first term of (6.16), the integrand cannot be negative, so the minimum of J occurs when this first term vanishes, i.e.

$$y_k(\mathbf{x};\mathbf{w}_{\text{opt}}) = \langle y_{dk}|\mathbf{x}\rangle, \tag{6.17}$$

where $\mathbf{w}_{\text{opt}}$ denotes the weights at the minimum of J. This is a very important result as it shows that the model output is simply the conditional mean of the target data. Thus in the limit of an infinite number of observations in the dataset, and with the use of a flexible enough model, the model output y_k for a given input $\mathbf{x}$ is the conditional mean of the target data at $\mathbf{x}$, as illustrated in Fig. 6.1. Also the derivation of this result is quite general, as it does not actually require the model mapping $y_k(\mathbf{x};\mathbf{w})$ to be restricted to NN models. This result also shows that in nonlinear regression problems, in the limit of infinite sample size, overfitting cannot occur, as the model output converges to the conditional mean of the target data. In practice, in the absence of outliers, overfitting ceases to be a problem when the number of independent observations is much larger than the number of model parameters.

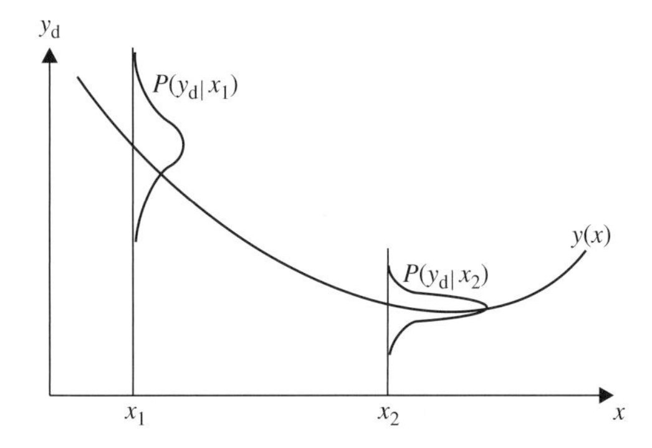

Fig. 6.1 Showing the model output y as the conditional mean of the target data y_d, with the conditional probability distribution $p(y_d|x)$ displayed at x_1 and at x_2.

Next, we turn to the MAE objective function (6.8). Under infinite N, (6.8) becomes

$$J = \sum_k \int \int |y_k(\mathbf{x}; \mathbf{w}) - y_{dk}| \; p(y_{dk}|\mathbf{x})\, p(\mathbf{x})\, \mathrm{d}y_{dk}\, \mathrm{d}\mathbf{x}. \tag{6.18}$$

This can be rewritten as

$$J = \sum_k \int \tilde{J}_k(\mathbf{x}) \; p(\mathbf{x})\, \mathrm{d}\mathbf{x}, \tag{6.19}$$

where

$$\tilde{J}_k(\mathbf{x}) \equiv \int |y_k(\mathbf{x}; \mathbf{w}) - y_{dk}| \; p(y_{dk}|\mathbf{x})\, \mathrm{d}y_{dk}. \tag{6.20}$$

$\tilde{J}_k(\mathbf{x}) \geq 0$ since the integrand of (6.20) is non-negative. Also J in (6.19) is minimized when $\tilde{J}_k(\mathbf{x})$ is minimized. To minimize $\tilde{J}_k(\mathbf{x})$ with respect to the model output y_k, we set

$$\frac{\partial \tilde{J}_k}{\partial y_k} = \int \mathrm{sgn}(y_k(\mathbf{x}; \mathbf{w}) - y_{dk}) \; p(y_{dk}|\mathbf{x})\, \mathrm{d}y_{dk} = 0, \tag{6.21}$$

where the function sgn(z) gives $+1$ or -1 depending on the sign of z. For this integral to vanish, the equivalent condition is

$$\int_{-\infty}^{y_k} p(y_{dk}|\mathbf{x})\, \mathrm{d}y_{dk} - \int_{y_k}^{\infty} p(y_{dk}|\mathbf{x})\, \mathrm{d}y_{dk} = 0, \tag{6.22}$$

which means that $y_k(\mathbf{x}; \mathbf{w})$ has to be the conditional *median*, so that the conditional probability density integrated to the left of y_k equals that integrated to the right of y_k. In statistics, it is well known that the median is robust to outliers whereas the mean is not. For instance, the mean price of a house in a small city can be raised considerably by the sale of a single palatial home in a given month, whereas the median, which is the price where there are equal numbers of sales above and below this price, is not affected. Thus in the presence of outliers, the MSE objective function can produce solutions which are strongly influenced by outliers, whereas the MAE objective function can largely eliminate this undesirable property (in practice an infinite N is not attainable, therefore using MAE does not completely eliminate this problem). However, a disadvantage of the MAE objective function is that it is less sensitive than the MSE objective function, so it may not fit the data as closely.

The error function of Huber (1964) combines the desirable properties of the MSE and the MAE functions. Let $z = y_k - y_{dk}$, then the *Huber function* is

$$h(z) = \begin{cases} \frac{1}{2}z^2 & \text{for } |z| \leq 2\gamma \\ 2\gamma(|z| - \gamma) & \text{otherwise,} \end{cases} \tag{6.23}$$

where γ is a positive constant. When the error z is not larger than 2γ, h behaves similarly to the MSE, whereas for larger errors, h behaves similarly to the MAE, thereby avoiding heavily weighting the target data lying at the distant tails of $p(y_{dk}|\mathbf{x})$ (Fig. 6.2).

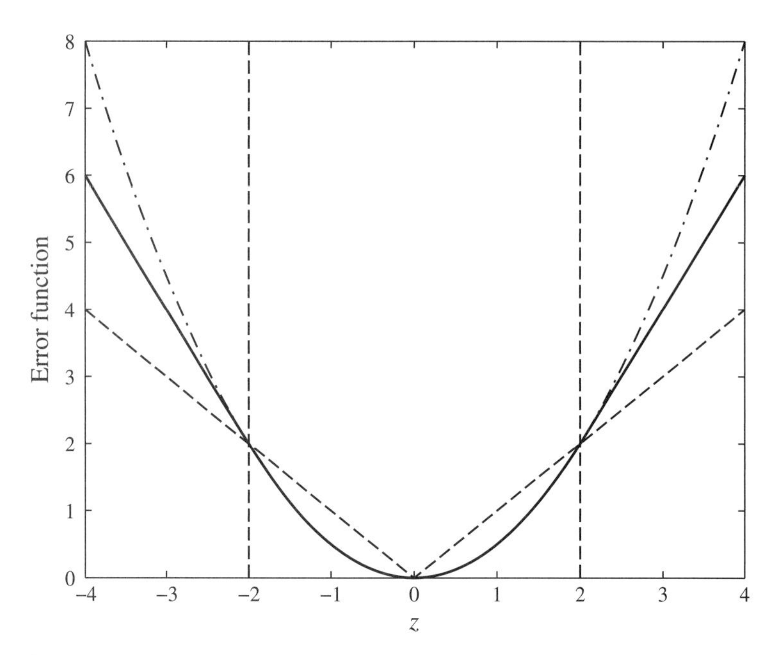

Fig. 6.2 The Huber error function (with parameter $\gamma = 1$) (solid curve) plotted versus z, the difference between the model output and the target data. The MSE (dot-dash) and MAE (dashed) error functions are also shown for comparison. The dashed vertical lines (at $\pm 2\gamma$) indicate where the Huber function changes from behaving like the MSE to behaving like the MAE.

6.3 Variance and bias errors

It is important to distinguish between two types of error when fitting a model to a dataset – namely variance error and bias error. To simplify the discussion, we will assume that the model output is a single variable $y = f(\mathbf{x})$. The true relation is $y_T = f_T(\mathbf{x})$. The model was trained over a dataset D. Let $\mathcal{E}[\cdot]$ denote the expectation or ensemble average over all datasets D. Note that $\mathcal{E}[\cdot]$ is not the expectation E$[\cdot]$ over $\mathbf{x}$ (see Section 1.1), so $\mathcal{E}[y] \equiv \bar{y}$ is still a function of $\mathbf{x}$. Thus the error of y is

$$
\begin{aligned}
\mathcal{E}[(y - y_T)^2] &= \mathcal{E}[(y - \bar{y} + \bar{y} - y_T)^2] \\
&= \mathcal{E}[(y - \bar{y})^2] + \mathcal{E}[(\bar{y} - y_T)^2] + 2\mathcal{E}[(y - \bar{y})(\bar{y} - y_T)] \\
&= \mathcal{E}[(y - \bar{y})^2] + (\bar{y} - y_T)^2 + 2(\bar{y} - y_T)\,\mathcal{E}[y - \bar{y}] \\
&= \mathcal{E}[(y - \bar{y})^2] + (\bar{y} - y_T)^2, \qquad (6.24)
\end{aligned}
$$

since $\mathcal{E}[y - \bar{y}] = 0$. The first term, $\mathcal{E}[(y - \bar{y})^2]$, is the *variance error*, as it measures the departure of y from its expectation $\bar{y}$. The second term, $(\bar{y} - y_T)^2$, is the *bias error*, as it measures the departure of $\bar{y}$ from the true value y_T. The variance error tells us how much the y estimated from a given dataset D can be expected to fluctuate about $\bar{y}$, the expectation over all datasets D. Even with this fluctuation caused by sampling for a particular dataset D removed, one has the bias error indicating the departure of the model expectation from the true value.

If one uses a model with few adjustable parameters, then the model may have trouble fitting to the true underlying relation accurately, resulting in a large bias error, as illustrated by the linear model fit in Fig. 6.3a. The variance error in this case is small, since the model is not flexible enough to fit to the noise in the data. In contrast, if one uses a model with many adjustable parameters, the model will fit to the noise closely, resulting in a large variance error (Fig. 6.3b), but the bias error

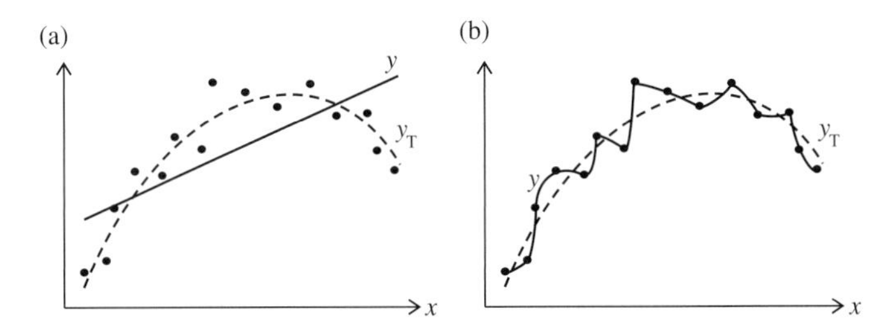

Fig. 6.3 Illustrating the results from using a model with (a) few adjustable parameters and (b) many adjustable parameters to fit the data. The model fit y is shown by a solid line and the true relation y_T by the dashed line. In (a), the bias error is large as y from a linear model is a poor approximation of y_T, but the variance error is small. In (b), the bias error is small but the variance error is large, since the model is fitting to the noise in the data.

will be small. The art of machine learning hinges on a balanced trade-off between variance error and bias error.

6.4 Reserving data for validation

In Chapter 4, we have mentioned the *early stopping* method, where a portion of the dataset is set aside for validation, and the remainder for training. As the number of training epochs increases, the objective function evaluated over the training data decreases, but the objective function evaluated over the validation data often decreases to a minimum and then increases due to overfitting (Fig. 4.6). The minimum gives an indication as to when to stop the training, as additional training epochs only contribute to overfitting.

What fraction of the data should one reserve for validation? Since more data for validation means fewer data for training the NN model, there is clearly an optimal fraction for validation. Using more than the optimal fraction for validation results in the training process generalizing less reliably from the smaller training dataset, while using a smaller fraction than the optimal could lead to overfitting.

A theory to estimate this optimal fraction was provided by Amari *et al.* (1996), assuming a large number of observations N. Of the N observations, fN are used for validation, and $(1 - f)N$ for training. If the number of weight and offset (i.e. bias) parameters in the NN model is N_p, then in the case of $N < 30\,N_p$, the optimal value for f is

$$f_{opt} = \frac{\sqrt{2N_p - 1} - 1}{2(N_p - 1)}, \tag{6.25}$$

$$\text{hence } f_{opt} \approx \frac{1}{\sqrt{2N_p}}, \quad \text{for large } N_p. \tag{6.26}$$

For example, if $N_p = 100$, (6.26) gives $f_{opt} \approx 0.0707$, i.e. only 7% of the data should be set aside for validation, with the remaining 93% used for training.

For the case $N > 30\,N_p$, Amari *et al.* (1996) showed that there is negligible difference between using the optimal fraction for validation, and not using early stopping at all (i.e. using the whole dataset for training till convergence). The reason is that in this case there are so many data relative to the number of model parameters that overfitting is not a problem. In fact, using early stopping with $f > f_{opt}$, leads to poorer results than not using early stopping, as found in numerical experiments by Amari *et al.* (1996). Thus when $N > 30\,N_p$, overfitting is not a problem, and reserving considerable validation data for early stopping is a poor approach.

6.5 Regularization

To prevent overfitting, the most common approach is via *regularization* of the objective function, i.e. by adding *weight penalty* (also known as *weight decay*) terms to the objective function. The objective function (4.18) now becomes

$$J = \frac{1}{N}\sum_{n=1}^{N}\left\{\frac{1}{2}\sum_{k}\left[y_k^{(n)} - y_{\mathrm{d}k}^{(n)}\right]^2\right\} + P\,\frac{1}{2}\sum_{j} w_j^2, \tag{6.27}$$

where w_j represents all the (weight and offset) parameters in the model, and P, a positive constant, is the *weight penalty parameter* or *regularization parameter*. Note that P is also referred to as a *hyperparameter* as it exerts control over the weight and offset parameters – during nonlinear optimization of the objective function, P is held constant while the optimal values of the other parameters are being computed. With a positive P, the selection of larger $|w_j|$ during optimization would increase the value of J, so larger $|w_j|$ values are penalized. Thus choosing a larger P will more strongly suppress the selection of larger $|w_j|$ by the optimization algorithm.

For sigmoidal activation functions such as tanh, the effect of weight penalty can be illustrated as follows: For $|wx| \ll 1$, the leading term of the Taylor expansion gives

$$y = \tanh(wx) \approx wx, \tag{6.28}$$

i.e. the nonlinear activation function tanh is approximated by a linear activation function when the weight $|w|$ is penalized to be small and x is reasonably scaled. Hence, using a relatively large P to penalize weights would diminish the nonlinear modelling capability of the model, thereby avoiding overfitting.

With the weight penalty term in (6.27), it is essential that the input variables have been scaled to similar magnitudes. The reason is that if for example the first input variable is much larger in magnitude than the second, then the weights multiplied to the second input variable will have to be much larger in magnitude than those multiplied to the first variable, in order for the second input to exert comparable influence on the output. However, the same weight penalty parameter P acts on both sets of weights, thereby greatly reducing the influence of the second input variable since the associated weights are not allowed to take on large values. Similarly, if there are multiple output variables, the target data for different variables should be scaled to similar magnitudes. Hence, when dealing with real unbounded variables, it is common to *standardize* the data first, i.e. each variable has its mean value subtracted, and then is divided by its standard deviation. After the NN model has been applied to the standardized variables, each output variable is rescaled to the original dimension, i.e. multiply by the original standard deviation and add back

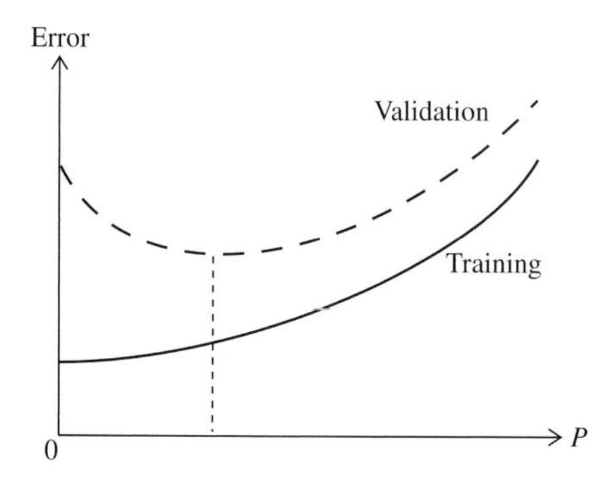

Fig. 6.4 Illustrating the model error (e.g. the MSE) for the training data (solid curve) and for the validation data (dashed curve) as a function of the weight penalty parameter P. The minimum in the dashed curve gives the optimal P value (as marked by the vertical dotted line).

the original mean value. As a cautionary tale, a graduate student of mine was once trying to compare NN with linear regression (LR). No matter how hard he tried, he was able to show only that LR was outperforming NN on test data. Only when he standardized the variables, was he finally able to show that NN was better than LR.

What value should one choose for the weight penalty parameter? A common way to select P is by *validation*. The dataset is divided into training data and validation data. Models are trained using the training data for a variety of P values, e.g. $P = 3, 1, 0.3, 0.1, 0.03, 0.01, \ldots$, or $P = 2^{-p}$ $(p = -1, 0, 1, 2, \ldots)$. Model performance over the validation data is then used to select the optimal P value. The model error (e.g. the MSE) over the training data generally drops as P drops (Fig. 6.4); however, the model error over the validation data eventually rises for small enough P, as the excessively nonlinear model has begun overfitting the training data.

For an MLP NN model with a single hidden layer, where there are m_1 inputs, m_2 hidden neurons and m_3 output neurons, we have assumed that m_2 is large enough so that the model has enough flexibility accurately to capture the underlying relation in the dataset. In practice, we may not know what m_2 value to use. Hence, instead of a single loop of model runs using a variety of P values, we may also need a second loop with $m_2 = 1, 2, 3, \ldots$. The run with the smallest validation error gives the best P and m_2 values.

6.6 Cross-validation

When there are plentiful data, reserving some data for validation poses no problem. Unfortunately, data are often not plentiful, and one can ill-afford setting aside a fair amount of data for validation, since this means fewer observations for model training. On the other hand, if one reserves few observations for validation so that one can use as many observations for training as possible, the validation error estimate may be very unreliable. Cross-validation is a technique which allows the entire dataset to be used for validation.

Given a data record, K-fold *cross-validation* involves dividing the record into K (approximately equal) segments. One segment is reserved as validation data, while the other $K-1$ segments are used for model training. This process is repeated K times, so that each segment of the data record has been used as validation data. Thus a validation error can be computed for the whole data record. A variety of models is run, with different numbers of model parameters and different weight penalties. Based on the lowest validation error over the whole data record, one can select the best model among the many runs.

For example, suppose the data record is 50 years long. In 5-fold cross-validation, the record is divided into 5 segments, i.e. years 1-10, 11-20, 21-30, 31-40, 41-50. First, we reserve years 1-10 for validation, and train the model using data from years 11-50. Next we reserve years 11-20 for validation, and train using data from years 1-10 and 21-50. This is repeated until the final segment of 41-50 is reserved for validation, with training done using the data from years 1-40. If one has more computing resources, one can try 10-fold cross-validation, where the 50 year record is divided into ten 5 year segments. If one has even more computing resources, one can try 50-fold cross-validation, with the record divided into fifty 1 year segments. At the extreme, one arrives at the *leave-one-out* cross-validation, where the validation segment consists of a single observation. For instance, if the 50 year record contains monthly values, then there are a total of 600 monthly observations, and a 600-fold cross-validation is the same as the leave-one-out approach.

With time series data, the neighbouring observations in the data record are often not independent of each other due to autocorrelation. If the dataset has a decorrelation time scale of 9 months, then leaving a single monthly observation out for independent validation would make little sense since it is well correlated with neighbouring observations already used for training. When there is autocorrelation in the time series data, the validation segments should be longer than the decorrelation time scale, i.e. in this example, the validation segments should exceed 9 months. Even then, at the boundary of a validation segment, there is still correlation between the data immediately to one side which are used for training and those to the other side used for validation. Thus under cross-validation, autocorrelation can lead to an underestimation of the model error over the validation data, especially when using small validation segments.

Because validation data are used for model selection, i.e. for choosing the best number of model parameters, weight penalty value, etc., the model error over the validation data cannot be considered an accurate model forecast error, since the validation data have already been involved in deriving the model. To assess the model forecast error accurately, the model error needs to be calculated over independent data not used in model training or model selection. Thus the data record needs to be divided into training data, validation data and 'testing' or verification data for measuring the true model forecast error. One then has to do a double cross-validation, which can be quite expensive computationally. Again consider the example of a 50-year data record, where we want to do a 10-fold cross-testing. We first reserve years 1-5 for testing data, and use years 6-50 for training and validation. We then implement 9-fold cross-validation over the data from years 6-50 to select the best model, which we use to forecast over the testing data. Next, we reserve years 6-10 for testing, and perform a 9-fold cross-validation over the data from years 1-5 and 11-50 to select the best model (which may have a different number of model parameters and different weight penalty value from the model selected in the previous cross-validation). The process is repeated until the model error is computed over test data covering the entire data record.

Generalized cross-validation (GCV) is an extension of the cross-validation method (Golub *et al.*, 1979), and has been applied to MLP NN models (Yuval, 2000) to determine the weight penalty parameter automatically. We will not pursue it further here, since the next section covers the widely used Bayesian approach to determining the weight penalty parameter(s).

6.7 Bayesian neural networks (BNN)

While cross-validation as presented in Section 6.6 allows one to find the weight penalty parameters which would give the model good generalization capability, separation of the data record into training and validation segments is cumbersome, and prevents the full data record from being used to train the model. Based on Bayes theorem (Section 1.5), MacKay (1992b,a) introduced a Bayesian neural network (BNN) approach which gives an estimate of the optimal weight penalty parameter(s) without the need of validation data. Foresee and Hagan (1997) applied this approach to the MLP NN model using the Levenberg–Marquardt optimization algorithm (Section 5.4), with their code implemented in the *Matlab* neural network toolbox as trainbr.m. BNN codes written in *Matlab* are also available in the package *Netlab* written by Nabney (2002).

The main objective of the training process is to minimize the sum of squared errors of the model output y (for simplicity, a single output is used here), over N observations. Let

$$E_{\mathrm{d}} = \frac{1}{2}\sum_{i=1}^{N}(y_i - y_{\mathrm{d}i})^2, \qquad (6.29)$$

where y_d is the target data from a dataset D. At the same time, regularization pushes for minimizing the magnitude of the weight parameters. Let

$$E_w = \frac{1}{2} \sum_{j=1}^{N_w} w_j^2, \tag{6.30}$$

where $\mathbf{w}$ contains all the weight (and offset) parameters of the MLP NN model, with a total of N_w parameters. The objective function is then

$$J(\mathbf{w}) = \beta E_d + \alpha E_w, \tag{6.31}$$

where α is the weight penalty hyperparameter. Previously, β was simply 1, but here in the Bayesian approach, the hyperparameter β is used to describe the strength of the Gaussian noise in the target data. The Bayesian approach will automatically determine the values for the two hyperparameters α and β. If $\alpha \ll \beta$, then the model will fit the data closely; but if $\alpha \gg \beta$, then the model will strive for small weights instead of close fit, resulting in a smooth fit to the data.

If we assume that the noise in the dataset D is Gaussian, then the likelihood function

$$p(D|\mathbf{w}, \beta, M) = \frac{1}{Z_d(\beta)} \exp(-\beta E_d), \tag{6.32}$$

where the normalization factor $Z_d(\beta) = (2\pi/\beta)^{N/2}$, and M is the particular model used (e.g. different models may have different numbers of hidden neurons). The likelihood function is the probability of the dataset D occurring, given the model M with weight parameters $\mathbf{w}$ and noise level specified by β.

If we assume that the prior probability density distribution for $\mathbf{w}$ is also a Gaussian centred on the origin, then

$$p(\mathbf{w}|\alpha, M) = \frac{1}{Z_w(\alpha)} \exp(-\alpha E_w), \tag{6.33}$$

with $Z_w(\alpha) = (2\pi/\alpha)^{N_w/2}$.

From Bayes theorem (1.54), the posterior probability density

$$p(\mathbf{w}|D, \alpha, \beta, M) = \frac{p(D|\mathbf{w}, \beta, M)\, p(\mathbf{w}|\alpha, M)}{p(D|\alpha, \beta, M)}, \tag{6.34}$$

where $p(D|\alpha, \beta, M)$ is the normalization factor which ensures that the posterior probability density function integrates to 1.

Substituting (6.32) and (6.33) into (6.34) yields

$$\begin{aligned} p(\mathbf{w}|D, \alpha, \beta, M) &= \frac{\frac{1}{Z_w(\alpha)} \frac{1}{Z_d(\beta)} \exp(-(\beta E_d + \alpha E_w))}{\text{normalization factor}} \\ &= \frac{1}{Z(\alpha, \beta)} \exp(-J), \end{aligned} \tag{6.35}$$

where J is the objective function in (6.31), and $Z(\alpha, \beta)$ the normalization factor. Thus to find the optimal $\mathbf{w}$, we need to maximize the posterior probability density $p(\mathbf{w}|D, \alpha, \beta, M)$, which is equivalent to minimizing the objective function J. Note that here we are assuming α and β are given or somehow known.

6.7.1 Estimating the hyperparameters

Our next step is to find the optimal values for the hyperparameters α and β. Note that the Bayesian approach is *hierarchical*. After deriving the model parameters assuming that the controlling hyperparameters are given, one then derives the hyperparameters (which could themselves be controlled by an even higher level of hyperparameters, and so forth). We again turn to Bayes theorem:

$$p(\alpha, \beta|D, M) = \frac{p(D|\alpha, \beta, M)p(\alpha, \beta|M)}{p(D|M)}. \tag{6.36}$$

If the prior density $p(\alpha, \beta|M)$ is assumed to be uniformly distributed, then the posterior probability density $p(\alpha, \beta|D, M)$ is maximized by simply maximizing the likelihood function $p(D|\alpha, \beta, M)$ (also called the *evidence* for α and β). From (6.34), we get

$$p(D|\alpha, \beta, M) = \frac{p(D|\mathbf{w}, \beta, M)p(\mathbf{w}|\alpha, M)}{p(\mathbf{w}|D, \alpha, \beta, M)}. \tag{6.37}$$

Substituting in (6.32), (6.33) and (6.35), we obtain

$$p(D|\alpha, \beta, M) = \frac{Z(\alpha, \beta)}{Z_{\rm d}(\beta)Z_{\rm w}(\alpha)} \frac{\exp(-\beta E_{\rm d} - \alpha E_{\rm w})}{\exp(-J)} = \frac{Z(\alpha, \beta)}{Z_{\rm d}(\beta)Z_{\rm w}(\alpha)}. \tag{6.38}$$

Since the normalization constants $Z_{\rm d}(\beta)$ and $Z_{\rm w}(\alpha)$ from (6.32) and (6.33) are known, we need only to derive $Z(\alpha, \beta)$, which is simply the integral of $\exp(-J)$ over $\mathbf{w}$-space.

We approximate $J(\mathbf{w})$ by a Taylor expansion about its minimum point $\mathbf{w}_{\rm MP}$,

$$J(\mathbf{w}) \approx J(\mathbf{w}_{\rm MP}) + \frac{1}{2}\Delta\mathbf{w}^{\rm T}\mathbf{H}_{\rm MP}\Delta\mathbf{w}, \tag{6.39}$$

where $\Delta\mathbf{w} = \mathbf{w} - \mathbf{w}_{\rm MP}$, the linear term vanishes since it is evaluated at a minimum point, and terms of higher order than the quadratic have been omitted. Here $\mathbf{H}$ is the Hessian matrix , i.e.

$$\mathbf{H} \equiv \nabla\nabla J = \beta\nabla\nabla E_{\rm d} + \alpha\nabla\nabla E_{\rm w} = \beta\nabla\nabla E_{\rm d} + \alpha\mathbf{I}, \tag{6.40}$$

with $\mathbf{I}$ the identity matrix.

Substituting the approximation (6.39) for J into

$$Z = \int \exp(-J(\mathbf{w}))\mathrm{d}\mathbf{w}, \tag{6.41}$$

and evaluating the integral (Bishop, 1995, Appendix B) yields

$$Z = (2\pi)^{N_w/2}[\det(\mathbf{H}_{MP})]^{-1/2}\exp(-J(\mathbf{w}_{MP})). \tag{6.42}$$

Approximating J by the 2nd order Taylor expansion (6.39) is equivalent to approximating the posterior probability density function (6.35) by a Gaussian function, which is called the *Laplace approximation*. This approximation allows the integral for Z to be evaluated analytically. If the Laplace approximation is avoided, then integrals have to be evaluated numerically using Markov chain Monte Carlo methods (Neal, 1996).

Taking the natural logarithm of (6.38) yields

$$\ln p(D|\alpha, \beta, M) = \ln Z(\alpha, \beta) - \ln Z_d(\beta) - \ln Z_w(\alpha). \tag{6.43}$$

To locate the maximum point of $\ln p(D|\alpha, \beta, M)$, substitute in the expressions for $Z(\alpha, \beta)$, $Z_d(\beta)$ and $Z_w(\alpha)$, differentiate with respect to α and β separately, set the derivatives to zero, and after some algebra obtain (Foresee and Hagan, 1997; Bishop, 1995, Sect. 10.4)

$$\alpha_{MP} = \frac{\gamma}{2E_w(\mathbf{w}_{MP})} \quad \text{and} \quad \beta_{MP} = \frac{N-\gamma}{2E_d(\mathbf{w}_{MP})}, \tag{6.44}$$

where

$$\gamma = N_w - \alpha_{MP}\,\mathrm{tr}((\mathbf{H}_{MP})^{-1}), \tag{6.45}$$

with tr denoting the trace of the matrix (i.e. sum of the diagonal elements). A very useful interpretation of γ (Gull, 1989; Bishop, 1995) is that it represents the effective number of parameters, with $\gamma \leq N_w$, the total number of parameters in $\mathbf{w}$.

In Section 5.4, we have looked at using the Gauss–Newton approximation of the Hessian matrix in the Levenberg–Marquardt optimization algorithm, which is the algorithm used by Foresee and Hagan (1997) to find the minimum point $\mathbf{w}_{MP}$. After scaling the inputs and outputs to the range $[-1, 1]$, the steps in their procedure are as set out below.

(1) Choose initial values for α, β and $\mathbf{w}$, e.g. set $\alpha = 0$, $\beta = 2$, and $\mathbf{w}$ according to the initialization scheme of Nguyen and Widrow (1990).
(2) Advance one step of the Levenberg–Marquardt algorithm towards minimizing the objective function $J(\mathbf{w})$.
(3) Compute the effective number of parameters γ from (6.45), where the Gauss–Newton approximation has been used in the Hessian matrix available in the Levenberg–Marquardt algorithm (see Section 5.4), i.e.

$$\mathbf{H} \approx \beta\mathbf{J}^{\mathrm{T}}\mathbf{J} + \alpha\mathbf{I}, \tag{6.46}$$

with $\mathbf{J}$ the Jacobian matrix of the training set errors.

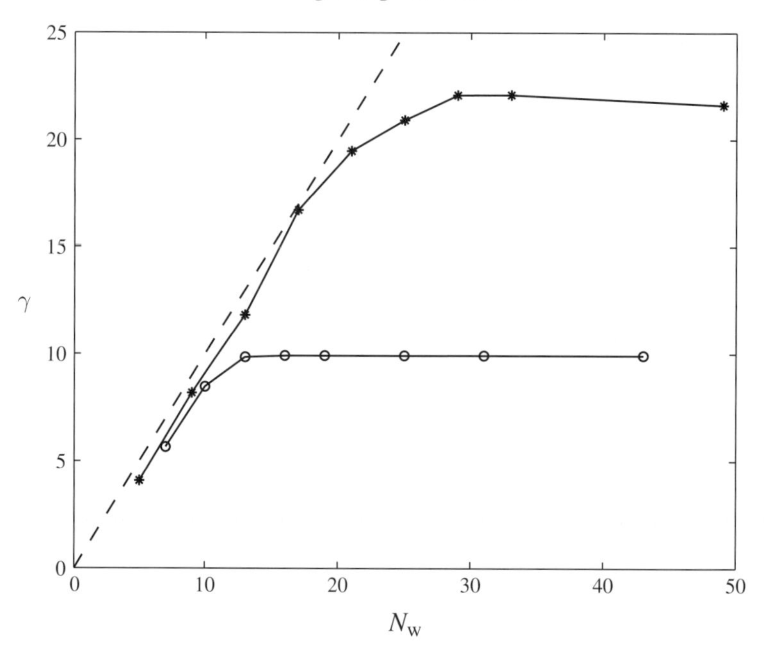

Fig. 6.5 Plot of γ, the effective number of parameters, as a function of N_w, the number of model parameters, for sample problem 1 (circles) and problem 3 (asterisks) from Foresee and Hagan (1997). The dashed line is the $\gamma = N_w$ line.

(4) Compute α and β from (6.44).
(5) Iterate steps (2)–(4) till convergence.
(6) With the optimal α and β now known, the objective function $J(\mathbf{w})$ in (6.31) can be minimized, yielding the optimal weights $\mathbf{w}_{MP}$ for the NN model.

Figure 6.5 shows the values of γ in the sample problems tested by Foresee and Hagan (1997) as N_w is varied by changing the number of neurons in the single hidden layer. In all cases, when N_w is relatively small, increasing N_w leads to an increase in γ. This indicates a need for more parameters to model the true function closely. However, γ eventually levels off, meaning that the addition of extra hidden neurons brings no advantage. Thus the optimal network size is when N_w is just large enough to cause γ to level off.

The Bayesian NN (BNN) model presented here has only one hyperparameter α controlling all the weights $\mathbf{w}$. One can conceive of situations where different weights would benefit from having different hyperparameters. Neal (1996) generalized the BNN model to have individual hyperparameters controlling groups of weights. For instance, a separate hyperparameter can be used to control all the weights originating from each individual input neuron. If a hyperparameter turned out to be huge, thereby causing all weights originating from a particular input neuron to be negligible in magnitude, then that input is irrelevant for predicting the output. This is known as *automatic relevance determination* (ARD).

The BNN model of MacKay (1992b,a) has only a single output variable. For multiple outputs, one has to assume the output variables to be independent of each other and apply the BNN approach to each output individually. For multiple outputs which are not independent, MacKay (1995) suggested the use of full covariance matrices to describe the output uncertainties, but did not elaborate on this idea. Noting the importance of uncertainty estimates for multi-dimensional mappings in atmospheric inverse problems, Aires (2004) and Aires *et al.* (2004a,b) extended the BNN model to multiple outputs using full covariance matrices.

6.7.2 Estimate of predictive uncertainty

A bonus of the Bayesian approach is that it naturally allows us to estimate the uncertainty in the model output. Suppose a new predictor $\tilde{\mathbf{x}}$ is available, we would like to know the uncertainty of the predictand $\tilde{y}$ through the conditional distribution $p(\tilde{y}|\,\tilde{\mathbf{x}}, D)$. (For notational brevity, we have not bothered to write $p(\tilde{y}|\,\tilde{\mathbf{x}}, D)$ as $p(\tilde{y}|\,\tilde{\mathbf{x}}, D, X)$, with X being the training dataset of predictors.) From basic rules of probability,

$$p(\tilde{y}|\,\tilde{\mathbf{x}}, D) = \int p(\tilde{y}|\,\tilde{\mathbf{x}}, \mathbf{w})\, p(\mathbf{w}|D)\mathrm{d}\mathbf{w}, \tag{6.47}$$

where $p(\tilde{y}|\,\tilde{\mathbf{x}}, \mathbf{w})$ is the distribution of the noisy target data for a given $\tilde{\mathbf{x}}$ and a given $\mathbf{w}$, and $p(\mathbf{w}|D)$ is the posterior distribution of the weights. Unlike traditional methods like maximum likelihood, which simply provides a single optimal estimate of the weights of a model, the Bayesian approach provides a posterior distribution of the weights. This means that when deriving $p(\tilde{y}|\,\tilde{\mathbf{x}}, D)$ in (6.47) we accept contributions from all possible NN models with various weights, with their contributions weighted by $p(\mathbf{w}|D)$.

As this integral cannot be evaluated analytically, we again approximate $p(\mathbf{w}|D)$ by a Gaussian distribution centred at $\mathbf{w}_{\mathrm{MP}}$ (see (6.35) and (6.39)), so

$$p(\tilde{y}|\,\tilde{\mathbf{x}}, D) \propto \int \exp\left(-\frac{\beta}{2}[\tilde{y} - y(\tilde{\mathbf{x}}; \mathbf{w})]^2\right) \exp\left(-\frac{1}{2}\Delta\mathbf{w}^{\mathrm{T}}\mathbf{H}_{\mathrm{MP}}\Delta\mathbf{w}\right) \mathrm{d}\mathbf{w}, \tag{6.48}$$

with constant factors ignored. We further assume that the width of the posterior distribution as governed by $\mathbf{H}_{\mathrm{MP}}$ is narrow enough to justify approximating the NN model by a first order Taylor expansion about $\mathbf{w}_{\mathrm{MP}}$, i.e.

$$y(\tilde{\mathbf{x}}; \mathbf{w}) \approx y(\tilde{\mathbf{x}}; \mathbf{w}_{\mathrm{MP}}) + \mathbf{g}^{\mathrm{T}}\Delta\mathbf{w}, \tag{6.49}$$

where the gradient

$$\mathbf{g}(\tilde{\mathbf{x}}) = \nabla_{\mathbf{w}}\, y(\tilde{\mathbf{x}}; \mathbf{w})|_{\mathbf{w}=\mathbf{w}_{\mathrm{MP}}}. \tag{6.50}$$

The integral can now be evaluated analytically (Bishop, 1995, p.399), yielding

$$p(\tilde{y}|\,\tilde{\mathbf{x}}, D) = \frac{1}{(2\pi\sigma^2)^{\frac{1}{2}}} \exp\left(-\frac{[\tilde{y} - y(\tilde{\boldsymbol{x}}; \mathbf{w}_{\mathrm{MP}})]^2}{2\sigma^2}\right), \tag{6.51}$$

which is a Gaussian distribution with mean $y(\tilde{\boldsymbol{x}}; \mathbf{w}_{\mathrm{MP}})$ and variance

$$\sigma^2 = \frac{1}{\beta} + \mathbf{g}^{\mathrm{T}}\mathbf{H}_{\mathrm{MP}}^{-1}\,\mathbf{g}. \tag{6.52}$$

There are two terms contributing to the variance of the predictand. The first term involving $1/\beta$ arises from the noise in the training target data, while the second term arises from the posterior distribution of the model weights $\mathbf{w}$. Note that through $\mathbf{g}(\tilde{\boldsymbol{x}})$, σ also varies with $\tilde{\boldsymbol{x}}$ – in regions where training data are sparse, the second term becomes large, i.e. increased uncertainty from the model weights $\mathbf{w}$.

Figure 6.6 illustrates a BNN fit to 16 training data points which consist of the signal $y = x\sin(\pi x)$ plus Gaussian noise. While the NN has a single input and single output, and six neurons in a single hidden layer (i.e. a total of 19 parameters), γ, the effective number of parameters, is found to be only about six. With Gaussian

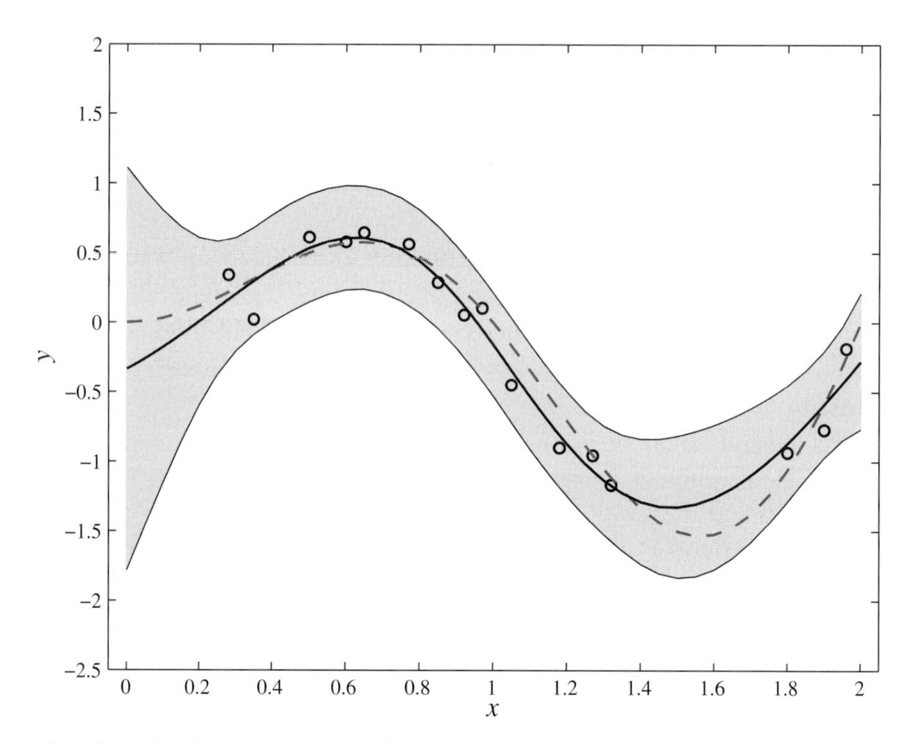

Fig. 6.6 Nonlinear regression fit by a Bayesian NN. The training data are indicated by circles, the BNN solution by the solid curve, and the signal $y = x\sin(\pi x)$ by the dashed curve. The shaded interval denotes ± 2 standard deviations.

noise, ± 2 standard deviations about the mean give the 95% prediction interval, which is seen to widen in regions with sparse or no data, e.g. for $x < 0.2$ and x around 1.6 in the figure.

For nonlinear regression problems, BNN codes written in the *Matlab* language are available in the *Matlab* Neural Network toolbox (based on Foresee and Hagan, 1997) and in the *Netlab* package (Nabney, 2002). Both are easy to use, and σ is provided by *Netlab*. However, the two codes, which differ somewhat in the way the hyperparameters are estimated, tend to give different values of γ for the same training dataset, with *Netlab* tending to yield a larger γ. In simple test problems run by the author, for some problems there are higher odds of underfitting from using the *Matlab* Neural Network toolbox, while for others there are higher odds of slight overfitting from the *Netlab* code. Of course, in most cases both codes yield comparable, good results. Since the final solution can be trapped in a local minimum, multiple runs using different random initial weights are recommended. These codes suffer from the disadvantage of making the Laplace approximation when estimating the hyperparameters, which can be avoided if Markov chain Monte Carlo methods are used (Neal, 1996). In Schlink *et al.* (2003), Bayesian MLP NN using the Laplace approximation underperformed MLP using early stopping (Fig. 4.6) in ground-level ozone predictions.

6.8 Ensemble of models

In weather forecasting, it is now standard practice to run a numerical weather prediction model multiple times from slightly perturbed initial conditions, giving an ensemble of model runs. The rationale is that the atmospheric models are very unstable to small perturbations in the initial conditions, i.e. a tiny error in the initial conditions would lead to a vastly different forecast a couple of weeks later. This behaviour was first noted by Lorenz (1963), which led to the discovery of the 'chaos' phenomenon. From this ensemble of model runs, the averaged forecast over the individual ensemble members is usually issued as the forecast, while the spread of the ensemble members provides information on the uncertainty of the forecast.

In NN applications, one usually trains a number of models for a variety of reasons, e.g. to deal with the multiple minima in the objective function, to experiment varying the number of model parameters, etc. One can test the models' skill over some validation data and simply select the best performer. However, model skill is dependent on the noise in the validation data, i.e. if a different validation dataset is used, a different model may be selected as the best performer. For this reason, it is common to retain a number of good models to form an *ensemble* of models, and use the ensemble average of their outputs as the desired output. In machine learning jargon, an ensemble of models is called a *committee*.

One way to generate multiple models is through *bagging* (abbreviated from Bootstrap AGGregatING) (Breiman, 1996), developed from the idea of *bootstrapping* (Efron, 1979; Efron and Tibshirani, 1993) in statistics. Under bootstrap resampling, data are drawn randomly from a dataset to form a new training dataset, which is to have the same number of data points as the original dataset. A data point in the original dataset can be drawn more than once into a training dataset. On average, 63.2% of the original data is drawn, while 36.8% is not drawn into a training dataset. This is repeated until a large number of training datasets are generated by this bootstrap procedure. During the random draws, predictor and predictand data pairs are of course drawn together. In the case of autocorrelated data, data segments about the length of the autocorrelation time scale are drawn instead of individual data points – i.e. if monthly data are found to be autocorrelated over the whole season, then one would draw an entire season of monthly data altogether. In the bagging approach, one model can be built from each training set, so from the large number of training sets, an ensemble of models is derived. By averaging the model output from all individual members of the ensemble, a final output is obtained. (If the problem is nonlinear classification instead of regression, the final output is chosen by *voting*, i.e. the class most widely selected by the individual members of the ensemble is chosen as the final output.)

Incidentally, the data not selected during the bootstrap resampling are not wasted, as they can be used as validation data. For instance, to avoid overfitting, these validation data can be used in the *early stopping* approach, i.e. NN model training is stopped when the model's error variance calculated using validation data starts to increase. Finally, from the distribution of the ensemble member solutions, statistical significance can be estimated easily – e.g. from the ensemble distribution, one can simply examine if at least 95% of the ensemble members give a value greater than zero, or less than zero, etc.

We next compare the error of the ensemble average to the average error of the individual models in the ensemble. Let $y_T(\mathbf{x})$ denote the true relation, $y_m(\mathbf{x})$ the mth model relation in an ensemble of M models, and $y^{(M)}(\mathbf{x})$ the ensemble average. The expected mean squared error of the ensemble average is

$$\begin{aligned}
\mathrm{E}[(y^{(M)} - y_T)^2] &= \mathrm{E}\left[\left(\frac{1}{M}\sum_{m=1}^{M} y_m - y_T\right)^2\right] \\
&= \mathrm{E}\left[\left(\frac{1}{M}\sum_{m}(y_m - y_T)\right)^2\right] = \frac{1}{M^2}\mathrm{E}\left[\left(\sum_{m}\epsilon_m\right)^2\right], \qquad (6.53)
\end{aligned}$$

where $\epsilon_m \equiv y_m - y_\mathrm{T}$ is the error of the mth model. From Cauchy inequality, we have

$$\left(\sum_{m=1}^{M} \epsilon_m\right)^2 \leq M \sum_{m=1}^{M} \epsilon_m^2, \tag{6.54}$$

hence

$$\mathrm{E}[(y^{(M)} - y_\mathrm{T})^2] \leq \frac{1}{M} \sum_{m=1}^{M} \mathrm{E}[\epsilon_m^2]. \tag{6.55}$$

This proves that the expected error of the ensemble average is less than or equal to the average expected error of the individual models in the ensemble, thereby providing the rationale for using ensemble averages instead of individual models. Note this is a general result, as it applies to an ensemble of dynamical models (e.g. general circulation models) as well as an ensemble of empirical models (e.g. NN models), or even to a mixture of completely different dynamical and empirical models. Perhaps this result is not so surprising, since in social systems we do find that, on average, democracy is better than the average dictatorship!

Next we restrict consideration to a single model for generating the ensemble members, e.g. by training the model with various bootstrapped resampled datasets, or performing nonlinear optimization with random initial parameters. We now repeat the variance and bias error calculation of Section 6.3 for an ensemble of models. Again, let $\mathcal{E}[\cdot]$ denote the expectation or ensemble average over all datasets D or over all random initial weights (as distinct from $\mathrm{E}[\cdot]$, the expectation over $\mathbf{x}$). Since all members of the ensemble were generated from a single model, we have for all the $m = 1, \ldots, M$ members,

$$\mathcal{E}[y_m] = \mathcal{E}[y] \equiv \bar{y}. \tag{6.56}$$

The expected square error of the ensemble average $y^{(M)}$ is

$$\begin{aligned}
\mathcal{E}[(y^{(M)} - y_\mathrm{T})^2] &= \mathcal{E}[(y^{(M)} - \bar{y} + \bar{y} - y_\mathrm{T})^2] \\
&= \mathcal{E}[(y^{(M)} - \bar{y})^2] + \mathcal{E}[(\bar{y} - y_\mathrm{T})^2] + 2\mathcal{E}[(y^{(M)} - \bar{y})(\bar{y} - y_\mathrm{T})] \\
&= \mathcal{E}[(y^{(M)} - \bar{y})^2] + (\bar{y} - y_\mathrm{T})^2 + 2(\bar{y} - y_\mathrm{T})\,\mathcal{E}[y^{(M)} - \bar{y}] \\
&= \mathcal{E}[(y^{(M)} - \bar{y})^2] + (\bar{y} - y_\mathrm{T})^2,
\end{aligned} \tag{6.57}$$

as $\mathcal{E}[y^{(M)} - \bar{y}] = 0$. The first term, $\mathcal{E}[(y^{(M)} - \bar{y})^2]$, is the variance error, while the second term, $(\bar{y} - y_\mathrm{T})^2$, is the bias error. Note that the variance error depends on M, the ensemble size, whereas the bias error does not.

Let us examine the variance error:

$$\mathcal{E}[(y^{(M)} - \bar{y})^2] = \mathcal{E}\left[\left(\frac{1}{M}\sum_{m=1}^{M} y_m - \bar{y}\right)^2\right] = \mathcal{E}\left[\left(\frac{1}{M}\sum_{m=1}^{M}(y_m - \bar{y})\right)^2\right]$$
$$= \frac{1}{M^2}\mathcal{E}\left[\left(\sum_m \delta_m\right)^2\right], \tag{6.58}$$

where $\delta_m = y_m - \bar{y}$. If the errors of the members are uncorrelated, i.e. $\mathcal{E}[\delta_m \delta_n] = 0$ if $m \neq n$, then

$$\frac{1}{M^2}\mathcal{E}\left[\left(\sum_m \delta_m\right)^2\right] = \frac{1}{M^2}\mathcal{E}\left[\sum_m \delta_m^2\right] = \frac{1}{M}\mathcal{E}[\delta^2], \tag{6.59}$$

as $\mathcal{E}[\delta_m^2] = \mathcal{E}[\delta^2]$, for all m. Thus if the member errors are uncorrelated, the variance error of the ensemble average is

$$\mathcal{E}[(y^{(M)} - \bar{y})^2] = \frac{1}{M}\mathcal{E}[(y - \bar{y})^2], \tag{6.60}$$

where the right hand side is simply the variance error of a single member divided by M. Hence, the variance error of the ensemble average $\to 0$ as $M \to \infty$. Of course, the decrease in the variance error of the ensemble average will not be as rapid as M^{-1} if the errors of the members are correlated.

In summary, for the ensemble average, the variance error can be decreased by increasing the ensemble size M, but the bias error is unchanged. This suggests that one should use models with small bias errors, and then rely on the ensemble averaging to reduce the variance error. In other words, one would prefer using models which overfit slightly to models which underfit, as ensemble averaging can alleviate the overfitting. Cannon and Whitfield (2002, Fig. 2 and 3) compared the performance of a single MLP NN model using the early stopping method and that of an ensemble of MLP models by bagging (without early stopping) using real hydrological data. They showed the ensemble approach to perform better than the single model, hence the ensemble method is an effective way to control overfitting.

So far, all the members are equally weighted in forming the ensemble average. As some members may be more skilful than others, one would like to weight them according to their skill, i.e.

$$y^{(M)} = \sum_m a_m y_m, \tag{6.61}$$

where the weights a_m are obtained from an optimization procedure. The procedure, described in Bishop (1995, Section 9.6), involves working with the error covariance matrix $\mathbf{C}$. Unfortunately, in most climate problems, the data records are probably too short to allow an accurate estimate of **C**. For problems where the data records are long enough to estimate **C** with accuracy, then this approach could improve on the equally-weighted ensemble averaging method.

An even more sophisticated way to form the ensemble average is to use an MLP NN model to perform nonlinear ensemble averaging. Here the MLP NN has M inputs and one output. The M inputs simply receive the outputs from the M trained ensemble members, while the output is trained towards the same target data used in training the individual ensemble members. Krasnopolsky (2007) used an ensemble of ten MLP NN models to emulate sea level height anomalies using state variables from an ocean model as input. The outputs of the ten ensemble members were then nonlinearly averaged by another MLP NN and compared with simple averaging (Fig. 6.7). The simple ensemble average has smaller error standard deviation than all ten individuals, but its bias is just the average of the bias of the individuals. In contrast, the nonlinear ensemble average by NN results in even smaller error standard deviation plus a considerable reduction in bias.

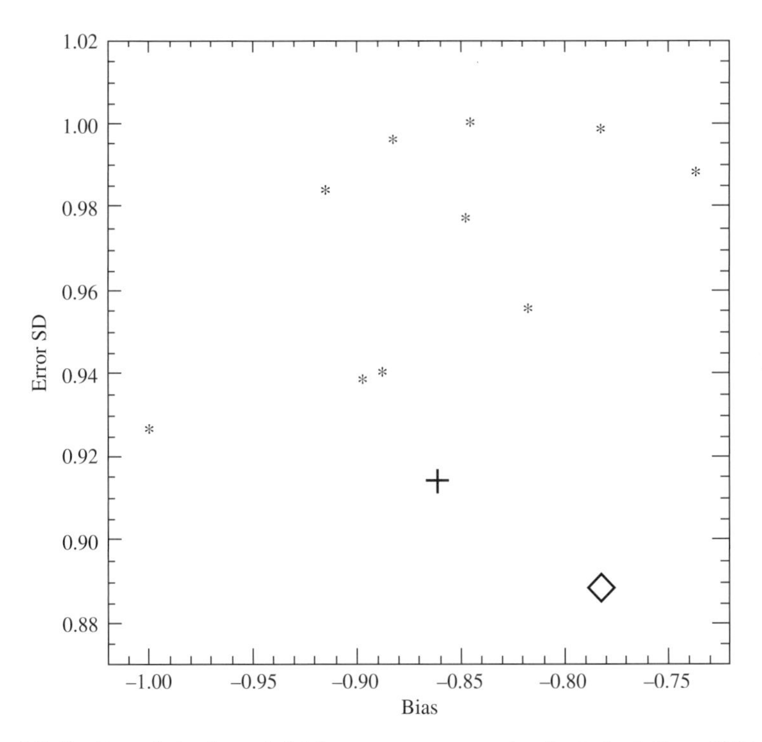

Fig. 6.7 Scatter plot of model bias versus error standard deviation (SD) for the ten individual ensemble members (asterisks), the simple ensemble average (cross) and the nonlinear ensemble average by NN (diamond). (Reproduced from Krasnopolsky (2007) with permission of the American Geophysical Union.)

Another ensemble/committee approach called *boosting* differs from other ensemble methods such as bagging in that the models in the ensemble are trained in sequence, with 'improvement' from one model to the next, and the final output of the ensemble being a weighted sum of the output from all its members. The most popular boosting algorithm is *AdaBoost* (Freund and Schapire, 1997), developed originally for classification problems, but also extended to regression problems. The key idea is that in the mth model there are some data points which are not well predicted, so when we train the next model, we increase the weighting of these difficult data points in our objective function. This type of learning approach is used by students, e.g. if a student does poorly in some courses, he will put more effort into the difficult courses in order to improve his overall grade. Since boosting tries to improve on predicting the difficult data points, we must ensure that the difficult data points are of sound quality, i.e. the data are not simply wrong!

The outline of the boosting approach is as follows: let $w_n^{(m)}$ denote the weight placed on the nth data point ($n = 1, \ldots, N$) in the objective function of the mth model ($m = 1, \ldots, M$). For the first model, we use uniform weight, i.e. $w_n^{(1)} = 1/N$. We next generate a sequence of models. For model m, the weights $w_n^{(m)}$ are increased relative to $w_n^{(m-1)}$ for a data point n if this point was poorly predicted by model $m - 1$. The final output of the M models is a weighted sum of the individual model outputs, with the weights a_m being larger for the better models.

6.9 Approaches to predictive uncertainty

The ability to provide an estimate of the predictive uncertainty to accompany the prediction of a single value is of great practical value (Cawley *et al.*, 2007). For instance, we dress very differently if on a particular day the temperature forecast has a large forecasted variance than if the same temperature forecast has a small variance. We have so far studied a variety of methods which would provide some measure of the predictive uncertainty, e.g. the NN modelling of conditional probability distribution $p(\mathbf{y}|\mathbf{x})$ in Section 4.7, the Bayesian neural network (BNN) model in Section 6.7, and the use of bootstrapping in Section 6.8. It is appropriate at this stage to compare how these methods differ in the way their predictive uncertainties are estimated.

From (6.47) (with the tilde now dropped for brevity), we have seen that the conditional probability distribution

$$p(y|\,\mathbf{x}, D) = \int p(y|\,\mathbf{x}, \mathbf{w})\, p(\mathbf{w}|D)\mathrm{d}\mathbf{w}, \qquad (6.62)$$

where $p(y|\,\mathbf{x}, \mathbf{w})$ is the distribution of the noisy target data for given $\mathbf{x}$ and $\mathbf{w}$, and $p(\mathbf{w}|D)$ is the posterior distribution of the weights given training data D. These two terms indicate that there are two separate contributions to predictive uncertainty, namely noise in the target data, and uncertainty in the weights $\mathbf{w}$ as these are estimated from a sample of finite size. Since the Bayesian approach derives an estimate of the posterior distribution $p(\mathbf{w}|D)$, it allows the uncertainty in the weights to be taken into account in the estimate of the predictive uncertainty as given by $p(y|\mathbf{x}, D)$.

In the NN modelling of the conditional probability distribution in Section 4.7, the principle of *maximum likelihood* was used to estimate the model weights. This more classical approach provides a single optimal estimate of $\mathbf{w}$, not a distribution $p(\mathbf{w}|D)$ as in the more modern Bayesian approach. Hence its predictive uncertainty estimate can only include the contribution from the noise in the target data, thereby omitting the contribution from the uncertainty of the weights estimated from a sample of limited size.

In the bootstrap approach of Section 6.8, we resampled the data record repeatedly to generate a large number of samples. Training the model on these samples led to an ensemble of models, each with different values of $\mathbf{w}$. The scatter in the predicted values y by the ensemble members provides an estimate of the predictive uncertainty, which has taken into account the uncertainty in the model weights, but not the noise in the target data.

Thus the maximum likelihood modelling of conditional distribution and the bootstrap modelling approach both provide incomplete estimates of the predictive uncertainty, as they can take into account only one of the two components. The former approach takes into account the noise in the target data while the latter takes into account the uncertainty in estimating $\mathbf{w}$ from a sample of limited size. The Bayesian approach incorporates both components in its predictive uncertainty estimate.

6.10 Linearization from time-averaging

Time-averaging is widely used to reduce noise in the data; however, it also linearizes the relations in the dataset. In a study of the nonlinear relation between the precipitation rate (the predictand) and ten other atmospheric variables (the predictors $\mathbf{x}$) in the NCEP/NCAR reanalysis data (Kalnay *et al.*, 1996), Yuval and Hsieh (2002) examined the daily, weekly and monthly averaged data by nonlinear multiple regression using the MLP NN model over three regions (Middle East, northeastern China and off the west coast of Canada), and discovered that the strongly nonlinear relations found in the daily data became dramatically reduced by time-averaging to the almost linear relations found in the monthly data. To measure the degree of nonlinearity in the NN relation, the ratio R was used, where

$$R = \frac{f(\mathbf{x}) + \nabla f(\mathbf{x}) \cdot \delta\mathbf{x}}{f(\mathbf{x} + \delta\mathbf{x})}, \tag{6.63}$$

with f the NN modelled predictand. If f is a linear function, then $R = 1$, hence the departure of R from 1 is a measure of nonlinearity. Figure 6.8 shows the precipitation at (50°N, 130°W), just offshore from British Columbia, Canada. The spread in R is seen to diminish dramatically as we move from daily data to monthly data, indicating a strong linearizing tendency. For the daily, weekly and monthly data, the ratio of the correlation skill of the NN model to that of LR (linear regression) was 1.10, 1.02 and 1.00, respectively, while the ratio of the RMSE (root mean squared error) of NN to that of LR was 0.88, 0.92 and 0.96 respectively. Hence the advantage of the nonlinear NN model over LR has almost vanished in the monthly data. Similar conclusions were found for data in the Middle East (33°N, 35°E) and northeastern China (50°N, 123°E).

To explain this surprising phenomenon, Yuval and Hsieh (2002) invoked the well-known central limit theorem from statistics. For simplicity, consider the relation between two variables x and y. If $y = f(x)$ is a nonlinear function, then even if x is a normally distributed random variable, y will in general not have a normal distribution. Now consider the effects of time-averaging on the (x, y) data. The bivariate central limit theorem (Bickel and Doksum, 1977, Theorem 1.4.3) says that if $(x_1, y_1), \ldots, (x_n, y_n)$ are independent and identically distributed random vectors with finite second moments, then (X, Y), obtained from averaging $(x_1, y_1), \ldots, (x_n, y_n)$, will, as $n \to \infty$, approach a bivariate normal distribution $N(\mu_1, \mu_2, \sigma_1^2, \sigma_2^2, \rho)$, where μ_1 and μ_2 are the mean of X and Y, respectively, σ_1^2 and σ_2^2 are the corresponding variance, and ρ the correlation between X and Y.

From the bivariate normal distribution, the conditional probability distribution of Y (given X) is also a normal distribution (Bickel and Doksum, 1977, Theorem 1.4.2), with mean

$$\mathrm{E}[Y|X] = \mu_2 + (X - \mu_1)\rho\sigma_2/\sigma_1. \tag{6.64}$$

This linear relation in X explains why time-averaging tends to linearize the relationship between the two variables. With more variables, the bivariate normal distribution readily generalizes to the multivariate normal distribution.

To visualize this effect, consider the synthetic dataset

$$y = x + x^2 + \epsilon, \tag{6.65}$$

where x is a Gaussian variable with unit standard deviation and ϵ is Gaussian noise with a standard deviation of 0.5. Averaging these 'daily' data over 7 days and over 30 days reveals a dramatic weakening of the nonlinear relation (Fig. 6.9), and the shifting of the y density distribution towards Gaussian with the time-averaging. With real data, there is autocorrelation in the time series, so the monthly data will

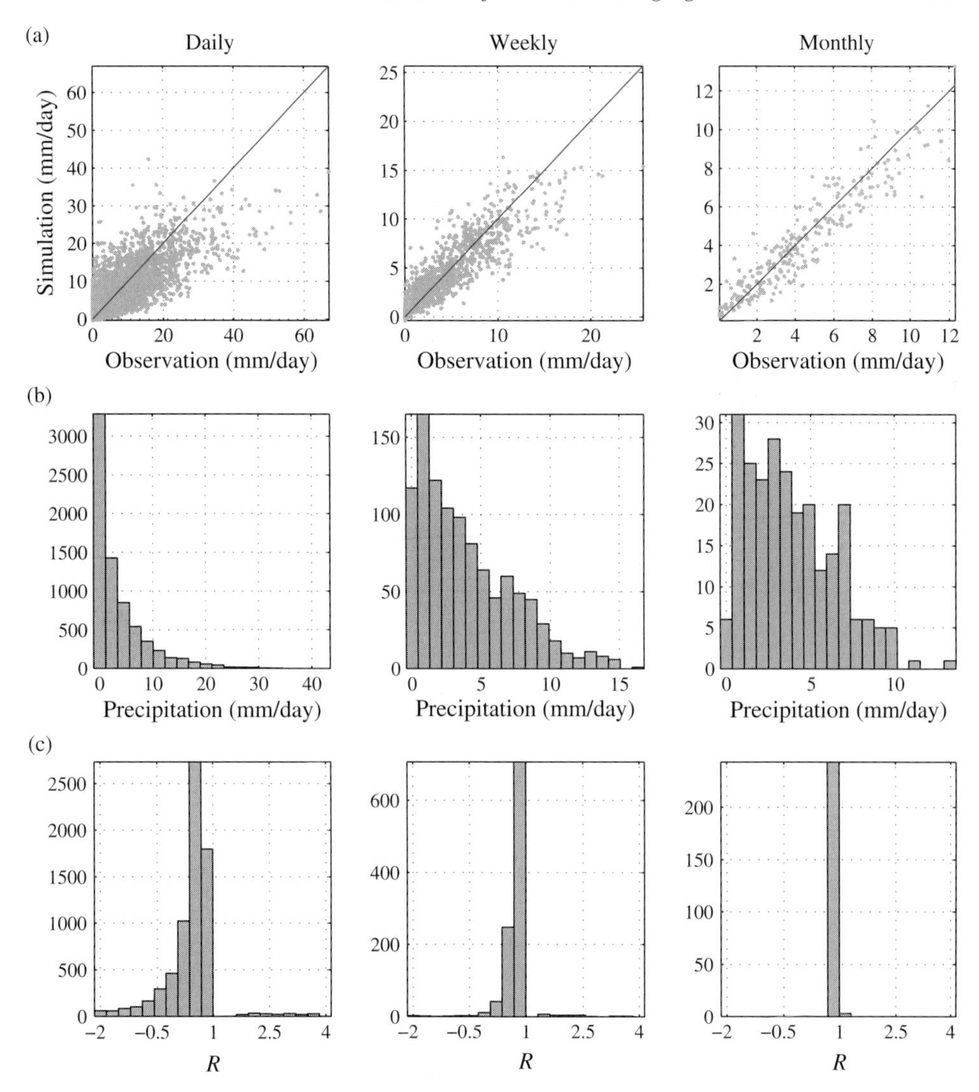

Fig. 6.8 Neural network simulated precipitation rate off the west coast of British Columbia, Canada. (a) Scatter plots of NN simulation of the daily, weekly and monthly precipitation rate versus the observed values, with the diagonal line indicating the ideal one-to-one relation. The weekly and monthly precipitation amounts were converted to mm/day units. (b) Histograms of the precipitation rate distribution. (c) The distributions of the ratio R defined in the text. The histograms are counts of the number of occurrences of precipitation rate and R values which fall within the limits of equally spaced bins covering the range of values. (Reproduced from Yuval and Hsieh (2002) with permission of the Royal Meteorological Society.)

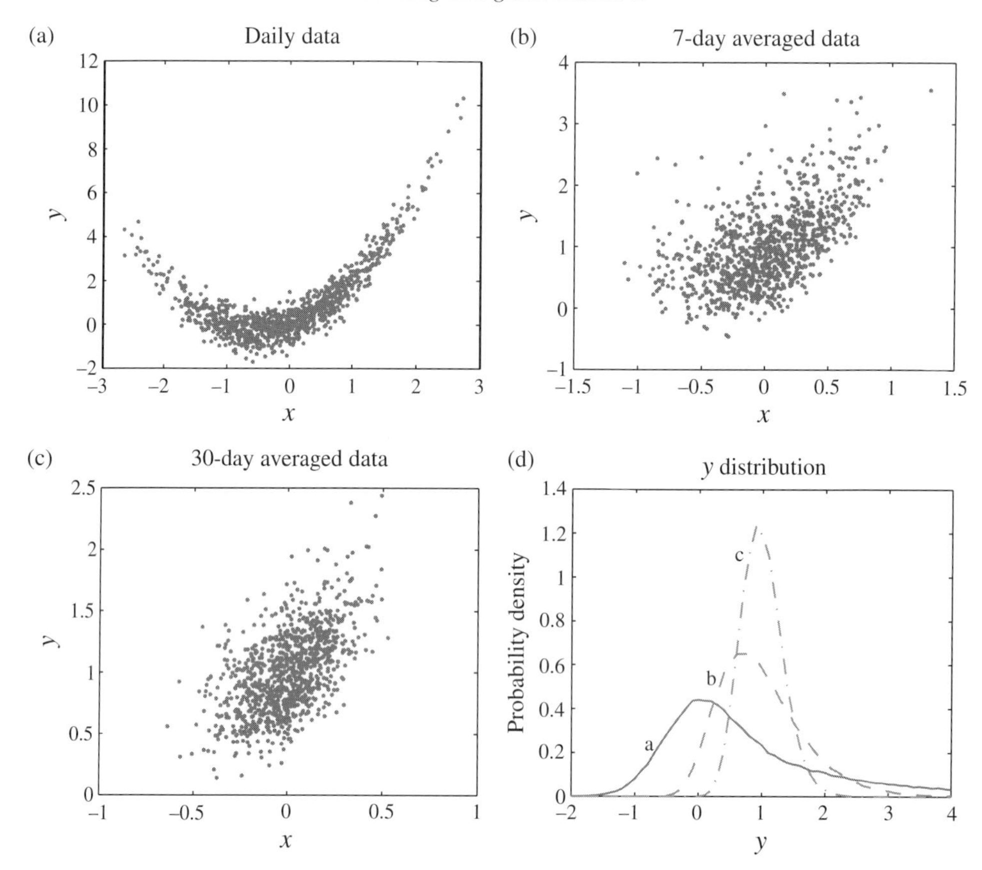

Fig. 6.9 Effects of time-averaging on a nonlinear relation. (a) Synthetic 'daily' data from a quadratic relation between x and y. (b) and (c) The data time-averaged over (b) 7 observations and (c) 30 observations. (d) The probability density distribution of y for cases (a), (b) and (c). (Reproduced from Hsieh and Cannon (2008) with permission of Springer.)

be effectively averaging over far fewer than 30 independent observations as done in this synthetic dataset.

If the data have strong autocorrelation, so that the integral time scale from the autocorrelation function is not small compared to the time-averaging window, then there are actually few independent observations used during the time-averaging, and the central limit theorem does not apply. For instance, the eastern equatorial Pacific sea surface temperatures have an integral time scale of about a year, hence nonlinear relations can be detected from monthly or seasonal data, as found by nonlinear principal component analysis (Monahan, 2001; Hsieh, 2001b) and nonlinear canonical correlation analysis (Hsieh, 2001a). In contrast, the mid-latitude weather variables have integral time scales of about 3–5 days, so monthly averaged data would have effectively averaged over about 6–10 independent observations, and seasonal data over 20–30 independent observations, so the influence of the central limit theorem cannot be ignored.

While time-averaging tends to reduce the nonlinear signal, it also smooths out the noise. Depending on the type of noise (and perhaps on the type of nonlinear signal), it is possible that time-averaging may nevertheless enhance detection of a nonlinear signal above the noise for some datasets. In short, researchers should be aware that time-averaging could have a major impact on their modelling or detection of nonlinear empirical relations, and that a nonlinear machine learning method often outperforms a linear method in weather applications, but fails to do so in climate applications.

Exercises

(6.1) Let $y = \sin(2\pi x)$ $(0 \leq x < 1)$ be the signal. The y data are generated by adding Gaussian noise to the y signal. Fit a multi-layer perceptron (MLP) NN model with one hidden layer and no regularization to the (x, y) data. Use the early stopping approach of Section 6.4, i.e. monitor the mean squared error (MSE) over training data and over validation data to determine when to stop the training, and the optimal number of hidden neurons to use.

(6.2) Let $y = \sin(2\pi x)$ $(0 \leq x < 1)$ be the signal. The y data are generated by adding Gaussian noise to the y signal. Fit an ensemble of MLP NN models with one hidden layer and no regularization to the (x, y) data. From independent data not used in the training, determine the average MSE of an individual model's prediction of y and the MSE of the ensemble averaged prediction of y. Vary the level of the Gaussian noise, the number of hidden neurons and the ensemble size to determine when the ensemble averaged prediction has the greatest advantage over a single individual model's prediction.

(6.3) For bootstrap resampling applied to a dataset with N observations (Section 6.8), derive an expression for the fraction of data in the original dataset drawn in an average bootstrap sample. What is this fraction as $N \to \infty$?

(6.4) In testing ensemble model forecasts, one compiles the following statistic:

N	1	2	5	10	20
$\mathrm{E}[(y^{(N)} - y_{\mathrm{T}})^2]$	0.40	0.27	0.19	0.17	0.15

where $y^{(N)}$ is the ensemble average prediction with N models used in the ensemble averaging, and y_{T} the true values from data. Assuming that the individual models in the ensemble are independent (i.e. errors are uncorrelated between models), estimate the bias error and $\mathrm{E}[(y - \bar{y})^2]$ (i.e. the variance error of a single model prediction y) from a simple graph. Is the assumption that individual models in the ensemble can be considered independent justified?

7

Kernel methods

Neural network methods became popular in the mid to late 1980s, but by the mid to late 1990s, kernel methods also became popular in the field of machine learning. The first kernel methods were nonlinear classifiers called *support vector machines* (SVM), which were then generalized to nonlinear regression (*support vector regression*, SVR). Soon kernel methods were further developed to nonlinearly generalize principal component analysis (PCA), canonical correlation analysis (CCA), etc. In this chapter, a basic introduction to the kernel approach is given, while further applications of the kernel method to nonlinear classification, regression, PCA, etc. are given in the following chapters. The kernel method has also been extended to probabilisitic models, e.g. *Gaussian processes* (GP).

Section 7.1 tries to bridge from NN to kernel methods. Sections 7.2–7.4 present the mathematical foundation of the kernel method. Since the mathematics behind kernel methods is more sophisticated than that for NN methods, Section 7.5 tries to summarize the main ideas behind kernel methods, as well as their advantages and disadvantages. The pre-image problem, a disadvantage of kernel methods, is discussed in Section 7.6.

7.1 From neural networks to kernel methods

First we recall linear regression, where

$$y_k = \sum_j w_{kj} x_j + b_k \,, \tag{7.1}$$

with $\mathbf{x}$ the predictors and $\mathbf{y}$ the predictands or response variables. When a multi-layer perceptron (MLP) NN is used for nonlinear regression, the mapping does not proceed directly from $\mathbf{x}$ to $\mathbf{y}$, but passes through an intermediate layer of variables $\mathbf{h}$, i.e. the hidden neurons,

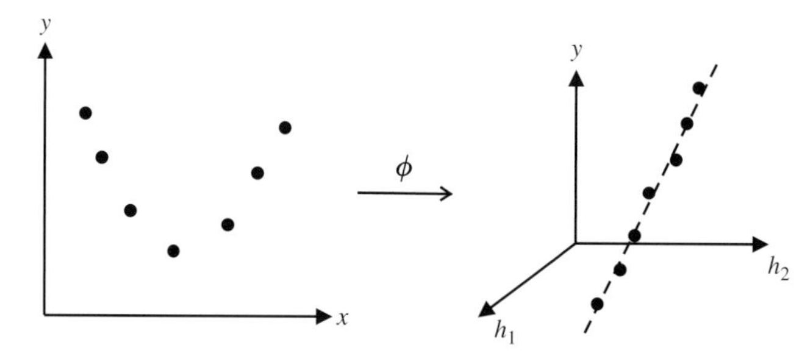

Fig. 7.1 Illustrating the effect of the nonlinear mapping $\boldsymbol{\phi}$ from the input space to the hidden space, where a nonlinear relation between the input x and the output y becomes a linear relation (dashed line) between the hidden variables $\mathbf{h}$ and y.

$$h_j = \tanh\Big(\sum_i w_{ji} x_i + b_j\Big), \tag{7.2}$$

where the mapping from $\mathbf{x}$ to $\mathbf{h}$ is nonlinear through an activation function like the hyperbolic tangent. The next stage is to map from $\mathbf{h}$ to $\mathbf{y}$, and is most commonly done via a linear mapping

$$y_k = \sum_j \tilde{w}_{kj} h_j + \tilde{b}_k \,. \tag{7.3}$$

Since (7.3) is formally identical to (7.1), one can think of the NN model as first mapping nonlinearly from the input space to a 'hidden' space (containing the hidden neurons), i.e. $\boldsymbol{\phi} : \mathbf{x} \to \mathbf{h}$, then performing linear regression between the hidden space variables and the output variables (Fig. 7.1). If the relation between $\mathbf{x}$ and $\mathbf{y}$ is highly nonlinear, then one must use more hidden neurons, i.e. increase the dimension of the hidden space, before one can find a good linear regression fit between the hidden space variables and the output variables. Of course, since the parameters w_{ji}, $\tilde{w}_{kj}$, b_j and $\tilde{b}_k$ are solved together, nonlinear optimization is involved, leading to local minima, which is the main disadvantage of the MLP NN.

The *kernel methods* follow a somewhat similar procedure: in the first stage, a nonlinear function $\boldsymbol{\phi}$ maps from the input space to a hidden space, called the *'feature' space*. In the second stage, one performs linear regression from the feature space to the output space. Instead of linear regression, one can also perform linear classification, PCA, CCA, etc. during the second stage. Like the radial basis function NN with non-adaptive basis functions (Section 4.6) (and unlike the MLP NN with adaptive basis functions), the optimization in stage two of a kernel method is independent of stage 1, so only linear optimization is involved, and there are no local minima – a main advantage over the MLP. (Some kernel methods, e.g. Gaussian processes, use nonlinear optimization to find the hyperparameters, and thus have the local minima problem.) However, the feature space can be of very high (or even infinite) dimension. This disadvantage is eliminated by the use of a kernel trick, which manages to avoid direct evaluation of the high dimensional function $\boldsymbol{\phi}$ altogether. Hence many methods which were previously buried due to the 'curse of dimensionality' (Section 4.3) have been revived in a kernel renaissance.

7.2 Primal and dual solutions for linear regression

To illustrate the kernel method, we will start with linear regression. We need to find a way of solving the regression problem which will compute efficiently under the kernel method. From Section 1.4.4, the multiple linear regression problem is

$$y_i = w_0 + \sum_{l=1}^{m} x_{il} w_l + e_i, \qquad i = 1, \ldots, n, \tag{7.4}$$

where there are n observations and m predictors x_l for the predictand y, with e_i the errors or residuals, and w_0 and w_l the regression coefficients. By introducing $x_{i0} \equiv 1$, we can place w_0 as the first element of the vector $\mathbf{w}$, which also contains w_l, $(l = 1, \ldots, m)$, so the regression problem can be expressed as

$$\mathbf{y} = \mathbf{X}\mathbf{w} + \mathbf{e}. \tag{7.5}$$

where $\mathbf{X}$ is an $n \times m$ matrix. The solution (see Section 1.4.4) is

$$\mathbf{w} = (\mathbf{X}^{\mathrm{T}}\mathbf{X})^{-1}\mathbf{X}^{\mathrm{T}}\mathbf{y}, \tag{7.6}$$

where $(\mathbf{X}^{\mathrm{T}}\mathbf{X})$ is an $m \times m$ matrix, so $(\mathbf{X}^{\mathrm{T}}\mathbf{X})^{-1}$ takes $\mathrm{O}(m^3)$ operations to compute. Let us rewrite the equation as

$$\mathbf{w} = (\mathbf{X}^{\mathrm{T}}\mathbf{X})(\mathbf{X}^{\mathrm{T}}\mathbf{X})^{-1}(\mathbf{X}^{\mathrm{T}}\mathbf{X})^{-1}\mathbf{X}^{\mathrm{T}}\mathbf{y}. \tag{7.7}$$

If we introduce *dual variables* $\boldsymbol{\alpha}$, with

$$\boldsymbol{\alpha} = \mathbf{X}(\mathbf{X}^{\mathrm{T}}\mathbf{X})^{-1}(\mathbf{X}^{\mathrm{T}}\mathbf{X})^{-1}\mathbf{X}^{\mathrm{T}}\mathbf{y}, \tag{7.8}$$

then

$$\mathbf{w} = \mathbf{X}^{\mathrm{T}}\boldsymbol{\alpha}. \tag{7.9}$$

Linear regression when performed with a weight penalty term is called *ridge regression*, where the objective function to be minimized becomes

$$J = \sum_{i=1}^{n}\Big(y_i - \sum_{l} x_{il} w_l\Big)^2 + p\|\mathbf{w}\|^2, \tag{7.10}$$

with p the weight penalty parameter. This can be rewritten as

$$J = (\mathbf{y} - \mathbf{X}\mathbf{w})^{\mathrm{T}}(\mathbf{y} - \mathbf{X}\mathbf{w}) + p\mathbf{w}^{\mathrm{T}}\mathbf{w}. \tag{7.11}$$

Setting the gradient of J with respect to $\mathbf{w}$ to zero, we have

$$-\mathbf{X}^{\mathrm{T}}(\mathbf{y}-\mathbf{X}\mathbf{w})+p\mathbf{w}=0, \tag{7.12}$$

i.e.

$$(\mathbf{X}^{\mathrm{T}}\mathbf{X}+p\mathbf{I})\mathbf{w}=\mathbf{X}^{\mathrm{T}}\mathbf{y}, \tag{7.13}$$

where $\mathbf{I}$ is the identity matrix. This yields

$$\mathbf{w}=(\mathbf{X}^{\mathrm{T}}\mathbf{X}+p\mathbf{I})^{-1}\,\mathbf{X}^{\mathrm{T}}\mathbf{y}, \tag{7.14}$$

which requires $\mathrm{O}(m^3)$ operations to solve.

From (7.12),

$$\mathbf{w}=p^{-1}\mathbf{X}^{\mathrm{T}}(\mathbf{y}-\mathbf{X}\mathbf{w}). \tag{7.15}$$

Invoking (7.9), we have

$$\boldsymbol{\alpha}=p^{-1}(\mathbf{y}-\mathbf{X}\mathbf{w})=p^{-1}(\mathbf{y}-\mathbf{X}\mathbf{X}^{\mathrm{T}}\boldsymbol{\alpha}). \tag{7.16}$$

Rewriting as

$$(\mathbf{X}\mathbf{X}^{\mathrm{T}}+p\mathbf{I})\boldsymbol{\alpha}=\mathbf{y}, \tag{7.17}$$

we obtain

$$\boldsymbol{\alpha}=(\mathbf{G}+p\mathbf{I})^{-1}\mathbf{y}, \tag{7.18}$$

with the $n\times n$ matrix $G\equiv\mathbf{X}\mathbf{X}^{\mathrm{T}}$ called the *Gram matrix*. Using this equation to solve for $\boldsymbol{\alpha}$ takes $\mathrm{O}(n^3)$ operations. If $n\ll m$, solving $\mathbf{w}$ from (7.9) via (7.18), called the *dual solution*, will be much faster than solving from (7.14), the *primal solution*.

If a new datum $\tilde{\mathbf{x}}$ becomes available, and if one wants to map from $\tilde{\mathbf{x}}$ to $\tilde{y}$, then using

$$\tilde{y}=\tilde{\mathbf{x}}^{\mathrm{T}}\mathbf{w} \tag{7.19}$$

requires $\mathrm{O}(m)$ operations if $\mathbf{w}$ from the primal solution (7.14) is used.

If the dual solution is used, we have

$$\tilde{y}=\mathbf{w}^{\mathrm{T}}\tilde{\mathbf{x}}=\boldsymbol{\alpha}^{\mathrm{T}}\mathbf{X}\tilde{\mathbf{x}}=\sum_{i=1}^{n}\alpha_i\,\mathbf{x}_i^{\mathrm{T}}\tilde{\mathbf{x}}, \tag{7.20}$$

where $\mathbf{x}_i^{\mathrm{T}}$ is the ith row of $\mathbf{X}$. With the dimensions of $\boldsymbol{\alpha}^{\mathrm{T}}$, $\mathbf{X}$ and $\tilde{\mathbf{x}}$ being $1\times n$, $n\times m$ and $m\times 1$, respectively, computing $\tilde{y}$ via the dual solution requires $\mathrm{O}(nm)$ operations, more expensive than via the primal solution. Writing $\mathbf{k}=\mathbf{X}\tilde{\mathbf{x}}$ (with $k_i=\mathbf{x}_i^{\mathrm{T}}\tilde{\mathbf{x}}$) and invoking (7.18), we have

$$\tilde{y}=\boldsymbol{\alpha}^{\mathrm{T}}\mathbf{k}=((\mathbf{G}+p\mathbf{I})^{-1}\mathbf{y})^{\mathrm{T}}\mathbf{k}=\mathbf{y}^{\mathrm{T}}(\mathbf{G}+p\mathbf{I})^{-1}\mathbf{k}, \tag{7.21}$$

where we have used the fact that $\mathbf{G}+p\mathbf{I}$ is symmetric.

7.3 Kernels

In the kernel method, a feature map $\boldsymbol{\phi}(\mathbf{x})$ maps from the input space X to the feature space F. For instance, if $\mathbf{x} \in X = \mathbb{R}^m$, then $\boldsymbol{\phi}(\mathbf{x}) \in F \subseteq \mathbb{R}^M$, where F is the feature space, and

$$\boldsymbol{\phi}^{\mathrm{T}}(\mathbf{x}) = [\phi_1(\mathbf{x}), \ldots, \phi_M(\mathbf{x})]. \tag{7.22}$$

In problems where it is convenient to incorporate a constant element $x_{i0} \equiv 1$ into the vector $\mathbf{x}_i$ (e.g. to deal with the constant weight w_0 in (7.4)), one also tends to add a constant element $\phi_0(\mathbf{x})$ ($\equiv 1$) to $\boldsymbol{\phi}(\mathbf{x})$. With appropriate choice of the nonlinear mapping functions $\boldsymbol{\phi}$, the relation in the high dimensional space F becomes linear.

In Section 7.2, we had $\mathbf{G} = \mathbf{X}\mathbf{X}^{\mathrm{T}}$ and $k_i = \mathbf{x}_i^{\mathrm{T}}\tilde{\mathbf{x}}$. Now, as we will be performing regression in the feature space, we need to work with

$$G_{ij} = \boldsymbol{\phi}^{\mathrm{T}}(\mathbf{x}_i)\boldsymbol{\phi}(\mathbf{x}_j), \tag{7.23}$$

$$k_i = \boldsymbol{\phi}^{\mathrm{T}}(\mathbf{x}_i)\boldsymbol{\phi}(\tilde{\mathbf{x}}). \tag{7.24}$$

If we assume $\boldsymbol{\phi}(\mathbf{x})$ requires $\mathrm{O}(M)$ operations, then $\boldsymbol{\phi}^{\mathrm{T}}(\mathbf{x}_i)\boldsymbol{\phi}(\mathbf{x}_j)$ is still of $\mathrm{O}(M)$, but (7.23) has to be done n^2 times as $\mathbf{G}$ is an $n \times n$ matrix, thus requiring a total of $\mathrm{O}(n^2 M)$ operations. To get $\boldsymbol{\alpha}$ from (7.18), computing the inverse of the $n \times n$ matrix $(\mathbf{G} + p\mathbf{I})$ takes $\mathrm{O}(n^3)$ operations, thus a total of $\mathrm{O}(n^2 M + n^3)$ operations is needed.

With new datum $\tilde{\mathbf{x}}$, (7.20) now becomes

$$\tilde{y} = \sum_{i=1}^{n} \alpha_i \, \boldsymbol{\phi}^{\mathrm{T}}(\mathbf{x}_i)\boldsymbol{\phi}(\tilde{\mathbf{x}}), \tag{7.25}$$

which requires $\mathrm{O}(nM)$ operations.

To save computation costs, instead of evaluating $\boldsymbol{\phi}$ explicitly, we introduce a *kernel function* K to evaluate the inner product (i.e. dot product) in the feature space,

$$K(\mathbf{x}, \mathbf{z}) \equiv \boldsymbol{\phi}^{\mathrm{T}}(\mathbf{x})\boldsymbol{\phi}(\mathbf{z}) = \sum_l \phi_l(\mathbf{x})\phi_l(\mathbf{z}), \tag{7.26}$$

for all $\mathbf{x}$, $\mathbf{z}$ in the input space. Since $K(\mathbf{x}, \mathbf{z}) = K(\mathbf{z}, \mathbf{x})$, K is a *symmetric function*. The key to the *kernel trick* is that if an algorithm in the input space can be formulated involving only inner products, then the algorithm can be solved in the feature space with the kernel function evaluating the inner products. Although the algorithm may only be solving for a linear problem in the feature space, it is equivalent to solving a nonlinear problem in the input space.

As an example, for $\mathbf{x} = (x_1, x_2) \in \mathbb{R}^2$, consider the feature map

$$\boldsymbol{\phi}(\mathbf{x}) = (x_1^2, \; x_2^2, \; \sqrt{2}\, x_1 x_2) \in \mathbb{R}^3. \tag{7.27}$$

Linear regression performed in F is then of the form

$$y = a_0 + a_1 x_1^2 + a_2 x_2^2 + a_3\sqrt{2}\,x_1 x_2, \qquad (7.28)$$

so quadratic relations with the inputs become linear relations in the feature space. For this $\boldsymbol{\phi}$,

$$\boldsymbol{\phi}^{\mathrm{T}}(\mathbf{x})\boldsymbol{\phi}(\mathbf{z}) = x_1^2 z_1^2 + x_2^2 z_2^2 + 2x_1 x_2 z_1 z_2 = (x_1 z_1 + x_2 z_2)^2 = (\mathbf{x}^{\mathrm{T}}\mathbf{z})^2. \qquad (7.29)$$

Hence

$$K(\mathbf{x}, \mathbf{z}) = (\mathbf{x}^{\mathrm{T}}\mathbf{z})^2. \qquad (7.30)$$

Although K is defined via $\boldsymbol{\phi}$ in (7.26), we can obtain K in (7.30) without explicitly involving $\boldsymbol{\phi}$. Note that for

$$\boldsymbol{\phi}(\mathbf{x}) = (x_1^2,\ x_2^2,\ x_1 x_2,\ x_2 x_1) \in \mathbb{R}^4, \qquad (7.31)$$

we would get the same K, i.e. a kernel function K is not uniquely associated with one feature map $\boldsymbol{\phi}$.

Next generalize to $\mathbf{x} \in \mathbb{R}^m$. Function K in (7.30) now corresponds to the feature map $\boldsymbol{\phi}(\mathbf{x})$ with elements $x_k x_l$, where $k = 1, \ldots, m$, $l = 1, \ldots, m$. Hence $\boldsymbol{\phi}(\mathbf{x}) \in \mathbb{R}^{m^2}$, i.e. the dimension of the feature space is $M = m^2$. Computing $\boldsymbol{\phi}^{\mathrm{T}}(\mathbf{x})\boldsymbol{\phi}(\mathbf{z})$ directly takes $\mathrm{O}(m^2)$ operations, but computing this inner product through $K(\mathbf{x}, \mathbf{z})$ using (7.30) requires only $\mathrm{O}(m)$ operations. This example clearly illustrates the advantage of using the kernel function to avoid direct computations with the high-dimensional feature map $\boldsymbol{\phi}$.

Given observations $\mathbf{x}_i$, $(i = 1, \ldots, n)$ in the input space, and a feature map $\boldsymbol{\phi}$, the $n \times n$ *kernel matrix* $\mathbf{K}$ is defined with elements

$$K_{ij} = K(\mathbf{x}_i, \mathbf{x}_j) = \boldsymbol{\phi}^{\mathrm{T}}(\mathbf{x}_i)\boldsymbol{\phi}(\mathbf{x}_j) \equiv G_{ij}, \qquad (7.32)$$

where (G_{ij}) is simply the Gram matrix.

A matrix $\mathbf{M}$ is *positive semi-definite* if for any vector $\mathbf{v} \in \mathbb{R}^n$, $\mathbf{v}^{\mathrm{T}}\mathbf{M}\mathbf{v} \geq 0$. A symmetric function $f : X \times X \longrightarrow \mathbb{R}$ is said to be a positive semi-definite function if any matrix $\mathbf{M}$ [with elements $M_{ij} = f(\mathbf{x}_i, \mathbf{x}_j)$], formed by restricting f to a finite subset of the input space X, is a positive semi-definite matrix.

A special case of *Mercer theorem* from functional analysis guarantees that a function $K : X \times X \longrightarrow \mathbb{R}$ is a kernel associated with a feature map $\boldsymbol{\phi}$ via

$$K(\mathbf{x}, \mathbf{z}) = \boldsymbol{\phi}^{\mathrm{T}}(\mathbf{x})\boldsymbol{\phi}(\mathbf{z}), \qquad (7.33)$$

if and only if K is a positive semi-definite function. To show that a kernel function K is positive semi-definite, consider any kernel matrix $\mathbf{K}$ derived from K, and any vector $\mathbf{v}$:

$$\mathbf{v}^{\mathrm{T}}\mathbf{K}\mathbf{v} = \sum_i \sum_j v_i v_j K_{ij} = \sum_i \sum_j v_i v_j \boldsymbol{\phi}(\mathbf{x}_i)^{\mathrm{T}} \boldsymbol{\phi}(\mathbf{x}_j)$$

$$= \Big(\sum_i v_i \boldsymbol{\phi}(\mathbf{x}_i)\Big)^{\mathrm{T}} \Big(\sum_j v_j \boldsymbol{\phi}(\mathbf{x}_j)\Big) = \Big\| \sum_i v_i \boldsymbol{\phi}(\mathbf{x}_i) \Big\|^2 \geq 0, \tag{7.34}$$

so K is indeed positive semi-definite. The proof for the converse is more complicated, and the reader is referred to Schölkopf and Smola (2002) or Shawe-Taylor and Cristianini (2004).

Let us now consider what operations can be performed on kernel functions which will yield new kernel functions. For instance, if we multiply a kernel function K_1 by a positive constant c, then $K = cK_1$ is a kernel function, since for any kernel matrix $\mathbf{K}_1$ derived from K_1 and any vector $\mathbf{v}$,

$$\mathbf{v}^{\mathrm{T}}(c\mathbf{K}_1)\mathbf{v} = c\,\mathbf{v}^{\mathrm{T}}\mathbf{K}_1\mathbf{v} \geq 0. \tag{7.35}$$

Next consider adding two kernel functions K_1 and K_2, which also yields a kernel since

$$\mathbf{v}^{\mathrm{T}}(\mathbf{K}_1 + \mathbf{K}_2)\mathbf{v} = \mathbf{v}^{\mathrm{T}}\mathbf{K}_1\mathbf{v} + \mathbf{v}^{\mathrm{T}}\mathbf{K}_2\mathbf{v} \geq 0. \tag{7.36}$$

Below is a list of operations on kernel functions (K_1 and K_2) which will yield new kernel functions K.

(1) $K(\mathbf{x}, \mathbf{z}) = cK_1(\mathbf{x}, \mathbf{z}), \quad \text{with } c \geq 0.$ (7.37)

(2) $K(\mathbf{x}, \mathbf{z}) = K_1(\mathbf{x}, \mathbf{z}) + K_2(\mathbf{x}, \mathbf{z}).$ (7.38)

(3) $K(\mathbf{x}, \mathbf{z}) = K_1(\mathbf{x}, \mathbf{z})K_2(\mathbf{x}, \mathbf{z}).$ (7.39)

(4) $K(\mathbf{x}, \mathbf{z}) = \dfrac{K_1(\mathbf{x}, \mathbf{z})}{\sqrt{K_1(\mathbf{x}, \mathbf{x})K_1(\mathbf{z}, \mathbf{z})}} = \dfrac{\boldsymbol{\phi}_1(\mathbf{x})^{\mathrm{T}}}{\|\boldsymbol{\phi}_1(\mathbf{x})\|} \dfrac{\boldsymbol{\phi}_1(\mathbf{z})}{\|\boldsymbol{\phi}_1(\mathbf{z})\|}.$ (7.40)

(5) $K(\mathbf{x}, \mathbf{z}) = K_1(\boldsymbol{\psi}(\mathbf{x}), \boldsymbol{\psi}(\mathbf{z})), \text{ with } \boldsymbol{\psi} \text{ a real function.}$ (7.41)

If f is a real function, $\mathbf{M}$ a symmetric positive semi-definite $m \times m$ matrix, and p a positive integer, we also obtain kernel functions from the following operations.

(6) $K(\mathbf{x}, \mathbf{z}) = f(\mathbf{x})f(\mathbf{z}).$ (7.42)

(7) $K(\mathbf{x}, \mathbf{z}) = \mathbf{x}^{\mathrm{T}}\mathbf{M}\mathbf{z}.$ (7.43)

(8) $K(\mathbf{x}, \mathbf{z}) = \big(K_1(\mathbf{x}, \mathbf{z})\big)^p,$ (7.44)
where a special case is the popular *polynomial kernel*

$$K(\mathbf{x}, \mathbf{z}) = \big(1 + \mathbf{x}^{\mathrm{T}}\mathbf{z}\big)^p. \tag{7.45}$$

For example, suppose the true relation is

$$y(x_1, x_2) = b_1 x_2 + b_2 x_1^2 + b_3 x_1 x_2. \tag{7.46}$$

With a $p = 2$ polynomial kernel, $\boldsymbol{\phi}(\mathbf{x})$ in the feature space contains the elements $x_1, x_2, x_1^2(\equiv x_3), x_1x_2(\equiv x_4), x_2^2(\equiv x_5)$. Linear regression in F is

$$y = a_0 + a_1 x_1 + a_2 x_2 + a_3 x_3 + a_1 x_4 + a_1 x_4 + a_5 x_5, \tag{7.47}$$

so the true nonlinear relation (7.46) can be extracted by a linear regression in F.

Kernels can also be obtained through exponentiation.

(9) $K(\mathbf{x}, \mathbf{z}) = \exp\left(K_1(\mathbf{x}, \mathbf{z})\right)$. (7.48)

(10) $K(\mathbf{x}, \mathbf{z}) = \exp\left(-\frac{\|\mathbf{x}-\mathbf{z})\|^2}{2\sigma^2}\right)$. (7.49)

The *Gaussian kernel* or *radial basis function (RBF) kernel* (7.49) is the most commonly used kernel. As the exponential function can be approximated arbitrarily accurately by a high-degree polynomial with positive coefficients, the exponential kernel is therefore a limit of kernels. To extend further from the exponential kernel (7.48) to the Gaussian kernel (7.49), we first note that $\exp(\mathbf{x}^{\mathrm{T}}\mathbf{z}/\sigma^2)$ is a kernel according to (7.48). Normalizing this kernel by (7.40) gives the new kernel

$$\begin{aligned}\frac{\exp(\mathbf{x}^{\mathrm{T}}\mathbf{z}/\sigma^2)}{\sqrt{\exp(\mathbf{x}^{\mathrm{T}}\mathbf{x}/\sigma^2)\exp(\mathbf{z}^{\mathrm{T}}\mathbf{z}/\sigma^2)}} &= \exp\left(\frac{\mathbf{x}^{\mathrm{T}}\mathbf{z}}{\sigma^2} - \frac{\mathbf{x}^{\mathrm{T}}\mathbf{x}}{2\sigma^2} - \frac{\mathbf{z}^{\mathrm{T}}\mathbf{z}}{2\sigma^2}\right) \\ &= \exp\left(\frac{-\mathbf{x}^{\mathrm{T}}\mathbf{x} + \mathbf{z}^{\mathrm{T}}\mathbf{x} + \mathbf{x}^{\mathrm{T}}\mathbf{z} - \mathbf{z}^{\mathrm{T}}\mathbf{z}}{2\sigma^2}\right) \\ &= \exp\left(\frac{-(\mathbf{x}-\mathbf{z})^{\mathrm{T}}(\mathbf{x}-\mathbf{z})}{2\sigma^2}\right) \\ &= \exp\left(\frac{-\|\mathbf{x}-\mathbf{z}\|^2}{2\sigma^2}\right). \qquad (7.50)\end{aligned}$$

Hence the Gaussian kernel is a normalized exponential kernel. For more details on operations on kernels, see Shawe-Taylor and Cristianini (2004).

7.4 Kernel ridge regression

For the first application of the kernel method, let us describe *kernel ridge regression* as a technique for performing nonlinear regression. We will use the Gaussian kernel K and ridge regression. The Gaussian kernel (7.49) has the parameter σ governing the width of the Gaussian function, while ridge regression (7.10) has a weight penalty parameter p. These two parameters are usually called *hyperparameters*, as we need to perform a search for their optimal values above a search for the basic model parameters like the regression coefficients.

The procedure for kernel regression is set out below.

(1) Set aside some data for validation later; use the remaining observations $\mathbf{x}_i$, y_i, $(i = 1, \ldots, n)$ for training.
(2) Choose values for σ and p.
(3) Calculate the kernel matrix $\mathbf{K}$ with $K_{ij} = K(\mathbf{x}_i, \mathbf{x}_j)$.
(4) Solve the dual problem for ridge regression in (7.18) with $\mathbf{K}$ replacing $\mathbf{G}$, i.e. $\boldsymbol{\alpha} = (\mathbf{K} + p\mathbf{I})^{-1}\mathbf{y}$.

(5) For validation data $\tilde{\mathbf{x}}$, compute $\tilde{y}$ using (7.21), i.e. $\tilde{y} = \boldsymbol{\alpha}^{\mathrm{T}}\mathbf{k}$, with $k_i = K(\mathbf{x}_i, \tilde{\mathbf{x}})$.
(6) Calculate the mean squared error (MSE) of $\tilde{y}$ over the validation data.
(7) Go to (2) and repeat the calculations with different values of σ and p.
(8) Choose the σ and p values with the smallest MSE over the validation data as the optimal solution.

7.5 Advantages and disadvantages

Since the mathematical formulation of the last three sections may be difficult for some readers, a summary of the main ideas of kernel methods and their advantages and disadvantages is presented in this section.

First consider the simple linear regression problem with a single predictand variable y,

$$y = \sum_i a_i\, x_i + a_0, \tag{7.51}$$

with x_i the predictor variables, and a_i and a_0 the regression parameters. In the MLP NN approach to nonlinear regression (Section 4.3), nonlinear *adaptive* basis functions h_j (also called hidden neurons) are introduced, so the linear regression is between y and h_j,

$$y = \sum_j a_j h_j(\mathbf{x}; \mathbf{w}) + a_0, \tag{7.52}$$

with typically

$$h_j(\mathbf{x}; \mathbf{w}) = \tanh(\sum_i w_{ji} x_i + b_j). \tag{7.53}$$

Since y depends on $\mathbf{w}$ nonlinearly (due to the nonlinear function tanh), the resulting optimization is nonlinear, with multiple minima in general.

What happens if instead of adaptive basis functions $h_j(\mathbf{x}; \mathbf{w})$, we use *non-adaptive* basis functions $\phi_j(\mathbf{x})$, i.e. the basis functions do not have adjustable parameters $\mathbf{w}$? In this situation,

$$y = \sum_j a_j\, \phi_j(\mathbf{x}) + a_0. \tag{7.54}$$

For instance, Taylor series expansion with two predictors (x_1, x_2) would have $\{\phi_j\} = x_1, x_2, x_1^2, x_1x_2, x_2^2, \ldots$ The advantage of using non-adaptive basis functions is that y does not depend on any parameter nonlinearly, so only linear optimization is involved with no multiple minima problem. The disadvantage is the curse of dimensionality (Section 4.3), i.e. one generally needs a huge number of non-adaptive basis functions compared to relatively few adaptive basis functions to model a nonlinear relation, hence the dominance of MLP NN models despite the local minima problem.

The curse of dimensionality is finally solved with the kernel trick, i.e. although $\boldsymbol{\phi}$ is a very high (or even infinite) dimensional vector function, as long as the solution of the problem can be formulated to involve only inner products like $\boldsymbol{\phi}^{\mathrm{T}}(\mathbf{x}')\boldsymbol{\phi}(\mathbf{x})$, then a kernel function K can be introduced

$$K(\mathbf{x}', \mathbf{x}) = \boldsymbol{\phi}^{\mathrm{T}}(\mathbf{x}')\boldsymbol{\phi}(\mathbf{x}). \tag{7.55}$$

The solution of the problem now involves working only with a very manageable kernel function $K(\mathbf{x}', \mathbf{x})$ instead of the unmanageable $\boldsymbol{\phi}$. From Section 7.4, the kernel ridge regression solution is

$$y = \sum_{k=1}^{n} \alpha_k K(\mathbf{x}_k, \mathbf{x}), \tag{7.56}$$

where there are $k = 1, \ldots, n$ data points $\mathbf{x}_k$ in the training dataset. If a Gaussian kernel function is used, then y is simply a linear combination of Gaussian functions. If a polynomial kernel function is used, then y is a linear combination of polynomial functions.

The kernel approach has an elegant architecture with simple modularity, which allows easy interchange of parts. The modularity, as illustrated in Fig. 7.2, is as follows: after the input data have been gathered in stage 1, stage 2 consists of choosing a kernel function K and calculating the kernel matrix $\mathbf{K}$, stage 3 consists of using a pattern analysis algorithm to perform regression, classification, principal component analysis (PCA), canonical correlation analysis (CCA), or other tasks in the feature space, whereby the extracted information is given as output in stage 4. For example, if at stage 3, one decides to switch from ridge regression to a different regression algorithm such as support vector regression, or even to CCA, the other stages require no major adjustments to accommodate this switch. Similarly, we can switch kernels in stage 2 from say the Gaussian kernel to a polynomial kernel without having to make significant modifications to the other stages. The pattern analysis algorithms in stage 3 may be limited to working with vectorial data, yet

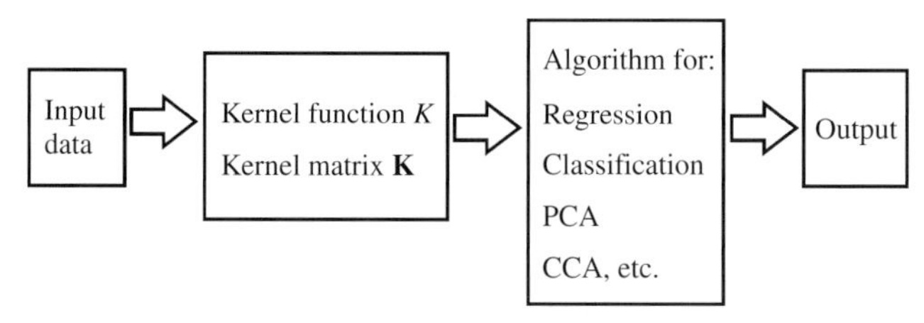

Fig. 7.2 The modular achitecture of the kernel method. (After Shawe-Taylor and Cristianini (2004).)

with cleverly designed kernels in stage 2, kernel methods have been used to analyze data with very different structures, e.g. strings (used in text and DNA analyses) and trees (Shawe-Taylor and Cristianini, 2004).

In summary, with the kernel method, one can analyze the structure in a high-dimensional feature space with only moderate computational costs thanks to the kernel function which gives the inner product in the feature space without having to evaluate the feature map $\boldsymbol{\phi}$ directly. The kernel method is applicable to all pattern analysis algorithms expressed only in terms of inner products of the inputs. In the feature space, linear pattern analysis algorithms are applied, with no local minima problems since only linear optimization is involved. Although only linear pattern analysis algorithms are used, the fully nonlinear patterns in the input space can be extracted due to the nonlinear feature map $\boldsymbol{\phi}$.

The main disadvantage of the kernel method is the lack of an easy way to map inversely from the feature space back to the input data space – a difficulty commonly referred to as the *pre-image* problem. This problem arises e.g. in kernel principal component analysis (kernel PCA, see Section 10.4), where one wants to find the pattern in the input space corresponding to a principal component in the feature space. In kernel PCA, where PCA is performed in the feature space, the eigenvectors are expressed as a linear combination of the data points in the feature space, i.e.

$$\mathbf{v} = \sum_{i=1}^{n} \alpha_i \boldsymbol{\phi}(\mathbf{x}_i). \tag{7.57}$$

As the feature space F is generally a much higher dimensional space than the input space X and the mapping function $\boldsymbol{\phi}$ is nonlinear, one may not be able to find an $\mathbf{x}$ (i.e. the 'pre-image') in the input space, such that $\boldsymbol{\phi}(\mathbf{x}) = \mathbf{v}$, as illustrated in Fig. 7.3. We will look at methods for finding an approximate pre-image in the next section.

This chapter has provided a basic introduction to the kernel approach, and further applications of the kernel method to nonlinear regression, classification, PCA, CCA, etc. are given in the following chapters. The kernel method has also been extended to probabilisitic models, e.g. Gaussian processes.

7.6 The pre-image problem

Since in general an exact pre-image may not exist (Fig. 7.3), various methods have been developed to find an approximate pre-image. In Fig. 7.4, we illustrate the situation where we want to find a pre-image for a point $\mathbf{p}(\boldsymbol{\phi}(\mathbf{x}))$. Here $\mathbf{p}$ can represent for instance the projection of $\boldsymbol{\phi}(\mathbf{x})$ onto the first PCA eigenvector in the feature space. Mika *et al.* (1999) proposed finding a pre-image $\mathbf{x}'$ by minimizing

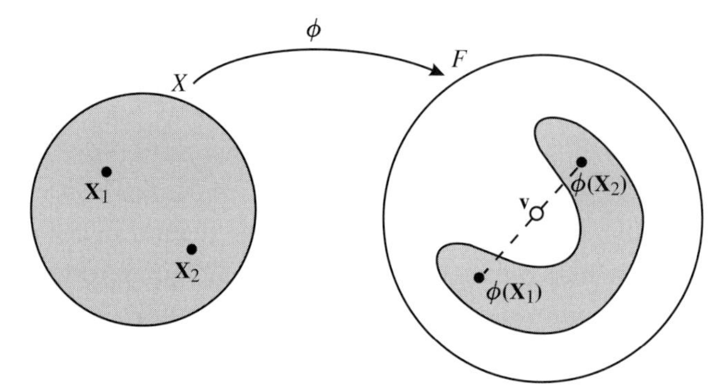

Fig. 7.3 Illustrating the pre-image problem in kernel methods. The input space X is mapped by $\boldsymbol{\phi}$ to the grey area in the much larger feature space F. Two data points $\mathbf{x}_1$ and $\mathbf{x}_2$ are mapped to $\boldsymbol{\phi}(\mathbf{x}_1)$ and $\boldsymbol{\phi}(\mathbf{x}_2)$, respectively, in F. Although $\mathbf{v}$ is a linear combination of $\boldsymbol{\phi}(\mathbf{x}_1)$ and $\boldsymbol{\phi}(\mathbf{x}_2)$, it lies outside the grey area in F, hence there is no 'pre-image' $\mathbf{x}$ in X, such that $\boldsymbol{\phi}(\mathbf{x}) = \mathbf{v}$.

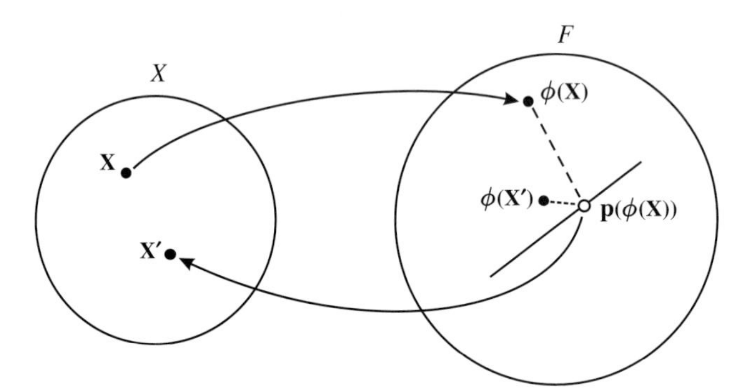

Fig. 7.4 Illustrating the approach used by Mika *et al.* (1999) to extract an approximate pre-image in the input space X for a point $\mathbf{p}(\boldsymbol{\phi}(\mathbf{x}))$ in the feature space F. Here, for example, $\mathbf{p}(\boldsymbol{\phi}(\mathbf{x}))$ is shown as the projection of $\boldsymbol{\phi}(\mathbf{x})$ onto the direction of the first PCA eigenvector (solid line). The optimization algorithm looks for $\mathbf{x}'$ in X which minimizes the squared distance between $\boldsymbol{\phi}(\mathbf{x}')$ and $\mathbf{p}(\boldsymbol{\phi}(\mathbf{x}))$.

the squared distance between $\boldsymbol{\phi}(\mathbf{x}')$ and $\mathbf{p}(\boldsymbol{\phi}(\mathbf{x}))$, i.e.

$$\mathbf{x}' = \arg\min_{\mathbf{x}'} \|\boldsymbol{\phi}(\mathbf{x}') - \mathbf{p}(\boldsymbol{\phi}(\mathbf{x}))\|^2. \tag{7.58}$$

This is a nonlinear optimization problem, hence susceptible to finding local minima. It turns out the method is indeed quite unstable, and alternatives have been proposed.

In the approach of Kwok and Tsang (2004), the distances between $\mathbf{p}(\boldsymbol{\phi}(\mathbf{x}))$ and its k nearest neighbours, $\boldsymbol{\phi}(\mathbf{x}_1) \ldots, \boldsymbol{\phi}(\mathbf{x}_k)$ in F are used. Noting that there is usually a simple relation between feature-space distance and input-space distance for

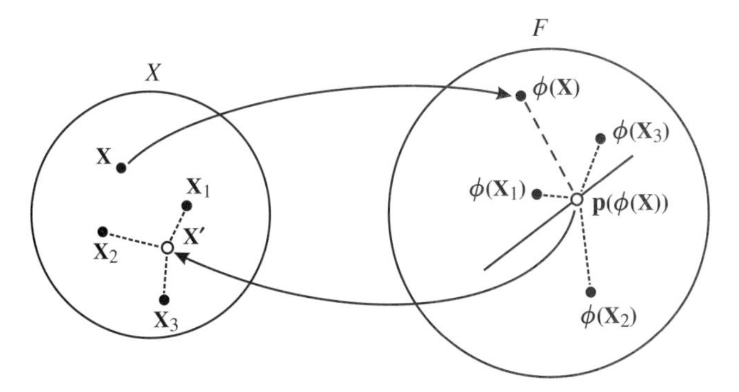

Fig. 7.5 Illustrating the approach used by Kwok and Tsang (2004) to extract an approximate pre-image in the input space X for a point $\mathbf{p}(\boldsymbol{\phi}(\mathbf{x}))$ in the feature space F. The distance information in F between $\mathbf{p}(\boldsymbol{\phi}(\mathbf{x}))$ and its several nearest neighbours (e.g. $\boldsymbol{\phi}(\mathbf{x}_1), \boldsymbol{\phi}(\mathbf{x}_2), \ldots$), and the relationship between distance in X and distance in F are exploited to allow $\mathbf{x}_1, \mathbf{x}_2, \ldots$ to pinpoint the desired approximate pre-image $\mathbf{x}'$ in X.

many commonly used kernels, and borrowing an idea from multi-dimensional scaling (MDS), they were able to use the corresponding distances among $\mathbf{x}_1, \ldots, \mathbf{x}_k$ to pinpoint the approximate pre-image $\mathbf{x}'$ for $\mathbf{p}(\boldsymbol{\phi}(\mathbf{x}))$ (Fig. 7.5), analogous to the use of global positioning system (GPS) satellites to pinpoint the location of an object. Since the solution required only linear algebra and is non-iterative, there are no numerical instability and local minima problems. Note that the method can be used to find the pre-image for any point $\tilde{\mathbf{p}}$ in F. Other applications beside kernel PCA include kernel clustering, e.g. $\tilde{\mathbf{p}}$ can be the mean position of a cluster in F obtained from a K-means clustering algorithm (Section 1.7). Alternative approaches to finding an approximate pre-image include Bakir *et al.* (2004) and Tang and Mazzoni (2006).

Exercises

(7.1) Prove (7.39), i.e. the product of two kernels is also a kernel.

(7.2) In the kernel method, a function $\boldsymbol{\phi}$ maps from the input $\mathbf{x}$ to a point $\boldsymbol{\phi}(\mathbf{x})$ in the feature space, with a kernel function $K(\mathbf{x}_i, \mathbf{x}_j) = \boldsymbol{\phi}(\mathbf{x}_i)^{\mathrm{T}}\boldsymbol{\phi}(\mathbf{x}_j)$. For some applications, we need to remove the mean $\bar{\boldsymbol{\phi}}$ from $\boldsymbol{\phi}$, i.e. we need to work with the centred kernel $\tilde{K}(\mathbf{x}_i, \mathbf{x}_j) = (\boldsymbol{\phi}(\mathbf{x}_i) - \bar{\boldsymbol{\phi}})^{\mathrm{T}}(\boldsymbol{\phi}(\mathbf{x}_j) - \bar{\boldsymbol{\phi}})$. Express this centred kernel $\tilde{K}$ in terms of the original kernel K.

8

Nonlinear classification

So far, we have used NN and kernel methods for nonlinear regression. However, when the output variables are discrete rather than continuous, the regression problem turns into a classification problem. For instance, instead of a prediction of the wind speed, the public may be more interested in a forecast of either 'storm' or 'no storm'. There can be more than two classes for the outcome – for seasonal temperature, forecasts are often issued as one of three classes, 'warm', 'normal' and 'cool' conditions. Issuing just a forecast for one of the three conditions (e.g. 'cool conditions for next season') is often not as informative to the public as issuing posterior probability forecasts. For instance, a '25% chance of warm, 30% normal and 45% cool' condition forecast, is quite different from a '5% warm, 10% normal and 85% cool' condition forecast, even though both are forecasting cool conditions. We will examine methods which choose one out of k classes, and methods which issue posterior probability over k classes. In cases where a class is not precisely defined, *clustering* methods are used to group the data points into clusters.

In Section 1.6, we introduced the two main approaches to classification – the first by discriminant functions, the second by posterior probability. A *discriminant function* is simply a function which tells us which class we should assign to a given predictor data point $\mathbf{x}$ (also called a *feature vector*). For instance, $\mathbf{x}$ can be the reading from several weather instruments at a given moment, and the discriminant function then tells us whether it is going to be 'storm' or 'no storm'. Discriminant functions provide *decision boundaries* in the $\mathbf{x}$-space, where data points on one side of a decision boundary are estimated to belong to one class, while those on the opposite side, to a different class. Figure 1.3 illustrates linear and nonlinear decision boundaries. Since linear discriminant analysis methods provide only hyperplane decision boundaries, neural network and kernel methods are used to model nonlinear decision boundaries.

In the posterior probability approach to classification, Bayes theorem (Section 1.5) is used to estimate the posterior probability $P(C_i|\mathbf{x})$ of belonging to class C_i given the observation $\mathbf{x}$, i.e.

$$P(C_i|\mathbf{x}) = \frac{p(\mathbf{x}|C_i)P(C_i)}{\sum_i p(\mathbf{x}|C_i)P(C_i)}, \quad i = 1, \ldots, k, \tag{8.1}$$

where $p(\mathbf{x}|C_i)$ is the likelihood and $P(C_i)$ the prior probability. The posterior probabilities can then be used to perform classification. For a given $\mathbf{x}$, suppose C_j is found to have the highest posterior probability, i.e.

$$P(C_j|\mathbf{x}) > P(C_i|\mathbf{x}), \quad \text{for all } i \neq j, \tag{8.2}$$

then C_j is chosen as the appropriate class.

There are pros and cons for the two different approaches. Under the framework of *statistical learning theory* (Vapnik, 1995), posterior probabilities are avoided in classification problems. The argument is that in terms of complexity, the three main types of problem can be ranked in the following order: classification (simplest), regression (harder) and estimating posterior probabilities (hardest). Hence the classical approach of estimating posterior probabilities before proceeding to classification amounts to solving a much harder problem first in order to solve a relatively simple classification problem (Cherkassky and Mulier, 1998). In this chapter, we will look at classification methods from both approaches – neural network (NN) and support vector machines (SVM). The NN outputs posterior probabilities, then uses them to perform classification (Section 8.1), whereas SVM performs classification without use of posterior probabilities (Section 8.4).

How to score the forecast performance of a classifier is examined in Section 8.5. How *unsupervised learning* (i.e. learning without target data) can classify input data into categories or clusters is presented in Section 8.6. Applications of machine learning methods for classification in the environmental sciences are reviewed in Chapter 12.

8.1 Multi-layer perceptron classifier

The multi-layer perceptron (MLP) NN model for nonlinear regression can easily be modified for classification problems. If there are only two classes C_1 and C_2, we can take the target data y_d to be a binary variable, with $y_d = 1$ denoting C_1 and $y_d = 0$ denoting C_2. Since the output is bounded in classification problems, instead of using a linear activation function in the output layer as in the case of MLP regression, we now use the logistic sigmoidal function (4.5). With s denoting the logistic sigmoidal activation function, the MLP network with one layer of hidden neurons h_j has the output y given by

$$y = s\left(\sum_j \tilde{w}_j h_j + \tilde{b}\right), \quad \text{with } h_j = s(\mathbf{w}_j \cdot \mathbf{x} + b_j), \tag{8.3}$$

where $\tilde{w}_j$ and $\mathbf{w}_j$ are weights and $\tilde{b}$ and b_j are offset or bias parameters.

In MLP regression problems, the objective function J minimizes the mean squared error (MSE), i.e.

$$J = \frac{1}{2N} \sum_{n=1}^{N} (y_n - y_{dn})^2 . \tag{8.4}$$

This objective function can still be used in classification problems, though there is an alternative objective function (the cross entropy function) to be presented later. Regularization (i.e. weight penalty or decay terms) (Section 6.5) can also be added to J to control overfitting.

To classify the MLP output y as either 0 or 1, we invoke the *indicator function* I, where

$$I(x) = \begin{cases} 1 & \text{if} \quad x > 0, \\ 0 & \text{if} \quad x \leq 0. \end{cases} \tag{8.5}$$

The classification is then given by

$$f(\mathbf{x}) = I[y(\mathbf{x}) - 0.5]. \tag{8.6}$$

The classification error can be defined by

$$E = \frac{1}{2N} \sum_{n=1}^{N} (f(\mathbf{x}_n) - y_{dn})^2 . \tag{8.7}$$

While posterior probabilities have not been invoked in this MLP classifier, y, the output from a logistic sigmoidal function, can be interpreted as a posterior probability. This was shown in Section 4.2 for the single-layer perceptron model. With the extra hidden layer in MLP, complicated, nonlinear decision boundaries can be modelled.

Figure 8.1 illustrates classification of noiseless data in the x_1-x_2 plane by an MLP with different values of m_h (the number of neurons in the hidden layer). The theoretical decision boundaries for the data are shown by the dashed curves (i.e. an ellipse and a rectangle). With m_h = 2 and 3, the model does not have enough complexity to model the correct decision boundaries. Classification of noisy data is shown in Fig. 8.2, where overfitting occurs when m_h becomes large, as regularization (weight penalty) is not used.

When one is interested only in classification and not in posterior probabilities, there is an alternative coding of target data which speeds up the computation. Instead of the target data being coded as 0 or 1, the data are coded as 0.1 or 0.9. The reason is that 0 and 1 are the asymptotic values of the logistic sigmoidal function, and would require weights of very large magnitude to represent them, hence a computational disadvantage (Cherkassky and Mulier, 1998). However, with good optimization algorithms, the difference in computational speed between the two coding schemes is actually quite modest.

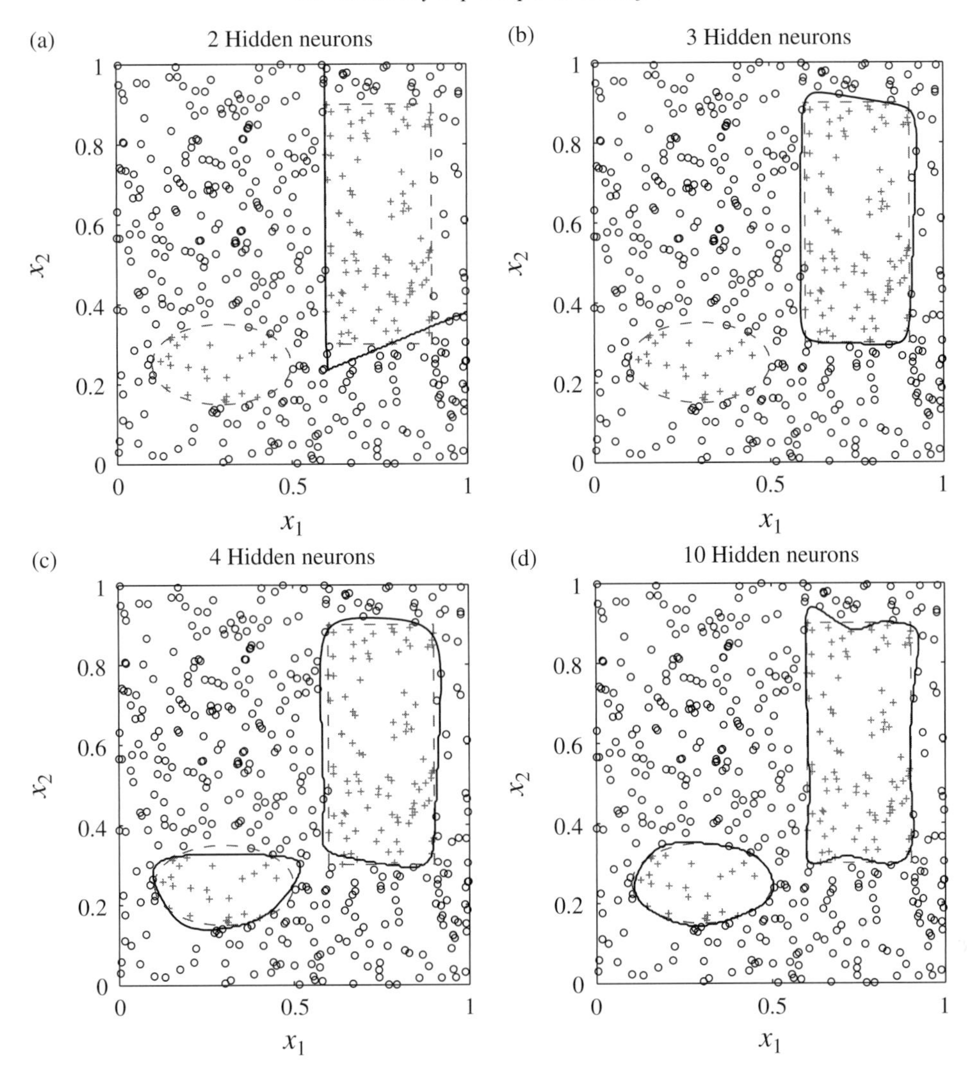

Fig. 8.1 Classification of noiseless data by an MLP model with the number of hidden neurons being (a) 2, (b) 3, (c) 4 and (d) 10. The two classes of data point are indicated by plus signs and circles, with the decision boundaries found by the MLP shown as solid curves, and the theoretical decision boundaries shown as dashed curves.

8.1.1 Cross entropy error function

An alternative to minimizing the mean squared error in the objective function is to minimize the cross entropy instead. In Section 6.1, the MSE objective function was found from the maximum likelihood principle assuming the target data to be continuous variables with Gaussian noise. For classification problems (with two classes), the targets are binary variables, and their noise, if present, is not well-described by Gaussian distributions. We therefore proceed to consider an alternative objective function designed specially for classification problems.

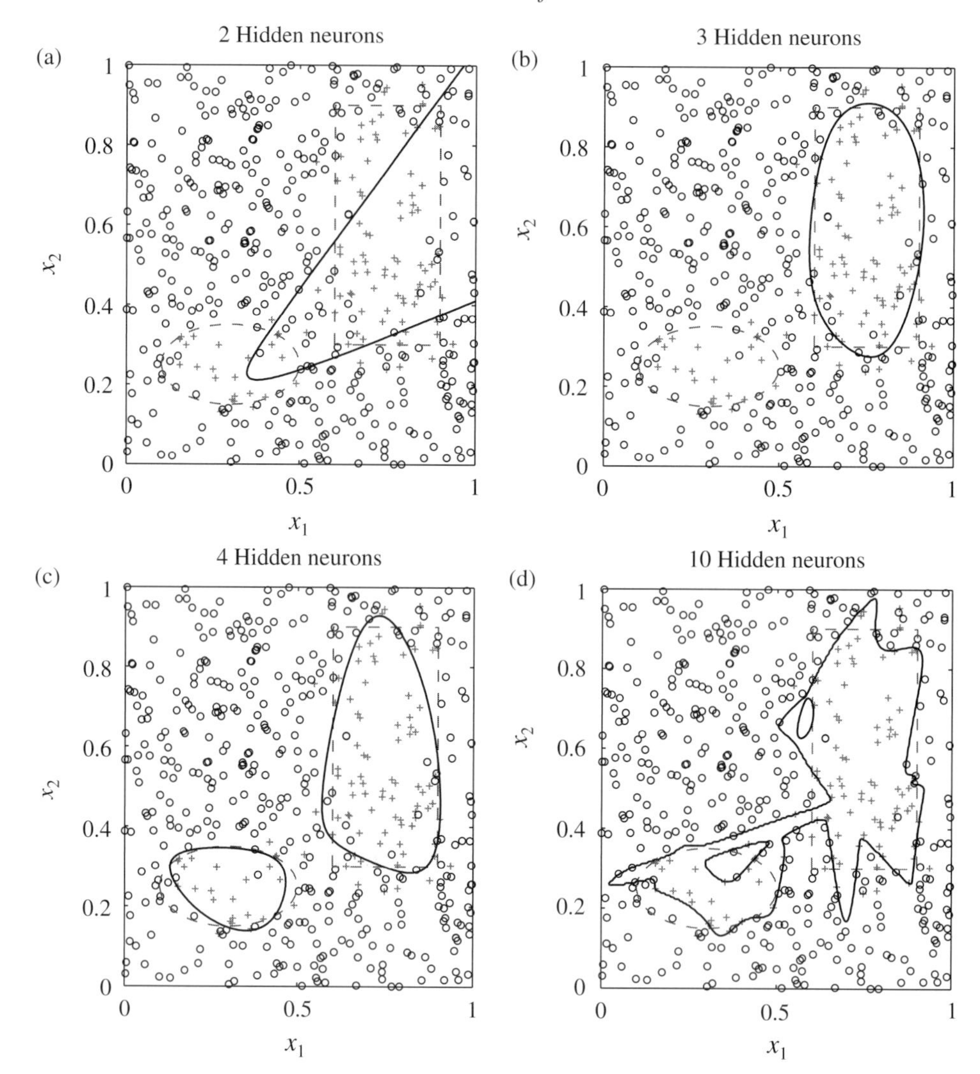

Fig. 8.2 Classification of noisy data by an MLP model with the number of hidden neurons being (a) 2, (b) 3, (c) 4 and (d) 10. Since regularization (weight penalty) is not used, overfitting occurred in (d).

For a two-class problem, we consider a network with a single output y modelling the posterior probability $P(C_1|\mathbf{x})$. The target data y_d equals 1 if the input $\mathbf{x}$ belongs to C_1, and equals 0 if C_2. We can combine the two expressions $P(C_1|\mathbf{x}) = y$ and $P(C_2|\mathbf{x}) = 1 - y$ into a single convenient expression

$$P(y_d|\mathbf{x}) = y^{y_d}(1 - y)^{1-y_d}, \tag{8.8}$$

since changing from $y_d = 1$ to $y_d = 0$ switches from the first expression to the second.

For a dataset with $n = 1, \ldots, N$ independent observations, the likelihood function is then

$$L = \prod_n y_n^{y_{dn}} (1 - y_n)^{1 - y_{dn}}. \tag{8.9}$$

As in Section 6.1, the objective function J is taken to be the negative logarithm of the likelihood function, i.e.

$$J = -\ln L = -\sum_n y_{dn} \ln y_n + (1 - y_{dn}) \ln(1 - y_n), \tag{8.10}$$

which is called the *cross entropy* error function. This error function also applies to the case where the target y_d is not a binary variable, but is a real variable in the interval [0, 1] representing the probability of the input $\mathbf{x}$ belonging to C_1 (Bishop, 1995).

8.2 Multi-class classification

We next turn to classification problems where there are now c classes (c being an integer > 2). For instance, the temperature is to be classified as warm, normal or cold. The target data typically use a 1-of-c coding scheme, e.g. warm is $(1, 0, 0)$, normal is $(0, 1, 0)$ and cold is $(0, 0, 1)$. This suggests that there should be three model outputs, one for each class. If we are interested in the outputs giving the posterior probability of each class, we will need a generalization of the logistic sigmoidal function. Since posterior probabilities are non-negative, one can model non-negative outputs by exponential functions like $\exp(a_k)$ for the kth model output. Next we require the posterior probabilites to sum to 1, so the kth model output y_k is

$$y_k = \frac{\exp(a_k)}{\sum_{k'} \exp(a_{k'})}, \tag{8.11}$$

which satistifes $\sum_k y_k = 1$. This normalized exponential activation function is commonly referred to as the *softmax* activation function. The name softmax comes about because the function is a smooth form of a 'maximum' function – e.g. if $a_j \gg a_k$, for all $k \neq j$, then $y_j \approx 1$ and all other $y_k \approx 0$.

That the softmax activation function is a generalization of the logistic sigmoidal function can be shown readily. Let us try to express y_k in the logistic sigmoidal form

$$y_k = \frac{1}{1 + \exp(-\alpha_k)}. \tag{8.12}$$

Equating the right hand side of (8.11) and (8.12), we then cross multiply to get

$$\sum_{k'} \exp(a_{k'}) = [1 + \exp(-\alpha_k)] \exp(a_k). \tag{8.13}$$

Solving for α_k yields

$$\alpha_k = a_k - \ln\left[\sum_{k' \neq k} \exp(a_{k'})\right]. \tag{8.14}$$

Next, we derive the cross entropy error function for multi-class classification. Assume that the model has one output y_k for each class C_k, and the target data y_{dk} are coded using the 1-of-c binary coding, i.e. $y_{dk} = 1$ if the input $\mathbf{x}$ belongs to C_k, and 0 otherwise. The output y_k is to represent the posterior probability $P(C_k|\mathbf{x})$. We can combine all $k = 1, \ldots, c$ expressions of $y_k = P(C_k|\mathbf{x})$ into a single expression

$$P(\mathbf{y}|\mathbf{x}) = \prod_{k=1}^{c} y_k^{\,y_{dk}}. \tag{8.15}$$

With $n = 1, \ldots, N$ independent observations, the likelihood function is then

$$L = \prod_{n=1}^{N}\prod_{k=1}^{c} y_{kn}^{\,y_{dkn}}. \tag{8.16}$$

Again, the objective function J is taken to be the negative logarithm of the likelihood function, i.e.

$$J = -\ln L = -\sum_n \sum_k y_{dkn} \ln(y_{kn}), \tag{8.17}$$

giving the cross entropy error function. This error function also applies to the case where the target y_{dk} is not a binary variable, but is a real variable in the interval [0, 1] representing the probability of the input $\mathbf{x}$ belonging to C_k (Bishop, 1995).

8.3 Bayesian neural network (BNN) classifier

So far, the objective functions for classication problems do not contain weight penalty or regularization terms to suppress overfitting. We have studied the use of Bayesian neural networks (BNN) for nonlinear regression in Section 6.7, where the regularization (or weight penalty) hyperparameter is automatically determined via a Bayesian approach to avoid overfitting. Here we will see that BNN can be applied in a similar way to perform nonlinear classification. Consider the two-class problem, where the objective function is the cross entropy error function (8.10), i.e. the negative log likelihood. A regularization or weight penalty term is now added to the objective function, i.e.

$$J = -\sum_n y_{dn} \ln y_n + (1 - y_{dn}) \ln(1 - y_n) + \frac{\alpha}{2}\mathbf{w}^{\mathrm{T}}\mathbf{w}, \tag{8.18}$$

where α is the hyperparameter controlling the size of the weight penalty, with $\mathbf{w}$ containing all the model weights (including offset parameters). Note that unlike the case of BNN regression (Section 6.7), here for classification there is only a single hyperparameter α. There is not a second hyperparameter β controlling the amount of noise in the target data, since the target data y_d are assumed to be correctly labelled as either 0 or 1.

As in BNN regression, an isotropic Gaussian form (6.33) is assumed for the prior distribution over $\mathbf{w}$. To find α iteratively, an initial guess value is used for α, and (8.18) is minimized using nonlinear optimization, yielding $\mathbf{w} = \mathbf{w}_\mathrm{MP}$.

Next, we need to improve the estimate for α by iterations. Without β (and not mentioning the model M dependence explicitly), the marginal log likelihood for BNN regression in (6.43) simplifies here to

$$\ln p(D|\alpha) = \ln Z(\alpha) - \ln Z_\mathrm{w}(\alpha), \tag{8.19}$$

with Z given by (6.42) and Z_w from (6.33). After a little algebra, it follows that

$$\ln p(D|\alpha) \approx -J(\mathbf{w}_\mathrm{MP}) - \frac{1}{2}\ln[\det(\mathbf{H}_\mathrm{MP})] + \frac{1}{2}N_\mathrm{w}\ln\alpha + \text{constant}, \tag{8.20}$$

where N_w is the number of model parameters, $\mathbf{H}_\mathrm{MP}$ the Hessian matrix at $\mathbf{w}_\mathrm{MP}$, and

$$J(\mathbf{w}_\mathrm{MP}) = -\sum_n y_{\mathrm{d}n}\ln y_n + (1 - y_{\mathrm{d}n})\ln(1 - y_n) + \frac{\alpha}{2}\mathbf{w}_\mathrm{MP}^\mathrm{T}\mathbf{w}_\mathrm{MP}, \tag{8.21}$$

with $y_n = y(\mathbf{x}_n, \mathbf{w}_\mathrm{MP})$. Maximizing $\ln p(D|\alpha)$ with respect to α again leads to

$$\alpha_\mathrm{MP} = \frac{\gamma}{\mathbf{w}_\mathrm{MP}^\mathrm{T}\mathbf{w}_\mathrm{MP}} \quad \text{and} \quad \gamma = N_\mathrm{w} - \alpha_\mathrm{MP}\,\mathrm{tr}((\mathbf{H}_\mathrm{MP})^{-1}), \tag{8.22}$$

which provides a new estimate for α to continue the iterative process till convergence.

8.4 Support vector machine (SVM) classifier

Since the mid 1990s, *support vector machines* (SVM) have become popular in nonlinear classification and regression problems. A kernel-based method, SVM for classification provides the decision boundaries but not posterior probabilities. While various individual features of SVM have been discovered earlier, Boser *et al.* (1992) and Cortes and Vapnik (1995) were the first papers to assemble the various ideas into an elegant new method for classification. Vapnik (1995) then extended the method to the nonlinear regression problem. Books covering SVM include Vapnik (1998), Cherkassky and Mulier (1998), Cristianini and Shawe-Taylor (2000), Schölkopf and Smola (2002), Shawe-Taylor and Cristianini (2004), and Bishop (2006).

The development of the SVM classifier for the two-class problem proceeds naturally in three stages. The basic *maximum margin classifier* is first introduced to problems where the two classes can be separated by a linear decision boundary (a hyperplane). Next, the classifier is modified to tolerate misclassification in problems not separable by a linear decision boundary. Finally, the classifier is nonlinearly generalized by using the kernel method.

8.4.1 Linearly separable case

First we consider the two-class problem and assume the data from the two classes can be separated by a hyperplane decision boundary. The separating hyperplane (Fig. 8.3) is given by the equation

$$y(\mathbf{x}) = \mathbf{w}^{\mathrm{T}}\mathbf{x} + w_0 = 0. \tag{8.23}$$

Any two points $\mathbf{x}_1$ and $\mathbf{x}_2$ lying on the hyperplane $y = 0$ satisfy

$$\mathbf{w}^{\mathrm{T}}(\mathbf{x}_1 - \mathbf{x}_2) = 0, \tag{8.24}$$

hence the unit vector

$$\hat{\mathbf{w}} = \frac{\mathbf{w}}{\|\mathbf{w}\|}, \tag{8.25}$$

is normal to the $y = 0$ hyperplane surface.

Any point $\mathbf{x}_0$ on the $y = 0$ surface satisfies $\mathbf{w}^{\mathrm{T}}\mathbf{x}_0 = -w_0$. In Fig. 8.3, the component of the vector $\mathbf{x} - \mathbf{x}_0$ projected onto the $\hat{\mathbf{w}}$ direction is given by

$$\hat{\mathbf{w}}^{\mathrm{T}}(\mathbf{x} - \mathbf{x}_0) = \frac{\mathbf{w}^{\mathrm{T}}(\mathbf{x} - \mathbf{x}_0)}{\|\mathbf{w}\|} = \frac{\mathbf{w}^{\mathrm{T}}\mathbf{x} + w_0}{\|\mathbf{w}\|} = \frac{y(\mathbf{x})}{\|\mathbf{w}\|}. \tag{8.26}$$

Thus $y(\mathbf{x})$ is proportional to the normal distance between $\mathbf{x}$ and the hyperplane $y = 0$, and the sign of $y(\mathbf{x})$ indicates on which side of the hyperplane $\mathbf{x}$ lies.

Let the training dataset be composed of predictors $\mathbf{x}_n$ and target data y_{dn} ($n = 1, \ldots, N$). Since there are two classes, y_{dn} takes on the value of -1 or $+1$. When

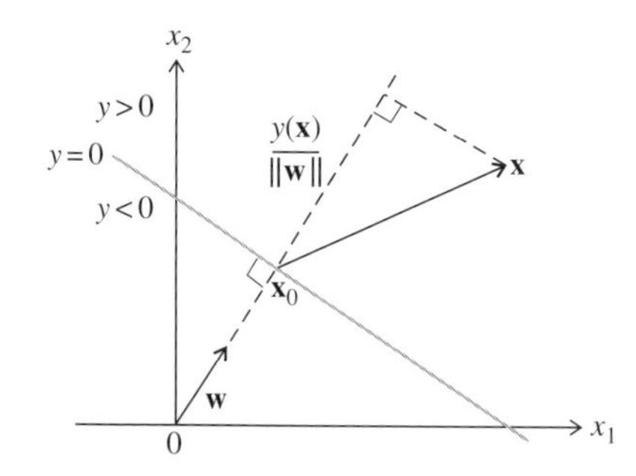

Fig. 8.3 The hyperplane $y = 0$, with the vector $\mathbf{w}$ normal to this hyperplane. The component of the vector $\mathbf{x} - \mathbf{x}_0$ projected onto the $\mathbf{w}$ direction is shown.

a new predictor vector $\mathbf{x}$ becomes available, it is classified according to the sign of the function

$$y(\mathbf{x}) = \mathbf{w}^{\mathrm{T}}\mathbf{x} + w_0. \tag{8.27}$$

The normal distance from a point $\mathbf{x}_n$ to the decision boundary $y = 0$ is, according to (8.26), given by

$$\frac{y_{dn}\, y(\mathbf{x}_n)}{\|\mathbf{w}\|} = \frac{y_{dn}\,(\mathbf{w}^{\mathrm{T}}\mathbf{x}_n + w_0)}{\|\mathbf{w}\|}, \tag{8.28}$$

where y_{dn} contributes the correct sign to ensure that the distance given by (8.28) is non-negative.

The margin (Fig. 8.4(a)) is given by the distance of the closest point(s) $\mathbf{x}_n$ in the dataset. A *maximum margin classifier* determines the decision boundary by maximizing the margin l, through searching for the optimal values of $\mathbf{w}$ and w_0. The optimization problem is then

$$\max_{\mathbf{w}, w_0} l \quad \text{subject to} \quad \frac{y_{dn}\,(\mathbf{w}^{\mathrm{T}}\mathbf{x}_n + w_0)}{\|\mathbf{w}\|} \geq l, \quad (n = 1, \ldots, N), \tag{8.29}$$

where the constraint simply ensures that no data points lie inside the margins. Obviously the margin is determined by relatively few points in the dataset, and these points circled in Fig. 8.4(a) are called *support vectors*.

Since the distance (8.28) is unchanged if we multiply $\mathbf{w}$ and w_0 by an arbitrary scale factor s, we are free to choose $\|\mathbf{w}\|$. If we choose $\|\mathbf{w}\| = l^{-1}$, then the constraint in (8.29) becomes

$$y_{dn}\,(\mathbf{w}^{\mathrm{T}}\mathbf{x}_n + w_0) \geq 1, \quad (n = 1, \ldots, N), \tag{8.30}$$

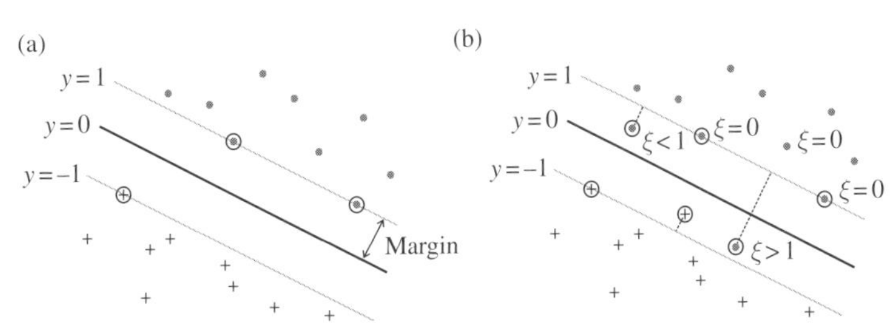

Fig. 8.4 (a) A dataset containing two classes (shown by round dots and plus signs) separable by a hyperplane decision boundary $y = 0$. The margin is maximized. Support vectors, i.e. data points used in determining the margins $y = \pm 1$, are circled. (b) A dataset not separable by a hyperplane boundary. Slack variables $\xi_n \geq 0$ are introduced, with $\xi_n = 0$ for data points lying on or within the correct margin, $\xi_n > 1$ for points lying to the wrong side of the decision boundary. Support vectors are circled.

and maximizing the margin l becomes equivalent to minimizing $\|\mathbf{w}\|$, which is equivalent to solving

$$\min_{\mathbf{w}, w_0} \frac{1}{2}\|\mathbf{w}\|^2, \tag{8.31}$$

subject to the constraint (8.30). This optimization of a quadratic function subject to constraints is referred to as a *quadratic programming* problem (Gill *et al.*, 1981).

Karush (1939) and Kuhn and Tucker (1951) (KKT) have solved this type of optimization with inequality constraint by a generalization of the Lagrange multiplier method. With a Lagrange multiplier $\lambda_n \geq 0$ introduced for each of the N constraints in (8.30) (see Appendix B), the Lagrange function L takes the form

$$L(\mathbf{w}, w_0, \boldsymbol{\lambda}) = \frac{1}{2}\|\mathbf{w}\|^2 - \sum_{n=1}^{N} \lambda_n \left[y_{dn}(\mathbf{w}^{\mathrm{T}}\mathbf{x}_n + w_0) - 1 \right], \tag{8.32}$$

with $\boldsymbol{\lambda} = (\lambda_1, \ldots, \lambda_N)^{\mathrm{T}}$. Setting the derivatives of L with respect to $\mathbf{w}$ and w_0 to zero yields, respectively,

$$\mathbf{w} = \sum_{n=1}^{N} \lambda_n y_{dn} \mathbf{x}_n, \tag{8.33}$$

$$0 = \sum_{n=1}^{N} \lambda_n y_{dn}. \tag{8.34}$$

Substituting these into (8.32) allows us to express L solely in terms of the Lagrange multipliers, i.e.

$$L_{\mathrm{D}}(\boldsymbol{\lambda}) = \sum_{n=1}^{N} \lambda_n - \frac{1}{2} \sum_{n=1}^{N} \sum_{j=1}^{N} \lambda_n \lambda_j y_{dn} y_{dj} \mathbf{x}_n^{\mathrm{T}} \mathbf{x}_j, \tag{8.35}$$

subject to $\lambda_n \geq 0$ and (8.34), with L_{D} referred to as the *dual Lagrangian*. If m is the dimension of $\mathbf{x}$ and $\mathbf{w}$, then the original primal optimization problem of (8.31) is a quadratic programming problem with about m variables. In contrast, the dual problem (8.35) is also a quadratic programming problem with N variables from $\boldsymbol{\lambda}$. The main reason we want to work with the dual problem instead of the primal problem is that, in the next stage, to generalize the linear classifier to a nonlinear classifier, SVM will invoke kernels (see Chapter 7), i.e. $\mathbf{x}$ will be replaced by $\boldsymbol{\phi}(\mathbf{x})$ in a feature space of dimension M, hence m will be replaced by M, which is usually much larger or even infinite.

Our constrained optimization problem satisfies the KKT conditions (Appendix B):

$$\lambda_n \geq 0, \tag{8.36}$$

$$y_{dn}(\mathbf{w}^T\mathbf{x}_n + w_0) - 1 \geq 0, \tag{8.37}$$

$$\lambda_n \,[\, y_{dn}(\mathbf{w}^T\mathbf{x}_n + w_0) - 1\,] = 0. \tag{8.38}$$

Combining (8.27) and (8.33), we have the following formula for classifying a new data point $\mathbf{x}$:

$$y(\mathbf{x}) = \sum_{n=1}^{N} \lambda_n y_{dn}\, \mathbf{x}_n^T \mathbf{x} + w_0, \tag{8.39}$$

where the class (+1 or −1) is decided by the sign of $y(\mathbf{x})$. The KKT condition (8.38) implies that for every data point $\mathbf{x}_n$, either $\lambda_n = 0$ or $y_{dn}y(\mathbf{x}_n) = 1$. But the data points with $\lambda_n = 0$ will be omitted in the summation in (8.39), hence will not contribute in the classification of a new data point $\mathbf{x}$. Data points with $\lambda_n > 0$ are the *support vectors* and they satisfy $y_{dn}y(\mathbf{x}_n) = 1$, i.e. they lie exactly on the maximum margins (Fig. 8.4a). Although only support vectors determine the margins and the decision boundary, the entire training dataset was used to locate the support vectors. However, once the model has been trained, classification of new data by (8.39) requires only the support vectors, and the remaining data (usually the majority of data points) can be ignored. The desirable property of being able to ignore the majority of data points is called *sparseness*, which greatly reduces computational costs.

To obtain the value for w_0, we use the fact that for any support vector $\mathbf{x}_j$, $y_{dj}y(\mathbf{x}_j) = 1$, i.e.

$$y_{dj}\left(\sum_{n=1}^{N} \lambda_n y_{dn}\, \mathbf{x}_n^T \mathbf{x}_j + w_0\right) = 1. \tag{8.40}$$

As $y_{dj} = \pm 1$, $y_{dj}^{-1} = y_{dj}$, so

$$w_0 = y_{dj} - \sum_{n=1}^{N} \lambda_n y_{dn}\, \mathbf{x}_n^T \mathbf{x}_j, \tag{8.41}$$

(where the summation need only be over n for which λ_n is non-zero). Since each support vector $\mathbf{x}_j$ gives an estimate of w_0, all the estimates are usually averaged together to provide the most accurate estimate of w_0.

8.4.2 Linearly non-separable case

So far, the maximum margin classifier has only been applied to the case where the two classes are separable by a linear decision boundary (a hyperplane). Next, we move to the case where the two classes are not separable by a hyperplane, i.e. no matter how hard we try, one or more data points will end up on the wrong side of the hyperplane and end up being misclassified. Hence our classifier needs to be generalized to tolerate misclassification.

We introduce *slack variables*, $\xi_n \geq 0, n = 1, \ldots, N$, i.e. each training data point $\mathbf{x}_n$ is assigned a slack variable ξ_n. The slack variables are defined by $\xi_n = 0$ for all $\mathbf{x}_n$ lying on or within the correct margin boundary, and $\xi_n = |y_{dn} - y(\mathbf{x}_n)|$ for all other points. Hence the slack variable measures the distance a data point protrudes beyond the correct margin (Fig. 8.4b). A data point which lies right on the decision boundary $y(\mathbf{x}_n) = 1$ will have $\xi_n = 1$, while $\xi_n > 1$ corresponds to points lying to the wrong side of the decision boundary, i.e. misclassified. Points with $0 < \xi_n \leq 1$ protrude beyond the correct margin but not enough to cross the decision boundary to be misclassified.

The constraint (8.30) is modified to allow for data points extending beyond the correct margin, i.e.

$$y_{dn}(\mathbf{w}^{\mathrm{T}}\mathbf{x}_n + w_0) \geq 1 - \xi_n, \quad (n = 1, \ldots, N). \tag{8.42}$$

The minimization problem (8.31) is modified to

$$\min_{\mathbf{w}, w_0} \left(\frac{1}{2}\|\mathbf{w}\|^2 + C \sum_{n=1}^{N} \xi_n \right). \tag{8.43}$$

If we divide this expression by C, the second term can be viewed as an error term, while the first term can be viewed as a weight penalty term, with C^{-1} as the weight penalty parameter – analogous to the regularization of NN models in Section 6.5 where we have a mean squared error (MSE) term plus a weight penalty term in the objective function. The effect of misclassification on the objective function is only linearly related to ξ_n in (8.43), in contrast to the MSE term which is quadratic. Since any misclassified point has $\xi_n > 1$, $\sum_n \xi_n$ can be viewed as providing an upper bound on the number of misclassified points.

To optimize (8.43) subject to constraint (8.42), we again turn to the method of Lagrange multipliers (Appendix B), where the Lagrange function is now

$$L = \frac{1}{2}\|\mathbf{w}\|^2 + C \sum_{n=1}^{N} \xi_n - \sum_{n=1}^{N} \lambda_n \left[y_{dn}(\mathbf{w}^{\mathrm{T}}\mathbf{x}_n + w_0) - 1 + \xi_n \right] - \sum_{n=1}^{N} \mu_n \xi_n, \tag{8.44}$$

with $\lambda_n \geq 0$ and $\mu_n \geq 0$ $(n = 1, \ldots, N)$ the Lagrange multipliers.

Setting the derivatives of L with respect to $\mathbf{w}$, w_0 and ξ_n to 0 yields, respectively,

$$\mathbf{w} = \sum_{n=1}^{N} \lambda_n y_{dn} \mathbf{x}_n, \tag{8.45}$$

$$0 = \sum_{n=1}^{N} \lambda_n y_{dn}, \tag{8.46}$$

$$\lambda_n = C - \mu_n. \tag{8.47}$$

Substituting these into (8.44) again allows L to be expressed solely in terms of the Lagrange multipliers $\boldsymbol{\lambda}$, i.e.

$$L_D(\boldsymbol{\lambda}) = \sum_{n=1}^{N} \lambda_n - \frac{1}{2} \sum_{n=1}^{N} \sum_{j=1}^{N} \lambda_n \lambda_j y_{dn} y_{dj} \mathbf{x}_n^T \mathbf{x}_j, \tag{8.48}$$

with L_D the dual Lagrangian. Note that L_D has the same form as that in the separable case, but the constraints are somewhat changed. As $\lambda_n \geq 0$ and $\mu_n \geq 0$, (8.47) implies

$$0 \leq \lambda_n \leq C. \tag{8.49}$$

Furthermore, there are the constraints from the KKT conditions (Appendix B):

$$y_{dn}(\mathbf{w}^T \mathbf{x}_n + w_0) - 1 + \xi_n \geq 0, \tag{8.50}$$

$$\lambda_n \left[\, y_{dn}(\mathbf{w}^T \mathbf{x}_n + w_0) - 1 + \xi_n \,\right] = 0, \tag{8.51}$$

$$\mu_n \, \xi_n = 0, \tag{8.52}$$

for $n = 1, \ldots, N$.

In the computation for $\mathbf{w}$ in (8.45), only data points with $\lambda_n \neq 0$ contributed in the summation, so these points (with $\lambda_n > 0$) are the *support vectors*. Some support vectors lie exactly on the margin (i.e. $\xi_n = 0$), while others may protrude beyond the margin ($\xi_n > 0$) (Fig. 8.4b). The value for w_0 can be obtained from (8.51) using any of the support vectors lying on the margin ($\xi_n = 0$), i.e. (8.41) again follows, and usually the values obtained from the individual vectors are averaged together to give the best estimate.

Optimizing the dual problem (8.48) is a simpler quadratic programming problem than the primal (8.44), and can be readily solved using standard methods (see e.g. Gill *et al.*, 1981). The parameter C is not solved by the optimization. Instead, one usually reserves some validation data to test the performance of various models trained with different values of C to determine the best value to use for C. With new data $\mathbf{x}$, the classification is, as before, based on the sign of $y(\mathbf{x})$ from (8.39).

8.4.3 Nonlinear classification by SVM

As our classifier is still restricted to linearly non-separable problems, the final step is to extend the classifier from being linear to nonlinear. This is achieved by performing linear classification not with the input $\mathbf{x}$ data but with the $\boldsymbol{\phi}(\mathbf{x})$ data in a feature space, where $\boldsymbol{\phi}$ is the nonlinear function mapping from the original input space to the feature space (Section 7.3), i.e.

$$y(\mathbf{x}) = \mathbf{w}^{\mathrm{T}}\boldsymbol{\phi}(\mathbf{x}) + w_0. \tag{8.53}$$

The dual Lagrangian L_{D} for this new problem is obtained from (8.48) simply through replacing $\mathbf{x}$ by $\boldsymbol{\phi}(\mathbf{x})$, giving

$$L_{\mathrm{D}}(\boldsymbol{\lambda}) = \sum_{n=1}^{N} \lambda_n - \frac{1}{2}\sum_{n=1}^{N}\sum_{j=1}^{N} \lambda_n \lambda_j y_{\mathrm{d}n} y_{\mathrm{d}j} \boldsymbol{\phi}^{\mathrm{T}}(\mathbf{x}_n)\boldsymbol{\phi}(\mathbf{x}_j). \tag{8.54}$$

Classification is based on the sign of $y(\mathbf{x})$, with $y(\mathbf{x})$ modified from (8.39) to

$$y(\mathbf{x}) = \sum_{n=1}^{N} \lambda_n y_{\mathrm{d}n}\, \boldsymbol{\phi}^{\mathrm{T}}(\mathbf{x}_n)\boldsymbol{\phi}(\mathbf{x}) + w_0. \tag{8.55}$$

Since the dimension of the feature space can be very high or even infinite, computations involving the inner product $\boldsymbol{\phi}^{\mathrm{T}}(\mathbf{x})\boldsymbol{\phi}(\mathbf{x}')$ are only practicable because of the *kernel trick* (see Section 7.3), where a *kernel function* K

$$K(\mathbf{x}, \mathbf{x}') \equiv \boldsymbol{\phi}^{\mathrm{T}}(\mathbf{x})\boldsymbol{\phi}(\mathbf{x}'), \tag{8.56}$$

is introduced to obviate direct computation of the inner product. Commonly used kernel functions include the *polynomial kernel* of degree p,

$$K(\mathbf{x}, \mathbf{x}') = \left(1 + \mathbf{x}^{\mathrm{T}}\mathbf{x}'\right)^p, \tag{8.57}$$

and the *Gaussian or radial basis function* (RBF) *kernel*

$$K(\mathbf{x}, \mathbf{x}') = \exp\left(-\frac{\|\mathbf{x} - \mathbf{x}')\|^2}{2\sigma^2}\right). \tag{8.58}$$

Under the kernel approach, (8.55) becomes

$$y(\mathbf{x}) = \sum_{n=1}^{N} \lambda_n y_{\mathrm{d}n}\, K(\mathbf{x}_n, \mathbf{x}) + w_0, \tag{8.59}$$

and (8.54) becomes

$$L_{\mathrm{D}}(\boldsymbol{\lambda}) = \sum_{n=1}^{N} \lambda_n - \frac{1}{2}\sum_{n=1}^{N}\sum_{j=1}^{N} \lambda_n \lambda_j y_{\mathrm{d}n} y_{\mathrm{d}j} K(\mathbf{x}_n, \mathbf{x}_j). \tag{8.60}$$

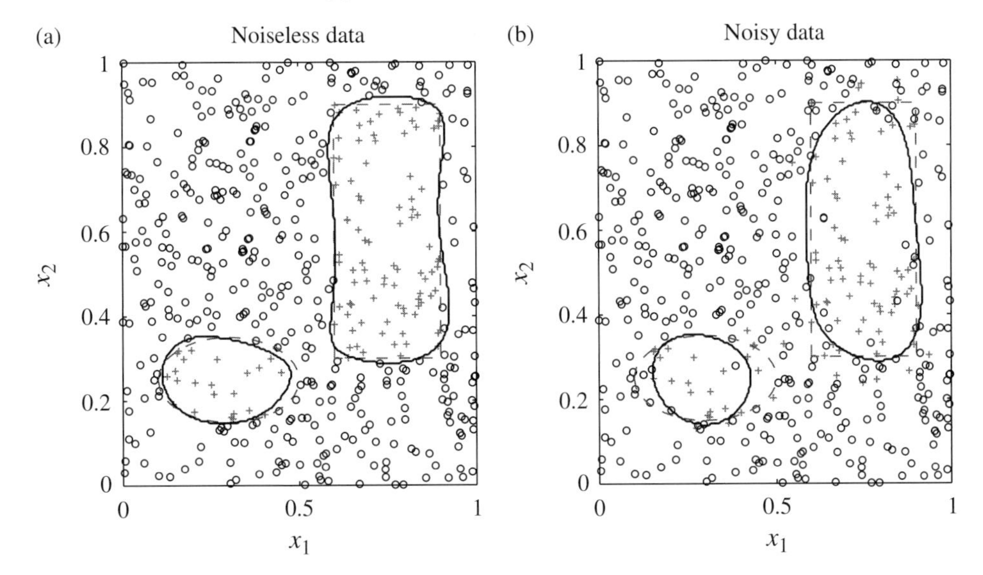

Fig. 8.5 The SVM classifier applied to (a) the noiseless dataset of Fig. 8.1 and (b) the moderately noisy dataset of Fig. 8.2. The two classes of data point are indicated by the plus signs and the circles, with the decision boundaries found by SVM shown as solid curves, and the theoretical decision boundaries for noiseless data shown as dashed curves.

Optimizing the dual problem (8.60) is again a quadratic programming problem. Since the objective function is only quadratic, and with the constraints being linear, there is no local minima problem, i.e. unlike NN models, the minimum found by SVM is the *global minimum*. The parameter C and the parameter σ (assuming the RBF kernel (8.58) is used) are not obtained from the optimization. To determine the values of these two hyperparameters, one usually trains multiple models with various values of C and σ, and from their classification performance over validation data, determines the best values for C and σ. Figure 8.5 illustrates classification of noiseless data and moderately noisy data by SVM, with the SVM code from LIBSVM (Chang and Lin, 2001). Note that SVM with the Gaussian kernel is similar in structure to the RBF NN (Section 4.6), but SVM appears to perform better (Schölkopf *et al.*, 1997).

This formulation of the SVM is sometimes called the C-SVM, as there is an alternative formulation, the ν-SVM, by Schölkopf *et al.* (2000). In ν-SVM, L_{D} in (8.60) is simplified to

$$L_{\mathrm{D}}(\boldsymbol{\lambda}) = -\frac{1}{2}\sum_{n=1}^{N}\sum_{j=1}^{N} \lambda_n \lambda_j y_{\mathrm{d}n} y_{\mathrm{d}j} K(\mathbf{x}_n, \mathbf{x}_j). \tag{8.61}$$

The constraints are

$$0 \leq \lambda_n \leq 1/N, \tag{8.62}$$

$$\sum_{n=1}^{N} \lambda_n y_{dn} = 0, \tag{8.63}$$

$$\sum_{n=1}^{N} \lambda_n \geq \nu, \tag{8.64}$$

where the parameter ν has replaced the inverse penalty parameter C. Relative to C, ν has more useful interpretations: (i) ν is an upper bound on the fraction of *margin errors*, i.e. fraction of points which lie on the wrong side of the margin ($\xi_n > 0$) (and may or may not be misclassified); (ii) ν is a lower bound on the fraction of support vectors ($\xi_n \geq 0$) (Schölkopf and Smola, 2002). While C ranges from 0 to ∞, ν ranges between $\nu_{\min}$ and $\nu_{\max}$, with $0 \leq \nu_{\min} \leq \nu_{\max} \leq 1$ (Chang and Lin, 2001).

8.4.4 Multi-class classification by SVM

Support vector machine was originally developed for two-class (binary) problems. The simplest extension to multi-class problems is via the *one-versus-the-rest* approach: i.e. given c classes, we train c SVM binary classifiers $y^{(k)}(\mathbf{x})$ ($k = 1, \ldots, c$) where the target data for class k are set to $+1$ and for all other classes to -1. New input data $\mathbf{x}$ are then classified as belonging to class j if $y^{(j)}(\mathbf{x})$ attains the highest value among $y^{(k)}(\mathbf{x})$ ($k = 1, \ldots, c$).

There are two criticisms of this heuristic approach: (a) as the c binary classifiers were trained on different problems, there is no guarantee that the outputs $y^{(k)}$ are on the same scale for fair comparison between the c binary classifiers; (b) if there are many classes, then the $+1$ target data can be greatly outnumbered by the -1 target data.

An alternative approach is *pairwise classification* or *one-versus-one*. A binary classifier is trained for each possible pair of classes, so there is a total of $c(c-1)/2$ binary classifiers, with each classifier using training data from only class k and class l ($k \neq l$). With new data $\mathbf{x}$, all the binary classifiers are applied, and if class j gets the most 'yes votes' from the classifiers, then $\mathbf{x}$ is classified as belonging to class j.

Recently, based on graph theory, Liu and Jin (2008) proposed the LATTICESVM method, which can significantly reduce storage and computational complexity for multi-class SVM problems, especially when c is large.

8.5 Forecast verification

In environmental sciences, forecasting has great socio-economic benefits for society. Forecasts for hurricanes, tornados, Arctic blizzards, floods, tsunami and earthquakes save lives and properties, while forecasts for climate variability such as the El Niño-Southern Oscillation bring economic benefits to farming and fishing. Hence, once a forecast model has been built, it is important that we evaluate the quality of its forecasts, a process known as *forecast verification* or forecast evaluation. A considerable number of statistical measures have been developed to evaluate how accurate forecasts are compared with observations (Jolliffe and Stephenson, 2003).

Let us start with forecasts for two classes or categories, where class 1 is for an event (e.g. tornado) and class 2 for a non-event (e.g. non-tornado). Model forecasts and observed data can be compared and arranged in a 2×2 *contingency table* (Fig. 8.6). The number of events forecast and indeed observed are called 'hits' and are placed in entry a of the table. In our tornado example, this would correspond to forecasts for tornados which turned out to be correct. Entry b is the number of false alarms, i.e. tornados forecast but which never materialized. Entry c is the number of misses, i.e. tornados appeared in spite of non-tornado forecasts, while entry d is the number of correct negatives, i.e. non-tornado forecasts which turned out to be correct. Marginal totals are also listed in the table, e.g. the top row sums to $a + b$, the total number of tornados forecast, whereas the first column sums to $a + c$, the total number of tornados observed. Finally, the total number of cases N is given by $N = a + b + c + d$.

The simplest measure of accuracy of binary forecasts is the *fraction correct* (FC) or *hit rate*, i.e. the number of correct forecasts divided by the total number of forecasts,

$$\mathrm{FC} = \frac{a + d}{N} = \frac{a + d}{a + b + c + d}, \tag{8.65}$$

		Observed		
		Yes	No	Total
Forecast	Yes	a = hits	b = false alarms	$a + b$ = forecast yes
Forecast	No	c = misses	d = correct negatives	$c + d$ = forecast no
	Total	$a + c$ = observed yes	$b + d$ = observed no	$a + b + c + d$ = total

Fig. 8.6 Illustrating a 2×2 contingency table used in the forecast verification of a two-class problem. The number of forecast 'yes' and 'no', and the number of observed 'yes' and 'no' are the entries in the table. The marginal totals are obtained by summing over the rows or the columns.

where FC ranges between 0 and 1, with 1 being the perfect score. Unfortunately, this measure becomes very misleading if the number of non-events vastly outnumbers the number of events. For instance if $d \gg a, b, c$, then (8.65) yields FC ≈ 1. For instance, in Marzban and Stumpf (1996), one NN tornado forecast model has $a = 41$, $b = 31$, $c = 39$ and $d = 1002$, since the vast majority of days has no tornado forecast and none observed. While a, b and c are comparable in magnitude, the overwhelming size of d lifts FC to a lofty value of 0.937.

In such situations, where the non-events vastly outnumber the events, including d in the score is rather misleading. Dropping d in both the numerator and the denominator in (8.65) gives the *threat score* (TS) or *critical success index* (CSI),

$$\mathrm{TS} = \mathrm{CSI} = \frac{a}{a+b+c}, \tag{8.66}$$

which is a much better measure of forecast accuracy than FC in such situations. The worst TS is 0 and the best TS is 1.

To see what fraction of the observed events ('yes') were correctly forecast, we compute the *probability of detection* (POD)

$$\mathrm{POD} = \frac{\text{hits}}{\text{hits} + \text{misses}} = \frac{a}{a+c}, \tag{8.67}$$

with the worst POD score being 0 and the best score being 1.

Besides measures of forecast accuracy, we also want to know if there is forecast *bias* or *frequency bias*:

$$B = \frac{\text{total 'yes' forecast}}{\text{total 'yes' observed}} = \frac{a+b}{a+c}. \tag{8.68}$$

For instance, it is easy to increase the POD if we simply issue many more forecasts of events ('yes'), despite most of them being false alarms. Increasing B would raise concern that the model is forecasting far too many events compared to the number of observed events.

To see what fraction of the forecast events ('yes') never materialized, we compute the *false alarm ratio* (FAR)

$$\mathrm{FAR} = \frac{\text{false alarms}}{\text{hits} + \text{false alarms}} = \frac{b}{a+b}, \tag{8.69}$$

with the worst FAR score being 0 and the best score being 1.

Be careful not to confuse the false alarm ratio (FAR) with the *false alarm rate* (F), also known as the *probability of false detection* (POFD). The value of F measures the fraction of the observed 'no' events which were incorrectly forecast as 'yes', i.e.

$$F = \mathrm{POFD} = \frac{\text{false alarms}}{\text{false alarms} + \text{correct negatives}} = \frac{b}{b+d}, \tag{8.70}$$

with the worst F score being 0 and the best score being 1. While F is not as commonly given as FAR and POD, it is one of the axes in the *relative operating characteristic* (ROC) diagram, used widely in probabilistic forecasts (Marzban, 2004; Kharin and Zwiers, 2003).

In an ROC diagram (Fig. 8.7), F is the abscissa and POD, the ordinate. Although our model may be issuing probabilistic forecasts in a two-class problem, we are actually free to choose the decision threshold used in the classification, i.e. instead of using a posterior probability of 0.5 as the threshold for deciding whether to issue a 'yes' forecast, we may want to use 0.7 as the threshold if we want fewer false alarms (i.e. lower F) (at the expense of a lower POD), or 0.3 if we want to increase our POD (at the expense of increasing F as well). The result of varying the threshold generates a curve in the ROC diagram. The choice of the threshold hinges on the cost associated with missing an event and that with issuing a false alarm. For instance, if we miss forecasting a powerful hurricane hitting a vulnerable coastal city, the cost may be far higher than that from issuing a false alarm, so we would want a low threshold value to increase the POD. If we have a second model, where for a given F, it has a higher POD than the first model, then its ROC curve (dashed curve in Fig. 8.7) lies above that from the first model. A model with zero skill (POD = F) is shown by the diagonal line in the ROC diagram. A real ROC diagram is shown later in Chapter 12 (Fig. 12.7).

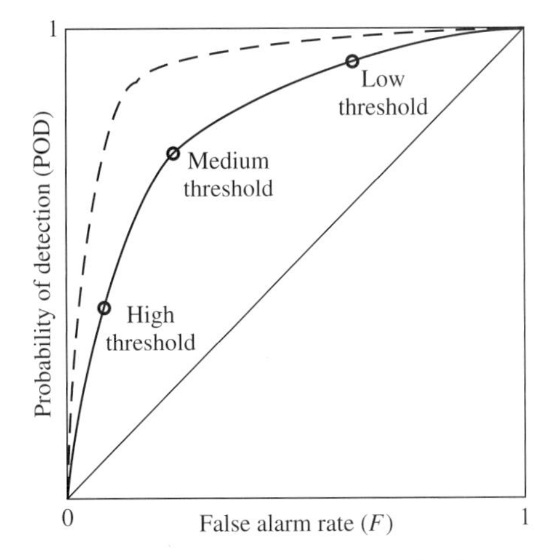

Fig. 8.7 A schematic relative operating characteristic (ROC) diagram illustrating the trade-off between the false alarm rate (F) and the probability of detection (POD) as the classification decision threshold is varied for a given model (solid curve). The dashed curve shows the ROC of a better model while the diagonal line (POD = F) indicates a model with zero skill.

8.5.1 Skill scores

Various *skill scores* have been designed to measure the relative accuracy of a set of forecasts, with respect to a set of *reference* or control forecasts. Choices for the reference forecasts include persistence, climatology, random forecasts and forecasts from a standard model. Persistence forecasts simply carry forward the anomalies to the future (e.g. tomorrow's weather is forecast to be the same as today's). Climatological forecasts simply issue the climatological mean value in the forecast. In random forecasts, events are forecast randomly but in accordance with the observed frequency of such events. For instance, if tornados are observed only 2% of the time, then random forecasts also only forecast tornados 2% of the time. Finally the reference model can be a standard model, as the researcher is trying to show that the new model is better.

For a particular measure of accuracy A, the skill score (SS) is defined generically by

$$\mathrm{SS} = \frac{A - A_{\mathrm{ref}}}{A_{\mathrm{perfect}} - A_{\mathrm{ref}}}, \tag{8.71}$$

where A_{perfect} is the value of A for a set of perfect forecasts, and A_{ref} is the value of A computed over the set of reference forecasts. Note that if we define $A' = -A$, then SS is unchanged if computed using A' instead of A. This shows that SS is unaffected by whether A is positively or negatively oriented (i.e. whether better accuracy is indicated by a higher or lower value of A).

The *Heidke skill score* (HSS) (Heidke, 1926) is the skill score (8.71) using the fraction correct (FC) for A and random forecasts as the reference, i.e.

$$\mathrm{HSS} = \frac{\mathrm{FC} - \mathrm{FC}_{\mathrm{random}}}{\mathrm{FC}_{\mathrm{perfect}} - \mathrm{FC}_{\mathrm{random}}}. \tag{8.72}$$

Hence, if the forecasts are perfect, HSS = 1; if they are only as good as random forecasts, HSS = 0; and if they are worse than random forecasts, HSS is negative. From (8.65), FC can be interpreted as the fraction of hits (a/N) plus the fraction of correct negatives (d/N). For $\mathrm{FC}_{\mathrm{random}}$ obtained from random forecasts, the fraction of hits is the product of two probabilities, P('yes' forecast) and P('yes' observed), i.e. $(a + b)/N$ and $(a + c)/N$, respectively. Similarly, the fraction of correct negatives from random forecasts is the product of P('no' forecast) and

P('no' observed), i.e. $(c + d)/N$ and $(b + d)/N$. With FC_{random} being the fraction of hits plus the fraction of correct negatives, we have

$$FC_{random} = \left(\frac{a+b}{N}\right)\left(\frac{a+c}{N}\right) + \left(\frac{c+d}{N}\right)\left(\frac{b+d}{N}\right). \tag{8.73}$$

Substituting this into (8.72) and invoking (8.65) and $FC_{perfect} = 1$, HSS is then given by

$$HSS = \frac{(a+d)/N - [(a+b)(a+c) + (b+d)(c+d)]/N^2}{1 - [(a+b)(a+c) + (b+d)(c+d)]/N^2}, \tag{8.74}$$

which can be simplified to

$$HSS = \frac{2(ad - bc)}{(a+c)(c+d) + (a+b)(b+d)}. \tag{8.75}$$

The *Peirce skill score* (PSS) (Peirce, 1884), also called the *Hansen and Kuipers' score*, or the *true skill statistics* (TSS) is similar to the HSS, except that the reference used in the denominator of (8.72) is unbiased, i.e. P('yes' forecast) is set to equal P('yes' observed), and P('no' forecast) to P('no' observed) for FC_{random} in the denominator of (8.72), whence

$$FC_{random} = \left(\frac{a+c}{N}\right)^2 + \left(\frac{b+d}{N}\right)^2. \tag{8.76}$$

The PSS is computed from

$$PSS = \frac{(a+d)/N - [(a+b)(a+c) + (b+d)(c+d)]/N^2}{1 - [(a+c)^2 + (b+d)^2]/N^2}, \tag{8.77}$$

which simplifies to

$$PSS = \frac{ad - bc}{(a+c)(b+d)}. \tag{8.78}$$

PSS can also be expressed as

$$PSS = \frac{a}{a+c} - \frac{b}{b+d} = POD - F, \tag{8.79}$$

upon invoking (8.67) and (8.70). Again, if the forecasts are perfect, PSS = 1; if they are only as good as random forecasts, PSS = 0; and if they are worse than random forecasts, PSS is negative.

8.5.2 Multiple classes

Next consider the forecast verification problem with c classes, where c is an integer > 2. The contingency table is then a $c \times c$ matrix, with the ith diagonal element giving the number of correct forecasts for class C_i. Fraction correct (FC) in (8.65) generalizes easily, as FC is simply the sum of all the diagonal elements divided by the sum of all elements of the matrix.

Other measures such as POD, FAR, etc. do not generalize naturally to higher dimensions. Instead the way to use them is to collapse the $c \times c$ matrix to a 2×2 matrix. For instance, if the forecast classes are ‘cold’, ‘normal’ and ‘warm’, we can put ‘normal’ and ‘warm’ together to form the class of ‘non-cold’ events. Then we are back to two classes, namely ‘cold’ and ‘non-cold’, and measures such as POD, FAR, etc. can be easily applied. Similarly, we can collapse to only ‘warm’ and ‘non-warm’ events, or to ‘normal’ and ‘non-normal’ events.

For multi-classes, HSS in (8.72) and (8.74) generalizes to

$$\mathrm{HSS} = \frac{\sum_{i=1}^{c} P(f_i, o_i) - \sum_{i=1}^{c} P(f_i)P(o_i)}{1 - \sum_{i=1}^{c} P(f_i)P(o_i)}, \tag{8.80}$$

where f_i denotes class C_i forecast, o_i denotes C_i observed, $P(f_i, o_i)$ the joint probability distribution of forecasts and observations, $P(f_i)$ the marginal distribution of forecasts and $P(o_i)$ the marginal distribution of observations. It is easy to see that (8.80) reduces to (8.74) when there are only two classes.

Note that PSS in (8.77) also generalizes to

$$\mathrm{PSS} = \frac{\sum_{i=1}^{c} P(f_i, o_i) - \sum_{i=1}^{c} P(f_i)P(o_i)}{1 - \sum_{i=1}^{c} [P(o_i)]^2}. \tag{8.81}$$

8.5.3 Probabilistic forecasts

In probabilistic forecasts, one can issue forecasts for binary events based on the posterior probability, then compute skill scores for the classification forecasts. Alternatively, one can apply skill scores directly to the probabilistic forecasts. The most widely used score for the probabilistic forecasts of an event is the *Brier score* (BS) (Brier, 1950). Formally, this score resembles the MSE, i.e.

$$\mathrm{BS} = \frac{1}{N} \sum_{n=1}^{N} (f_n - o_n)^2, \tag{8.82}$$

where there is a total of N pairs of forecasts f_n and observations o_n. While f_n is a continuous variable within [0, 1], o_n is a binary variable, being 1 if the event occurred and 0 if it did not occur. BS is negatively oriented, i.e. the lower the better. Since $|f_n - o_n|$ is bounded between 0 and 1 for each n, BS is also bounded between 0 and 1, with 0 being the perfect score.

From (8.71), the Brier skill score (BSS) is then

$$\mathrm{BSS} = \frac{\mathrm{BS} - \mathrm{BS}_{\mathrm{ref}}}{0 - \mathrm{BS}_{\mathrm{ref}}} = 1 - \frac{\mathrm{BS}}{\mathrm{BS}_{\mathrm{ref}}}, \tag{8.83}$$

where the reference forecasts are often taken to be random forecasts based on climatological probabilities. Unlike BS, BSS is positively oriented, with 1 being the perfect score and 0 meaning no skill relative to the reference forecasts.

8.6 Unsupervised competitive learning

In Section 1.7 we mentioned that in machine learning there are two general approaches, *supervised learning* and *unsupervised learning*. An analogy for the former is students in a French class where the teacher demonstrates the correct French pronunciation. An analogy for the latter is students working on a team project without supervision. When we use an MLP NN model to perform nonlinear regression or classification, the model output is fitted to the given target data, similar to students trying to imitate the French accent of their teacher, hence the learning is supervised. In unsupervised learning (e.g. clustering), the students are provided with learning rules, but must rely on self-organization to arrive at a solution, without the benefit of being able to learn from a teacher's demonstration.

In this section, we show how an NN model under unsupervised learning can classify input data into categories or clusters. The network consists only of an input layer and an output layer with no hidden layers between them. There are m units (i.e. neurons) in the input layer and k units in the output layer. Excitation at the ith output neuron is

$$e_i = \sum_{j=1}^{m} w_{ij} x_j = \mathbf{w}_i \cdot \mathbf{x}, \tag{8.84}$$

where $\mathbf{w}_i$ are the weight vectors to be determined.

Unsupervised learning is further divided into *reinforced* and *competitive learning*. In reinforced learning, all students are rewarded, whereas in competitive learning, only the best student is rewarded. In our network, under unsupervised competitive learning, the output neuron receiving the largest excitation is selected as the winner and is activated, while the others remain inactive. The output units are binary, i.e. assigned the value 0 if inactive, and 1 if active.

The input data and the weights are normalized, i.e. $\|\mathbf{x}\| = 1$ and $\|\mathbf{w}_i\| = 1$ for all i. The winner among the output units is the one with the largest excitation, i.e. the i' unit where

$$\mathbf{w}_{i'} \cdot \mathbf{x} \geq \mathbf{w}_i \cdot \mathbf{x}, \quad \text{for all } i. \tag{8.85}$$

From a geometric point of view, this means that the unit vector $\mathbf{w}_{i'}$ is oriented most closely in the direction of the current input unit vector $\mathbf{x}(t)$ (where t is either the time or some other label for the input data vectors), i.e.

$$\|\mathbf{w}_{i'} - \mathbf{x}\| \leq \|\mathbf{w}_i - \mathbf{x}\|, \quad \text{for all } i. \tag{8.86}$$

Set the output $y_{i'} = 1$, and all other $y_i = 0$. Only weights of the winner are updated by

$$\Delta w_{i'j} = \eta\,(x_j(t) - w_{i'j}), \tag{8.87}$$

where η, the *learning rate* parameter, is a (small) positive constant. Iterate the weights by feeding the network with the inputs $\{\mathbf{x}(t)\}$ repeatedly until the weights converge. This network will classify the inputs into k categories or clusters as determined by the weight vectors $\mathbf{w}_i$ $(i = 1, \ldots, k)$. Figure 8.8 illustrates the classification of 80 data points in a 2-dimensional input data space using either three or four output neurons.

The disadvantage of an unsupervised method is that the categories which emerge from the self-organization process may be rather unexpected. For instance, if pictures of human faces are separated into two categories by this method, depending on the inputs, the categories may turn out to be men and women, or persons with and without beards.

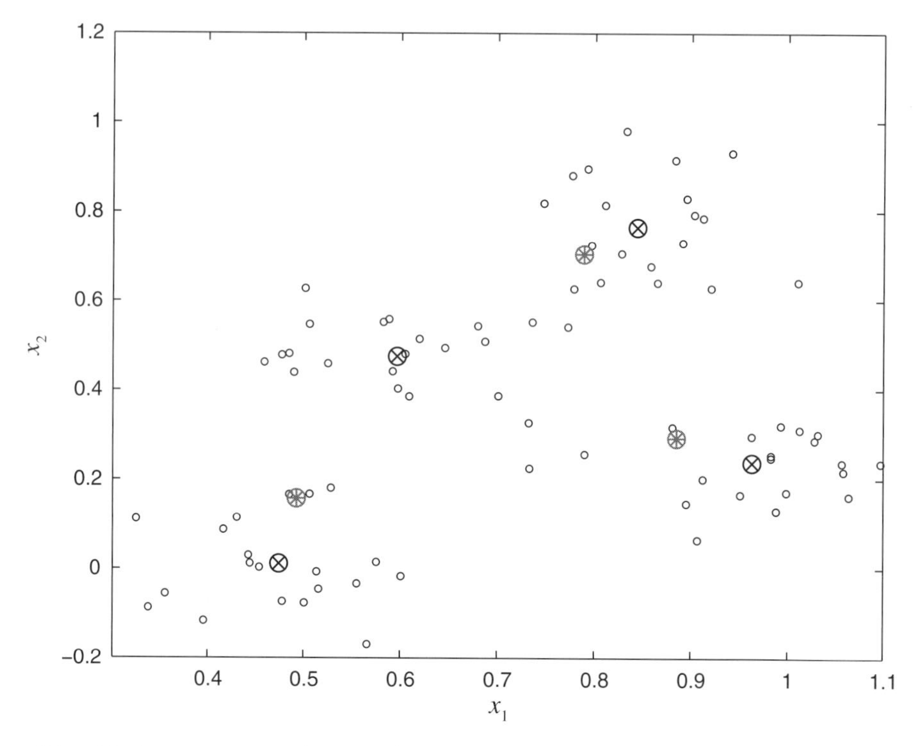

Fig. 8.8 An unsupervised competitive learning neural network clustering data. Small circles denote the data points, while the weight vectors $\mathbf{w}_i$ solved using three output neurons are marked by the asterisks, and four output neurons by the crosses, which mark the cluster centres.

Another disadvantage is that units with $\mathbf{w}_i$ far from any input vector $\mathbf{x}(t)$ will never have their weights updated, i.e. they never learn and are simpy dead units. One possibility is to update the weights of the losers with a much smaller η than the winner, so the losers will still learn slowly and avoid being dead. This approach is known as *leaky learning*.

Another popular unsupervised learning tool is the *self-organizing map* (SOM), which will be presented in Section 10.3, since it is a form of discrete nonlinear principal component analysis. The SOM tool has been applied to many environmental sciences problems.

Exercises

(8.1) Apply nonlinear classifiers (e.g. MLP, SVM) to solve the XOR problem (see Fig. 4.3(b)). To generate training data of about 50 data points, add Gaussian noise so that there is scatter around the four centres $(0, 0)$, $(1, 0)$, $(0, 1)$ and $(1, 1)$ in the (x, y) space. Vary the amount of Gaussian noise. For SVM, try both the linear kernel and nonlinear kernels.

(8.2) Derive expression (8.75) for the Heidke skill score (HSS).

(8.3) Derive expression (8.79) for the Peirce skill score (PSS).

(8.4) Denote the reference Brier score using climatological forecasts by B_{ref}. If we define s as the climatological probability for the occurrence of the event (i.e. using observed data, count the number of occurrences of the event, then divide by the total number of observations), show that $B_{\text{ref}} = s(1 - s)$.

(8.5) Using the data file provided in the book website, which contains two predictors x_1 and x_2 and predictand y data (400 observations), develop a nonlinear classification model of y as a function of x_1 and x_2. Briefly describe the approach taken (e.g. MLP, SVM, etc.). Forecast y_{test} using the new test predictor data $x_{1\text{test}}$ and $x_{2\text{test}}$ provided.

9

Nonlinear regression

The study of nonlinear regression started with neural network models in Chapter 4. The advent of kernel methods and tree methods has expanded the range of tools for nonlinear regression. In this chapter, we will examine two kernel methods, *support vector regression* (SVR) and *Gaussian process* (GP), as well as a tree method, *classification and regression tree* (CART).

9.1 Support vector regression (SVR)

In Chapter 8, we have studied support vector machines (SVM) for classification problems. We now extend SVM to regression problems. The reader should read Section 8.4 on SVM classification before proceeding to *support vector regression.*

In SVR, the objective function to be minimized is

$$J = C\sum_{n=1}^{N} E[y(\mathbf{x}_n) - y_{dn}] + \frac{1}{2}\|\mathbf{w}\|^2, \tag{9.1}$$

where C is the inverse weight penalty parameter, E is an error function and the second term is the weight penalty term. To retain the sparseness property of the SVM classifier, E is usually taken to be of the form

$$E_\epsilon(z) = \begin{cases} |z| - \epsilon, & \text{if } |z| > \epsilon, \\ 0, & \text{otherwise.} \end{cases} \tag{9.2}$$

This is an ϵ*-insensitive error function*, as it ignores errors of size smaller than ϵ (Fig. 9.1).

Slack variables are introduced as in SVM classification, except that for each data point $\mathbf{x}_n$, there are two slack variables $\xi_n \geq 0$ and $\xi_n' \geq 0$. We assign $\xi_n > 0$ only to data points lying above the ϵ-tube in Fig. 9.2, i.e. $y_{dn} > y(\mathbf{x}_n) + \epsilon$, and $\xi_n' > 0$ to data points lying below the ϵ-tube, i.e. $y_{dn} < y(\mathbf{x}_n) - \epsilon$.

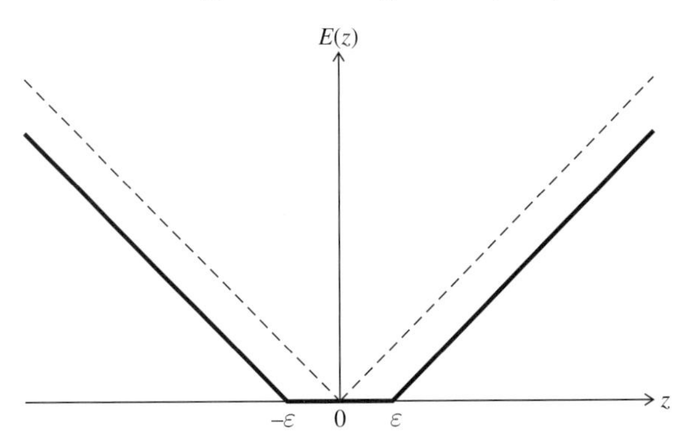

Fig. 9.1 The ϵ-insensitive error function $E_\epsilon(z)$. Dashed line shows the mean absolute error (MAE) function.

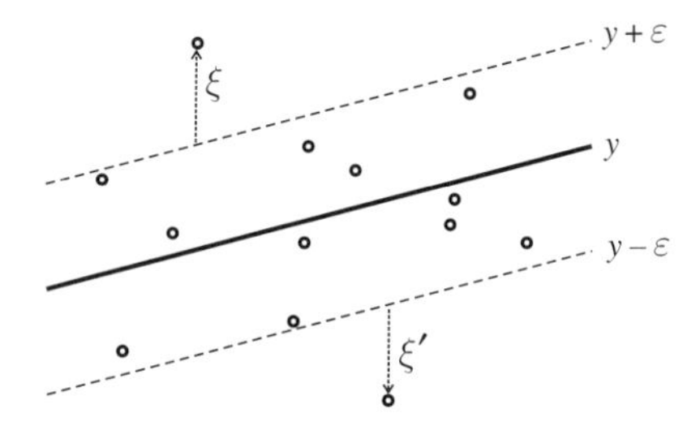

Fig. 9.2 Support vector regression (SVR). Data points lying within the 'ϵ tube' (i.e. between $y - \epsilon$ and $y + \epsilon$) are ignored by the ϵ-insensitive error function. Actually the commonly used term 'ϵ tube' is a misnomer since when the predictor is multi-dimensional, the 'tube' is actually a slab of thickness 2ϵ. For data points lying above and below the tube, their distances from the tube are given by the slack variables ξ and ξ', respectively. Data points lying inside (or right on) the tube have $\xi = 0 = \xi'$. Those lying above the tube have $\xi > 0$ and $\xi' = 0$, while those lying below have $\xi = 0$ and $\xi' > 0$.

The conditions for a data point to lie within the ϵ-tube, i.e. $y(\mathbf{x}_n) - \epsilon \le y_{dn} \le y(\mathbf{x}_n) + \epsilon$, can now be extended via the slack variables to allow for data points lying outside the tube, yielding the conditions

$$y_{dn} \le y(\mathbf{x}_n) + \epsilon + \xi_n, \tag{9.3}$$

$$y_{dn} \ge y(\mathbf{x}_n) - \epsilon - \xi'_n. \tag{9.4}$$

The objective function is then

$$J = C \sum_{n=1}^{N} (\xi_n + \xi'_n) + \frac{1}{2}\|\mathbf{w}\|^2, \tag{9.5}$$

to be minimized subject to the constraints $\xi_n \ge 0$, $\xi'_n \ge 0$, (9.3) and (9.4). To handle the constraints, Lagrange multipliers, $\lambda_n \ge 0$, $\lambda'_n \ge 0$, $\mu_n \ge 0$ and $\mu'_n \ge 0$, are introduced, and the Lagrangian function is

$$L = C\sum_{n=1}^{N}(\xi_n + \xi'_n) + \frac{1}{2}\|\mathbf{w}\|^2 - \sum_{n=1}^{N}(\mu_n\xi_n + \mu'_n\xi'_n)$$
$$- \sum_{n=1}^{N}\lambda_n[y(\mathbf{x}_n) + \epsilon + \xi_n - y_{dn}] - \sum_{n=1}^{N}\lambda'_n[y_{dn} - y(\mathbf{x}_n) + \epsilon + \xi'_n]. \quad (9.6)$$

Similarly to SVM classification (Section 8.4), the regression is performed in the feature space, i.e.

$$y(\mathbf{x}) = \mathbf{w}^{\mathrm{T}}\boldsymbol{\phi}(\mathbf{x}) + w_0, \quad (9.7)$$

where $\boldsymbol{\phi}$ is the feature map. Substituting this for $y(\mathbf{x}_n)$ in (9.6), then setting the derivatives of L with respect to $\mathbf{w}$, w_0, ξ_n and ξ'_n to zero, yields, respectively,

$$\mathbf{w} = \sum_{n=1}^{N}(\lambda_n - \lambda'_n)\boldsymbol{\phi}(\mathbf{x}_n), \quad (9.8)$$

$$\sum_{n=1}^{N}(\lambda_n - \lambda'_n) = 0, \quad (9.9)$$

$$\lambda_n + \mu_n = C, \quad (9.10)$$
$$\lambda'_n + \mu'_n = C. \quad (9.11)$$

Substituting these into (9.6), we obtain the dual Lagrangian

$$L_{\mathrm{D}}(\boldsymbol{\lambda}, \boldsymbol{\lambda}') = -\frac{1}{2}\sum_{n=1}^{N}\sum_{j=1}^{N}(\lambda_n - \lambda'_n)(\lambda_j - \lambda'_j)K(\mathbf{x}_n, \mathbf{x}_j)$$
$$- \epsilon\sum_{n=1}^{N}(\lambda_n + \lambda'_n) + \sum_{n=1}^{N}(\lambda_n - \lambda'_n)y_{dn}, \quad (9.12)$$

where the kernel function $K(\mathbf{x}, \mathbf{x}') = \boldsymbol{\phi}^{\mathrm{T}}(\mathbf{x})\boldsymbol{\phi}(\mathbf{x}')$. The optimization problem is now to maximize L_{D} subject to constraints. The Lagrange multipliers λ_n, λ'_n, μ_n and μ'_n are all ≥ 0 (Appendix B). Together with (9.10) and (9.11), we have the following constraints on λ_n and λ'_n

$$0 \leq \lambda_n \leq C$$
$$0 \leq \lambda'_n \leq C, \quad (9.13)$$

and (9.9).

The Karush–Kuhn–Tucker (KKT) conditions (Appendix B) require the product of each Lagrange multiplier and its associated constraint condition to vanish, yielding

$$\lambda_n \left[y(\mathbf{x}_n) + \epsilon + \xi_n - y_{dn}\right] = 0, \tag{9.14}$$

$$\lambda'_n \left[y_{dn} - y(\mathbf{x}_n) + \epsilon + \xi'_n\right] = 0, \tag{9.15}$$

$$\mu_n \xi_n = (C - \lambda_n)\xi_n = 0, \tag{9.16}$$

$$\mu'_n \xi'_n = (C - \lambda'_n)\xi'_n = 0. \tag{9.17}$$

For λ_n to be non-zero, $y(\mathbf{x}_n) + \epsilon + \xi_n - y_{dn} = 0$ must be satisfied, meaning that the data point y_{dn} must lie either above the ϵ-tube ($\xi_n > 0$) or exactly at the top boundary of the tube ($\xi_n = 0$). Similarly, for λ'_n to be non-zero, $y_{dn} - y(\mathbf{x}_n) + \epsilon + \xi'_n = 0$ must be satisfied, hence y_{dn} lies either below the ϵ-tube ($\xi'_n > 0$) or exactly at the bottom boundary of the tube ($\xi'_n = 0$).

Substituting (9.8) into (9.7) gives

$$y(\mathbf{x}) = \sum_{n=1}^{N} (\lambda_n - \lambda'_n) K(\mathbf{x}, \mathbf{x}_n) + w_0. \tag{9.18}$$

Support vectors are the data points which contribute in the above summation, i.e. either $\lambda_n \neq 0$ or $\lambda'_n \neq 0$, meaning that the data point must lie either outside the ϵ-tube or exactly at the boundary of the tube. All data points within the tube have $\lambda_n = \lambda'_n = 0$ and fail to contribute to the summation in (9.18), thus leading to a sparse solution. If K is the Gaussian or radial basis function (RBF) kernel (8.58), then (9.18) is simply a linear combination of radial basis functions centred at the training data points $\mathbf{x}_n$.

To solve for w_0, we look for a data point with $0 < \lambda_n < C$. From (9.16), $\xi_n = 0$, and from (9.14), $y(\mathbf{x}_n) + \epsilon - y_{dn} = 0$. Invoking (9.18), we get

$$w_0 = -\epsilon + y_{dn} - \sum_{j=1}^{N} (\lambda_j - \lambda'_j) K(\mathbf{x}_n, \mathbf{x}_j). \tag{9.19}$$

Similarly, for a data point with $0 < \lambda'_n < C$, we have $\xi'_n = 0$, $y_{dn} - y(\mathbf{x}_n) + \epsilon = 0$, and

$$w_0 = \epsilon + y_{dn} - \sum_{j=1}^{N} (\lambda_j - \lambda'_j) K(\mathbf{x}_n, \mathbf{x}_j). \tag{9.20}$$

These two equations for estimating w_0 can be applied to various data points satisfying $0 < \lambda_n < C$ or $0 < \lambda'_n < C$, then the various values of w_0 can be averaged together for better numerical accuracy.

If the Gaussian or radial basis function (RBF) kernel (8.58) is used, then there are three hyperparameters in SVR, namely C, ϵ and σ (controlling the width of

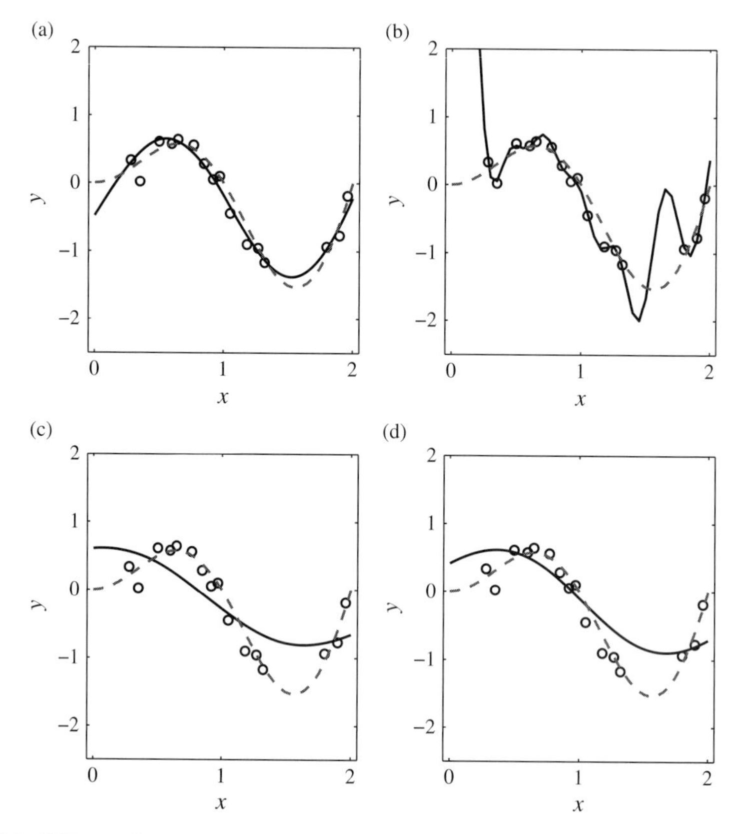

Fig. 9.3 SVR applied to a test problem. (a) Optimal values of the hyperparameters C, ϵ and σ obtained from validation are used. (b) A larger C (i.e. less weight penalty) and a smaller σ (i.e. narrower Gaussian functions) result in overfitting. Underfitting occurs when (c) a larger ϵ (wider ϵ-tube) or (d) a smaller C (larger weight penalty) is used. The training data are the circles, the SVR solution is the solid curve and the true relation is the dashed curve.

the Gaussian function in the kernel), in contrast to only two hyperparameters (C and σ) in the SVM classifier in the previous chapter. Multiple models are trained for various values of the three hyperparameters, and upon evaluating their performance over validation data, the best estimates of the hyperparameters are obtained. Figure 9.3 illustrates the SVR result, with the SVM code from LIBSVM (Chang and Lin, 2001). Note that SVR with the Gaussian kernel can be viewed as an extension of the RBF NN method of Section 4.6.

As in SVM classification, there is another popular alternative formulation of SVR (Schölkopf *et al.*, 2000) called ν-SVR. The new hyperparameter $0 \leq \nu \leq 1$ replaces ϵ (instead of C in SVM classification). As ν is an upper bound on the fraction of data points lying outside the ϵ-tube, it plays a similar role as ϵ (which by controlling the thickness of the ϵ-insensitive region, also affects the fraction of points lying outside the ϵ-tube). We now maximize

$$L_{\mathrm{D}}(\boldsymbol{\lambda}, \boldsymbol{\lambda}') = -\frac{1}{2}\sum_{n=1}^{N}\sum_{j=1}^{N}(\lambda_n - \lambda'_n)(\lambda_j - \lambda'_j)K(\mathbf{x}_n, \mathbf{x}_j) + \sum_{n=1}^{N}(\lambda_n - \lambda'_n)y_{\mathrm{d}n}, \tag{9.21}$$

subject to constraints

$$0 \le \lambda_n \le C/N, \tag{9.22}$$

$$0 \le \lambda'_n \le C/N, \tag{9.23}$$

$$\sum_{n=1}^{N}(\lambda_n - \lambda'_n) = 0, \tag{9.24}$$

$$\sum_{n=1}^{N}(\lambda_n + \lambda'_n) \le \nu C. \tag{9.25}$$

It turns out that ν provides not only an upper bound on the fraction of data points lying outside the ϵ-insensitive tube, but also a lower bound on the fraction of support vectors, i.e. there are at most νN data points lying outside the tube, while there are at least νN support vectors (i.e. data points lying either exactly on or outside the tube).

The advantages of SVR over MLP NN are: (a) SVR trains significantly faster since it only solves sets of linear equations, hence is better suited for high-dimensional datasets, (b) SVR avoids the local minima problem from nonlinear optimization, and (c) the ϵ-insensitive error norm used by SVR is more robust to outliers in the training data.

There have been further extensions of SVM methods. Rojo-Alvarez *et al.* (2003) developed SVR using an ϵ-insensitive Huber error function

$$h_\epsilon(z) = \begin{cases} 0, & \text{for } |z| \le \epsilon, \\ \frac{1}{2}(|z| - \epsilon)^2, & \text{for } \epsilon < |z| \le \epsilon + 2\gamma, \\ 2\gamma[(|z| - \epsilon) - \gamma], & \text{for } \epsilon + 2\gamma < |z|, \end{cases} \tag{9.26}$$

where γ and ϵ are parameters. When $\epsilon = 0$, this function is simply the Huber function (6.23). As $\gamma \to 0$, $h_\epsilon(z)$ becomes the standard ϵ-insensitive error function (9.2) multiplied by 2γ. In general, the error function $h_\epsilon(z)$ has the following form: it is zero for $|z| \le \epsilon$, then it grows quadratically for $|z|$ in the intermediate range, and finally grows only linearly for large $|z|$ to stay robust in the presence of outliers. A comparison of the performance of different error functions has been made by Camps-Valls *et al.* (2006) using satellite data for estimating ocean surface chorophyll concentration (see Fig. 12.1 in Section 12.1.1). Other applications of SVR in environmental sciences are described in Chapter 12.

Relevance Vector Machine (RVM) (Tipping, 2001) is a Bayesian generalization of SVM for regression and classification problems. Chu *et al.* (2004) developed a Bayesian SVR method using the ϵ-insensitive Huber error function.

9.2 Classification and regression trees (CART)

Tree-based methods partition the predictor $\mathbf{x}$-space into rectangular regions and fit a simple function $f(\mathbf{x})$ in each region. The most common tree-based method is *CART (classification and regression tree)* (Breiman *et al.*, 1984), which fits $f(\mathbf{x}) =$ constant in each region, so there is a step at the boundary between two regions. The CART method may seem crude but is useful for two main reasons: (i) it gives an intuitive display of how the predictand or response variable broadly depends on the predictors; (ii) when there are many predictors, it provides a computationally inexpensive way to select a smaller number of relevant predictors. These selected predictors can then be used in more accurate but computationally more expensive models like NN, SVR, etc., i.e. CART can be used to pre-screen predictors for more sophisticated models (Burrows, 1999).

While CART can be used for both nonlinear classification and regression problems, we will first focus on the regression problem, with y_d the predictand data. For simplicity, suppose there are only two predictor variables x_1 and x_2. We look for the partition point $x_1^{(1)}$ where the step function $f(\mathbf{x}) = c_1$ for $x_1 < x_1^{(1)}$, and $f(\mathbf{x}) = c_2$ for $x_1 \geq x_1^{(1)}$ gives the best fit to the predictand data. If the fit is determined by the mean squared error (MSE), then the constants c_1 and c_2 are simply given by the mean value of y_d over the two separate regions partitioned by $x_1^{(1)}$. A similar search for a partition point $x_2^{(1)}$ is performed in the x_2 direction. We decide on whether our first partition should be at $x_1^{(1)}$ or $x_2^{(1)}$ based on whichever partition yields the smaller MSE. In Fig. 9.4(a), the partition is made along $x_1^{(1)}$, and there are now two regions. The process is repeated, i.e. the next partition is made to one or both of the two regions. The partition process is repeated until some stopping criterion is met. In Fig. 9.4(b), the second partition is along $x_1 = x_1^{(2)}$, and the third partition along $x_2 = x_2^{(3)}$, resulting in the predictor space being partitioned into four regions, and the predictand described by the four constants over the four regions.

Let us illustrate CART with a dataset containing the daily maximum of the hourly-averaged ozone reading (in ppm) at Los Angeles, with high ozone level indicating poor air quality. The downloaded dataset was prepared by Leo Breiman, Department of Statistics, UC Berkeley, and was similar to that used in Breiman

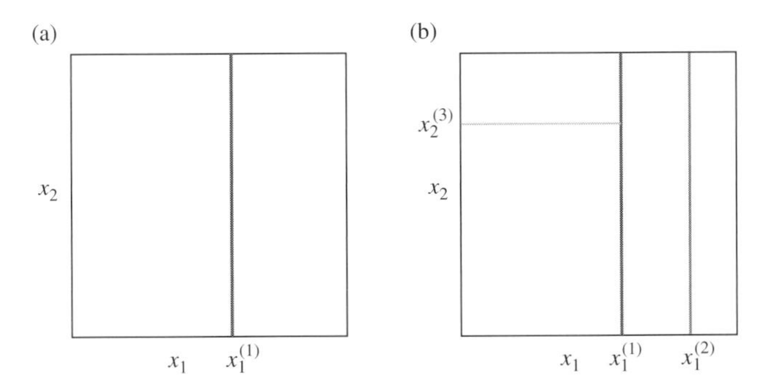

Fig. 9.4 Illustrating partitioning of the predictor **x**-space by CART. (a) First partition at $x_1 = x_1^{(1)}$ yields two regions, each with a constant value for the predictand y. (b) Second partition at $x_1 = x_1^{(2)}$ is followed by a third partition at $x_2 = x_2^{(3)}$, yielding four regions.

and Friedman (1985). The dataset also contained nine predictor variables for the ozone. Among the nine predictors, there are temperature measurements T_1 and T_2 (in °F) at two stations, visibility (in miles) measured at Los Angeles airport, and the pressure gradient (in mm Hg) between the airport and another station. In the 9-dimensional predictor space, CART made the first partition at $T_1 = 63.05$ °F, the second partition at $T_1 = 70.97$ °F and the third partition at $T_2 = 58.50$ °F. This sequence of partitions can be illustrated by the tree in Fig. 9.5(a). The partitioned regions are also shown schematically in Fig. 9.4(b), with x_1 and x_2 representing T_1 and T_2, respectively.

If one continues partitioning, the tree grows further. In Fig. 9.5(b), there is now a fourth partition at pressure gradient $= -13$ mm Hg, and a fifth partition at visibility = 75 miles. Hence CART tells us that the most important predictors, in decreasing order of importance, are T_1, T_2, pressure gradient and visibility. The tree now has six terminal or leaf nodes, denoting the six regions formed by the partitions. Each region is associated with a constant ozone value, the highest being 27.8 ppm (attained by the second leaf from the right). From this leaf node, one can then retrace the path towards the root, which tells us that this highest ozone leaf node was reached after satisfying first 63.05 °F $\leq T_1$, then 70.97 °F $\leq T_1$ and finally visibility < 75 miles, i.e. the highest ozone conditions tend to occur at high temperature and low visibility. The lowest ozone value of 5.61 ppm was attained by the leftmost leaf node, which satisfies $T_1 < 63.05$ °F and $T_2 < 58.5$ °F, indicating that the lowest ozone values tend to occur when both stations record low temperatures. The CART method also gives the number of data points in each partitioned region, e.g. 15 points belong to the highest ozone leaf node versus 88 to the lowest ozone node, out of a total of 203 data points. After training is done, when a new value of the predictor **x** becomes available, one proceeds along the tree till a leaf node

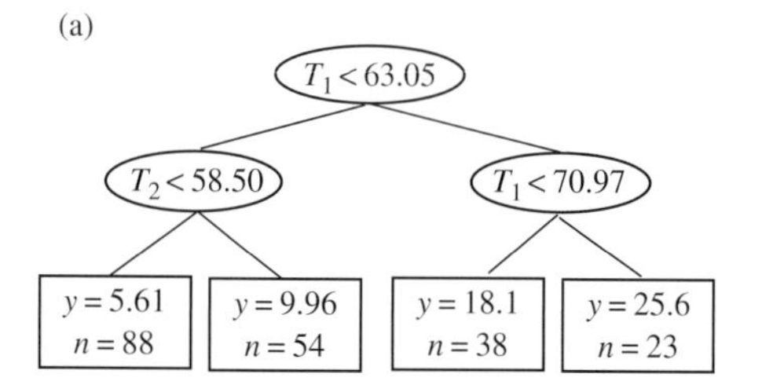

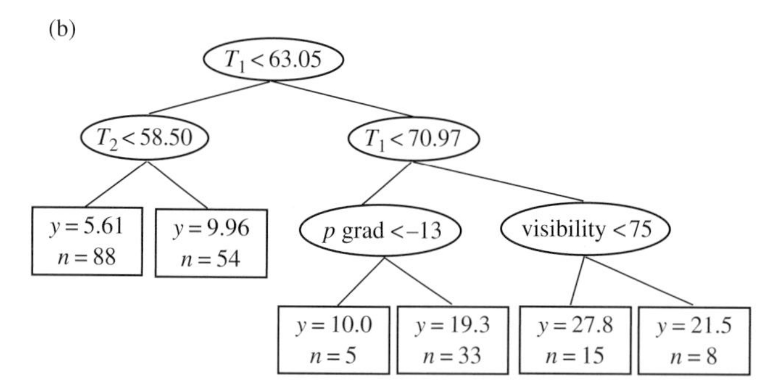

Fig. 9.5 Regression tree from CART where the predictand y is the Los Angeles ozone level (in ppm), and there are nine predictor variables. The 'tree' is plotted upside down, with the 'leaves' (i.e. terminal nodes) drawn as rectangular boxes at the bottom and the non-terminal nodes as ellipses. (a) The tree after three partitions has four leaf nodes. (b) The tree after five partitions has six leaf nodes. In each ellipse, a condition is given. Starting from the top ellipse, if the condition is satisfied, proceed along the left branch down to the next node; if not, proceed along the right branch; continue until a leaf node is reached. In each rectangular box, the constant value of y (computed from the mean of the data y_d) in the partitioned region associated with the particular leaf node is given, as well as n, the number of data points in that region. Among the nine predictor variables, the most relevant are the temperatures T_1 and T_2 (in °F) at two stations, p grad (pressure gradient in mm Hg) and visibility (in miles). CART was calculated here using the function treefit.m from the *Matlab* statistical toolbox.

is reached, and the forecast for y is then simply the constant value associated with that leaf node.

The intuitive interpretation of the tree structure contributes to CART's popularity in the medical field. Browsing over my copy of the American Medical Association's *Encyclopedia of Medicine*, I can see many tree-structured flow charts for patient diagnosis. For instance, the questions asked are: Is the body temperature above normal? Is the patient feeling pain? Is the pain in the chest area? The terminal nodes are of course the likely diseases, e.g. influenza, heart attack, food poisoning, etc. Hence the tree-structured logic in CART is indeed the type of reasoning used by doctors.

How big a tree should one grow? It is not a good idea to stop the growing process after encountering a partition which gave little improvement in the MSE, because a further partition may lead to a large drop in the MSE. Instead, one lets the tree grow to a large size, and uses regularization (i.e. weight penalty) to prune the tree down to the optimal size. Suppose L is the number of leaf nodes. A regularized objective function is

$$J(L) = E(L) + PL, \tag{9.27}$$

where $E(L)$ is the MSE for the tree with L leaf nodes, and P is the weight penalty parameter, penalizing trees with excessive leaf nodes. The process to generate a sequence of trees with varying L is as follows: (a) start with the full tree, remove the internal node the demise of which leads to the smallest increase in MSE, and continue until the tree has only one internal node; (b) from this sequence of trees with a wide range of L values, one chooses the tree with the smallest $J(L)$, thus selecting the tree with the optimal size for a given P. The best value for P is determined from cross-validation, where multiple runs with different P values are made, and the run with the smallest cross-validated MSE is chosen as the best.

CART can also be used for *classification*. The constant value for y over a region is now given by the class k to which the largest number of y_{d} belong. With p_{lk} denoting the proportion of data in region l belonging to class k, the error E is no longer the MSE, but is usually either the cross entropy

$$E(L) = \sum_{l=1}^{L} \sum_{k=1}^{K} p_{lk} \ln p_{lk}, \tag{9.28}$$

or the *Gini index*

$$E(L) = \sum_{l=1}^{L} \sum_{k=1}^{K} p_{lk}(1 - p_{lk}). \tag{9.29}$$

Applications of CART in the environmental sciences include use of CART classification models to predict lake-effect snow (Burrows, 1991), CART regression to predict ground level ozone concentration (Burrows *et al.*, 1995), CART regression to predict ultraviolet radiation at the ground in the presence of cloud and other environmental factors (Burrows, 1997), CART to select large-scale atmospheric predictors for surface marine wind prediction by a neuro-fuzzy system (Faucher *et al.*, 1999), and CART regression to predict lightning probability over Canada and northern USA (Burrows *et al.*, 2005). In general, the predictand variables are not forecast by the numerical weather prediction models. Instead, the numerical weather prediction models provide the predictors for the CART models.

So far, the partitions are of the form $x_i < s$, thus restricting the partition boundaries to lie parallel to the axes in **x**-space. If a decision boundary in the x_1-x_2 plane is oriented at 45° to the x_1 axis, then it would take many parallel-axes partitions to approximate such a decision boundary. Some versions of CART allow partitions of the form $\sum_i a_i x_i < s$, which are not restricted to be parallel to the axes, but the easy interpretability of CART is lost.

The CART method uses piecewise constant functions to represent the nonlinear relation $y = f(\mathbf{x})$. To extend beyond a zero-order model like CART, first-order models, like the M5 tree model (or its earlier versions, e.g. M4.5) (Quinlan, 1993), use piecewise linear functions to represent the nonlinear relation. The M5 model has been used to predict runoff from precipitation (Solomatine and Dulal, 2003).

A boostrap ensemble of CART models is called a *random forest* (Breiman, 2001). If N is the number of training data points and M the number of predictor variables, one generates many bootstrap samples (by selecting N data points with replacement from the training dataset), then trains CART on each bootstrap sample using m randomly chosen predictors out of the original M predictors ($m \ll M$ if M is large). The trees are fully grown without pruning. With new predictor data, y is taken to be the mean of the ensemble output in regression problems, or the class k chosen by the largest number of ensemble members in classification problems.

9.3 Gaussian processes (GP)

The kernel trick (Chapter 7) can also be applied to probabilistic models, leading to a new class of kernel methods known as *Gaussian processes* (GP) in the late 1990s. Historically, regression using Gaussian processes has actually been known for a long time in geostatistics as *kriging* (after D. G. Krige, the South African mining engineer), which evolved from the need to perform spatial interpolation of mining exploration data. The new GP method is on a Bayesian framework and has gained considerable popularity in recent years. Among Bayesian methods (e.g. Bayesian NN), the Bayesian inference generally requires either approximations or extensive numerical computations using Markov chain Monte Carlo methods; however, GP regression has the advantage that by assuming Gaussian distributions, the Bayesian inference at the first level (i.e. obtaining the posterior distribution of the model parameters) is analytically tractable and hence exact. The Gaussian process method is covered in texts such as Rasmussen and Williams (2006), Bishop (2006), MacKay (2003) and Nabney (2002).

In GP regression, the predictand or response variable y is a linear combination of M fixed basis functions $\phi_l(\mathbf{x})$, (analogous to the radial basis function NN in Section 4.6)

$$y(\mathbf{x}) = \sum_{l=1}^{M} w_l \phi_l(\mathbf{x}) = \boldsymbol{\phi}^{\mathrm{T}}(\mathbf{x})\,\mathbf{w}, \tag{9.30}$$

where $\mathbf{x}$ is the input vector and $\mathbf{w}$ the weight vector.

The Gaussian distribution over the M-dimensional space of $\mathbf{w}$ vectors is

$$N(\mathbf{w}|\boldsymbol{\mu}, \mathbf{C}) \equiv \frac{1}{(2\pi)^{M/2}|\mathbf{C}|^{1/2}} \exp\left[-\frac{1}{2}(\mathbf{w}-\boldsymbol{\mu})^{\mathrm{T}}\mathbf{C}^{-1}(\mathbf{w}-\boldsymbol{\mu})\right], \tag{9.31}$$

where $\boldsymbol{\mu}$ is the mean, $\mathbf{C}$ the $M \times M$ covariance matrix, and $|\mathbf{C}|$ the determinant of $\mathbf{C}$. In GP, we assume the prior distribution of $\mathbf{w}$ to be an isotropic Gaussian with zero mean and covariance $\mathbf{C} = \alpha^{-1}\mathbf{I}$ ($\mathbf{I}$ being the identity matrix), i.e.

$$p(\mathbf{w}) = N(\mathbf{w}|\mathbf{0}, \alpha^{-1}\mathbf{I}), \tag{9.32}$$

where α is a hyperparameter, with α^{-1} governing the variance of the distribution. This distribution in $\mathbf{w}$ leads to a probability distribution for the functions $y(\mathbf{x})$ through (9.30), with

$$\mathrm{E}[y(\mathbf{x})] = \boldsymbol{\phi}^{\mathrm{T}}(\mathbf{x})\,\mathrm{E}[\mathbf{w}] = 0, \tag{9.33}$$

$$\begin{aligned}\mathrm{cov}[y(\mathbf{x}_i), y(\mathbf{x}_j)] &= \mathrm{E}\left[\boldsymbol{\phi}^{\mathrm{T}}(\mathbf{x}_i)\,\mathbf{w}\,\mathbf{w}^{\mathrm{T}}\boldsymbol{\phi}(\mathbf{x}_j)\right] \\ &= \boldsymbol{\phi}^{\mathrm{T}}(\mathbf{x}_i)\,\mathrm{E}\left[\mathbf{w}\,\mathbf{w}^{\mathrm{T}}\right]\boldsymbol{\phi}(\mathbf{x}_j) \\ &= \alpha^{-1}\boldsymbol{\phi}(\mathbf{x}_i)^{\mathrm{T}}\boldsymbol{\phi}(\mathbf{x}_j) \equiv K_{ij} \equiv K(\mathbf{x}_i, \mathbf{x}_j),\end{aligned} \tag{9.34}$$

where $K(\mathbf{x}, \mathbf{x}')$ is the kernel function, and (given n training input data points $\mathbf{x}_1, \ldots, \mathbf{x}_n$), K_{ij} are the elements of the $n \times n$ covariance or kernel matrix $\mathbf{K}$.

To be a Gaussian process, the function $y(\mathbf{x})$ evaluated at $\mathbf{x}_1, \ldots, \mathbf{x}_n$ must have the probability distribution $p(y(\mathbf{x}_1), \ldots, y(\mathbf{x}_n))$ obeying a joint Gaussian distribution. Thus the joint distribution over $y(\mathbf{x}_1), \ldots, y(\mathbf{x}_n)$ is completely specified by second-order statistics (i.e. the mean and covariance). Since the mean is zero, the GP is completely specified by the covariance, i.e. the kernel function K. There are many choices for kernel functions (see Section 7.3). A simple and commonly used one is the (isotropic) Gaussian kernel function

$$K(\mathbf{x}, \mathbf{x}') = a \exp\left(-\frac{\|\mathbf{x}-\mathbf{x}'\|^2}{2\sigma_K^2}\right), \tag{9.35}$$

with two hyperparameters, a and σ_K.

When GP is used for regression, the target data y_{d} are the underlying relation $y(\mathbf{x})$ plus Gaussian noise with variance σ_{d}^2, i.e. the distribution of y_{d} conditional on y is a normal distribution N,

$$p(y_{\mathrm{d}}|y) = N(y_{\mathrm{d}}|\,y, \sigma_{\mathrm{d}}^2). \tag{9.36}$$

With n data points $\mathbf{x}_i$ ($i = 1, \ldots, n$), we write $y_i = y(\mathbf{x}_i)$, $\mathbf{y} = (y_1, \ldots, y_n)^{\mathrm{T}}$ and $\mathbf{y}_{\mathrm{d}} = (y_{\mathrm{d}1}, \ldots, y_{\mathrm{d}n})^{\mathrm{T}}$. Assuming that the noise is independent for each data point, the joint distribution of $\mathbf{y}_d$ conditional on $\mathbf{y}$ is given by an isotropic Gaussian distribution:

$$p(\mathbf{y}_d|\mathbf{y}) = N(\mathbf{y}_d|\,\mathbf{y}, \beta^{-1}\mathbf{I}), \tag{9.37}$$

where the hyperparameter $\beta = \sigma_d^{-2}$.

From (9.30), we note that y is a linear combination of the w_l variables, i.e. a linear combination of Gaussian distributed variables which again gives a Gaussian distributed variable, so upon invoking (9.33) and (9.34),

$$p(\mathbf{y}) = N(\mathbf{y}|\,\mathbf{0}, \mathbf{K}). \tag{9.38}$$

Since $p(\mathbf{y}_d|\mathbf{y})$ and $p(\mathbf{y})$ are both Gaussians, $p(\mathbf{y}_d)$ can be evaluated analytically (Bishop, 2006) to give another Gaussian distribution

$$p(\mathbf{y}_d) = \int p(\mathbf{y}_d|\mathbf{y})\, p(\mathbf{y})d\mathbf{y} = N(\mathbf{y}_d|\,\mathbf{0}, \mathbf{C}), \tag{9.39}$$

where the covariance matrix $\mathbf{C}$ has elements

$$C_{ij} = K(\mathbf{x}_i, \mathbf{x}_j) + \beta^{-1}\delta_{ij}, \tag{9.40}$$

with δ_{ij} the Kronecker delta function. These two terms indicate two sources of randomness, with the first term coming from $p(\mathbf{y})$, and the second from $p(\mathbf{y}_d|\mathbf{y})$ (i.e. from the noise in the target data).

Suppose we have built a GP regression model from the training dataset containing $\{\mathbf{x}_1, \ldots, \mathbf{x}_n\}$ and $\{y_{d1}, \ldots, y_{dn}\}$; now we want to make predictions with this model, i.e. given a new predictor point $\mathbf{x}_{n+1}$, what can we deduce about the distribution of the predictand or target variable $y_{d(n+1)}$?

To answer this, we need to find the conditional distribution $p(y_{d(n+1)}|\mathbf{y}_d)$, (where for notational brevity, the conditional dependence on the predictor data has been omitted). First we start with the joint distribution $p(\mathbf{y}_{d(n+1)})$, where $\mathbf{y}_{d(n+1)} = (\mathbf{y}_d^T, y_{d(n+1)})^T = (y_{d1}, \ldots, y_{dn}, y_{d(n+1)})^T$. From (9.39), it follows that

$$p(\mathbf{y}_{d(n+1)}) = N(\mathbf{y}_{d(n+1)}|\,\mathbf{0}, \mathbf{C}_{n+1}). \tag{9.41}$$

The $(n+1) \times (n+1)$ covariance matrix $\mathbf{C}_{n+1}$ given by (9.40) can be partitioned into

$$\mathbf{C}_{n+1} = \begin{bmatrix} \mathbf{C}_n & \mathbf{k} \\ \mathbf{k}^T & c \end{bmatrix}, \tag{9.42}$$

where $\mathbf{C}_n$ is the $n \times n$ covariance matrix, the column vector $\mathbf{k}$ has its ith element $(i = 1, \ldots, n)$ given by $K(\mathbf{x}_i, \mathbf{x}_{n+1})$, and the scalar $c = K(\mathbf{x}_{(n+1)}, \mathbf{x}_{(n+1)}) + \beta^{-1}$. Since the joint distribution $p(\mathbf{y}_{d(n+1)})$ is a Gaussian, it follows that the conditional distribution $p(y_{d(n+1)}|\mathbf{y}_d)$ is also a Gaussian (Bishop, 2006), with its mean and variance given by

$$\mu(\mathbf{x}_{n+1}) = \mathbf{k}^{\mathrm{T}}\mathbf{C}_n^{-1}\mathbf{y}_{\mathrm{d}}, \tag{9.43}$$

$$\sigma^2(\mathbf{x}_{n+1}) = c - \mathbf{k}^{\mathrm{T}}\mathbf{C}_n^{-1}\mathbf{k}. \tag{9.44}$$

The main computational burden in GP regression is the inversion of the $n \times n$ matrix $\mathbf{C}_n$, requiring $O(n^3)$ operations. The method becomes prohibitively costly for large sample size n. One can simply use a random subset of the original dataset to train the model, though this is wasteful of the data. A number of other approximations are given in Rasmussen and Williams (2006). An alternative is to solve the regression problem (9.30) with M basis functions directly without using the kernel trick. This would require $O(M^3)$ operations instead of $O(n^3)$ operations. The advantage of the kernel approach in GP regression is that it allows us to use kernel functions representing an infinite number of basis functions.

9.3.1 Learning the hyperparameters

Before we can compute the solutions in (9.43) and (9.44), we need to know the values of the hyperparameters, i.e. β, and a and σ_K if the Gaussian kernel (9.35) is used. Let $\boldsymbol{\theta}$ denote the vector of hyperparameters. Since parameters like β, and a and σ_K are all positive and we do not want to impose a bound on $\boldsymbol{\theta}$ during optimization, we can e.g. let $\boldsymbol{\theta} = [\ln\beta, \ln a, \ln\sigma_K]$.

A common way to find the optimal $\boldsymbol{\theta}$ is to maximize the likelihood function $p(\mathbf{y}_{\mathrm{d}}|\boldsymbol{\theta})$. In practice, the log of the likelihood function is maximized using a nonlinear optimization algorithm (Chapter 5), where

$$\ln p(\mathbf{y}_{\mathrm{d}}|\boldsymbol{\theta}) = -\frac{1}{2}\ln|\mathbf{C}_n| - \frac{1}{2}\mathbf{y}_{\mathrm{d}}^{\mathrm{T}}\mathbf{C}_n^{-1}\mathbf{y}_{\mathrm{d}} - \frac{n}{2}\ln(2\pi). \tag{9.45}$$

More efficient nonlinear optimization algorithms can be used if we supply the gradient information

$$\frac{\partial}{\partial\theta_i}\ln p(\mathbf{y}_{\mathrm{d}}|\boldsymbol{\theta}) = -\frac{1}{2}\mathrm{Tr}\left(\mathbf{C}_n^{-1}\frac{\partial\mathbf{C}_n}{\partial\theta_i}\right) + \frac{1}{2}\mathbf{y}_{\mathrm{d}}^{\mathrm{T}}\mathbf{C}_n^{-1}\frac{\partial\mathbf{C}_n}{\partial\theta_i}\mathbf{C}_n^{-1}\mathbf{y}_{\mathrm{d}}, \tag{9.46}$$

where Tr denotes the trace (i.e. the sum of the diagonal elements of a matrix). The nonlinear optimization problem will in general have multiple minima, hence it is best to run the optimization procedure multiple times from different initial conditions and choose the run with the lowest minimum. In contrast to MLP NN models where multiple minima tend to occur from the nonlinear optimization with respect to the model weight and offset parameters, multiple minima may occur in GP during nonlinear optimization of the hyperparameters, there being usually far fewer hyperparameters than the MLP weight and offset parameters.

The $\boldsymbol{\theta}$ values obtained from the above optimization can be used in (9.43) and (9.44) to give the GP regression solution. A full Bayesian treatment is also possible, where instead of using a single optimal $\boldsymbol{\theta}$ value, integration over all $\boldsymbol{\theta}$ values is performed. The integration cannot be done analytically, but can be computed using Markov chain Monte Carlo methods.

So far, the GP regression has been limited to a single predictand or output variable. If one has multiple predictands, one can apply the single predictand GP regression to each predictand separately, but this ignores any correlation between the predictands. Generalization to multiple predictands while incorporating correlation between the predictands is known as *co-kriging* in geostatistics (Cressie, 1993). Boyle and Frean (2005) have generalized GP to multiple dependent predictands.

In Fig. 9.6, we show examples of GP regression, for a univariate predictor x. The signal, as shown by the dashed curve, is

$$y = x\sin(\pi x), \quad 0 \le x < 2. \tag{9.47}$$

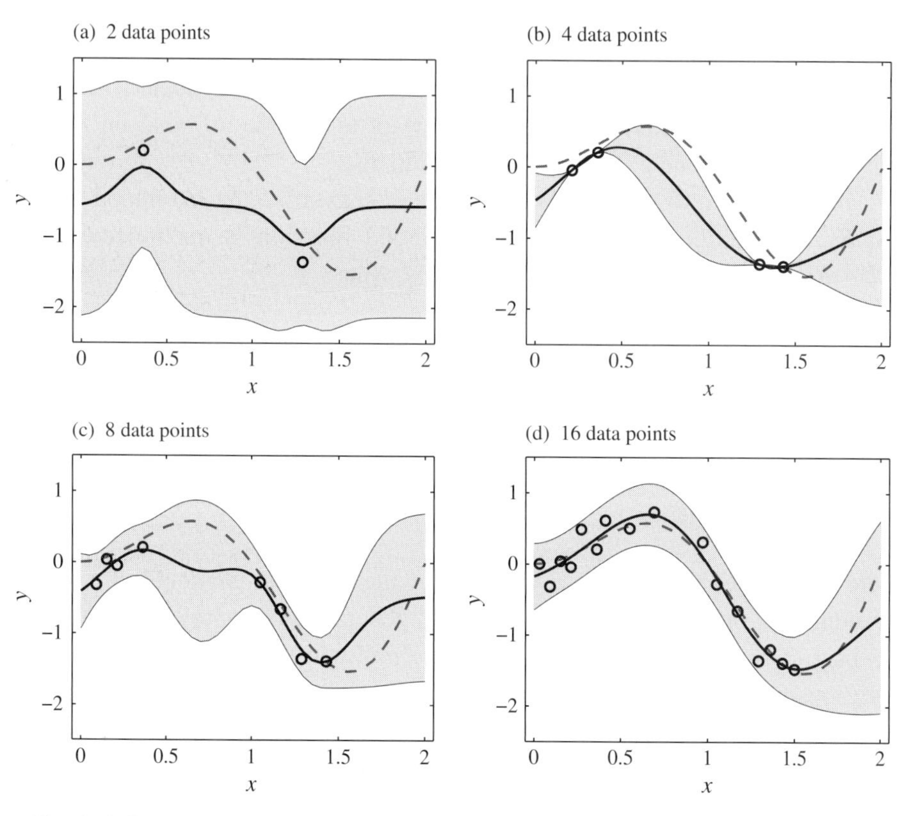

Fig. 9.6 GP regression using the isotropic Gaussian kernel, with the number of data points (small circles) being (a) 2, (b) 4, (c) 8 and (d) 16. The solid curve shows the predicted mean, with the two thin curves showing the 95% prediction interval (i.e. ±2 standard deviations). The true underlying signal is indicated by the dashed curve for comparison.

Gaussian noise with 1/3 the standard deviation of the signal is added to give the target data y_d. Using the isotropic Gaussian kernel (9.35), GP regression was performed with the number of data points varying from 2 to 16. The mean and variance from (9.43) and (9.44) are used to draw the solid curve and the two thin curves showing ± 2 standard deviations (i.e. the 95% prediction interval). In regions where data are lacking, the prediction interval widens.

For ground-level ozone concentration, Cai *et al.* (2008) found Bayesian models such as GP and BNN outperfoming non-Bayesian models when forecasting extreme events (i.e. high ozone events) (see Section 12.3.6).

9.3.2 Other common kernels

Besides the isotropic Gaussian kernel (9.35), there are a few other common kernels used in GP regression. A Gaussian kernel which allows *automatic relevance determination* (ARD) is

$$K(\mathbf{x}, \mathbf{x}') = a \exp\left(-\frac{1}{2} \sum_{l=1}^{m} \eta_l (x_l - x_l')^2 \right), \tag{9.48}$$

where the hyperparameters η_l govern the importance or relevance of each particular input x_l. Thus if a particular hyperparameter η_l turns out to be close to zero, then the corresponding input variable x_l is not a relevant predictor for y.

Another common class of kernels is the Matérn kernels (Rasmussen and Williams, 2006). With $r = \|\mathbf{x} - \mathbf{x}'\|$, and r_0 a hyperparameter, the two commonly used Matérn kernels are

$$K_{\nu=3/2}(\mathbf{x}, \mathbf{x}') = a \left(1 + \frac{\sqrt{3}\, r}{r_0} \right) \exp\left(-\frac{\sqrt{3}\, r}{r_0} \right), \tag{9.49}$$

$$K_{\nu=5/2}(\mathbf{x}, \mathbf{x}') = a \left(1 + \frac{\sqrt{5}\, r}{r_0} + \frac{5r^2}{3r_o^2} \right) \exp\left(-\frac{\sqrt{5}\, r}{r_0} \right), \tag{9.50}$$

which are the product of a polynomial kernel with an exponential kernel.

9.4 Probabilistic forecast scores

In Section 8.5.3, we encountered the Brier score for probabilistic forecast verification when the observed predictand is a binary variable. When the observed predictand y_d is a continuous variable, two well-designed scores for evaluating probabilistic forecasts are the continuous ranked probability score and the ignorance score (Gneiting *et al.*, 2005).

The continuous ranked probability score (CRPS) is the integral of the Brier scores at all possible threshold values y for the continuous predictand, and is defined as

$$\text{CRPS} = \frac{1}{N}\sum_{n=1}^{N} \text{crps}\,(F_n,\ y_{\text{d}n})$$
$$= \frac{1}{N}\sum_{n=1}^{N}\left(\int_{-\infty}^{\infty}\left[F_n(y) - H(y - y_{\text{d}n})\right]^2 \text{d}y\right), \tag{9.51}$$

where for the nth prediction, the cumulative probability $F_n(y) = p(Y \leq y)$, and $H(y - y_{\text{d}n})$ is the Heaviside step function which takes the value 0 when $y - y_{\text{d}n} < 0$, and 1 otherwise. For Gaussian distributions, an analytic form of the integral in (9.51) can be found (Gneiting *et al.*, 2005); for other distributions, numerical integration is needed.

The *ignorance score* (IGN) is the negative log predictive density, i.e.

$$\text{IGN} = \frac{1}{N}\sum_{n=1}^{N} \text{ign}\,(p_n,\ y_{\text{d}n}) = \frac{1}{N}\sum_{n=1}^{N}\left[-\ln\left(p_n\left(y_{\text{d}n}\right)\right)\right], \tag{9.52}$$

where p_n is the predictive density.

Both scores are negatively oriented. If the predictive distribution is Gaussian (with mean μ and standard deviation σ), the analytical forms of the scores can be derived (Gneiting *et al.*, 2005). For a Gaussian distribution, the key difference between these two scores is that CRPS grows linearly with the normalized prediction error $(y_{\text{d}n} - \mu)/\sigma$, but IGN grows quadratically. Note that CRPS can be interpreted as a generalized version of the MAE (Gneiting *et al.*, 2005). As the ignorance score assigns harsh penalties to particularly poor probabilistic forecasts, it can be exceedingly sensitive to outliers and extreme events. Hence Gneiting *et al.* (2005) preferred CRPS over IGN.

Exercises

(9.1) Let $y = \sin(2\pi x)$ $(0 \leq x < 1)$ be the signal. The y data are generated by adding Gaussian noise to the y signal. With these data, train a support vector regression (SVR) model, using both linear and nonlinear kernels. Vary the amount of Gaussian noise.

(9.2) Repeat Exercise 9.1 but use a Gaussian process model (GP) instead.

(9.3) Repeat Exercises 9.1 and 9.2, but turn some data points into outliers to check the robustness of the SVR and GP methods.

(9.4) Compare the prediction performance of CART and random forest (the ensemble version of CART as described in Section 9.2) with a real dataset.

(9.5) Using the data file provided in the book website, which contains two predictors x_1 and x_2 and predictand y data (80 observations), develop a nonlinear regression model of y as a function of x_1 and x_2. Briefly describe the approach taken (e.g. MLP, SVR, ensemble, etc.). Forecast y_{test} using the new test predictor data $x_{1\text{test}}$ and $x_{2\text{test}}$ provided.

10

Nonlinear principal component analysis

In Chapter 9, we have seen machine learning methods nonlinearly generalizing the linear regression method. In this chapter, we will examine ways to nonlinearly generalize principal component analysis (PCA) and related methods. Figure 10.1 illustrates the difference between linear regression, PCA, nonlinear regression, and nonlinear PCA.

Principal component analysis can be performed using neural network (NN) methods (Oja, 1982; Sanger, 1989). However, far more interesting is the nonlinear generalization of PCA, where the straight line in PCA is replaced by a curve which minimizes the mean squared error (MSE) (Fig. 10.1). Nonlinear PCA can be performed by a variety of methods, e.g. the auto-associative NN model using multi-layer perceptrons (MLP) (Kramer, 1991; Hsieh, 2001b), and the kernel PCA model (Schölkopf *et al.*, 1998). Nonlinear PCA belongs to the class of nonlinear dimensionality reduction techniques, which also includes principal curves (Hastie and Stuetzle, 1989), locally linear embedding (LLE) (Roweis and Saul, 2000) and isomap (Tenenbaum *et al.*, 2000). Self-organizing map (SOM) (Kohonen, 1982) can also be regarded as a discrete version of NLPCA. Dong and McAvoy (1996) combined the principal curve and MLP approaches, while Newbigging *et al.* (2003) used the principal curve projection concept to improve on the MLP approach. Another way to generalize PCA is via independent component analysis (ICA) (Comon, 1994; Hyvärinen *et al.*, 2001), which was developed from information theory, and has been applied to study the tropical Pacific sea surface temperature (SST) variability by Aires *et al.* (2000). Nonlinear complex PCA (Rattan and Hsieh, 2005) and nonlinear singular spectrum analysis (Hsieh and Wu, 2002) have also been developed.

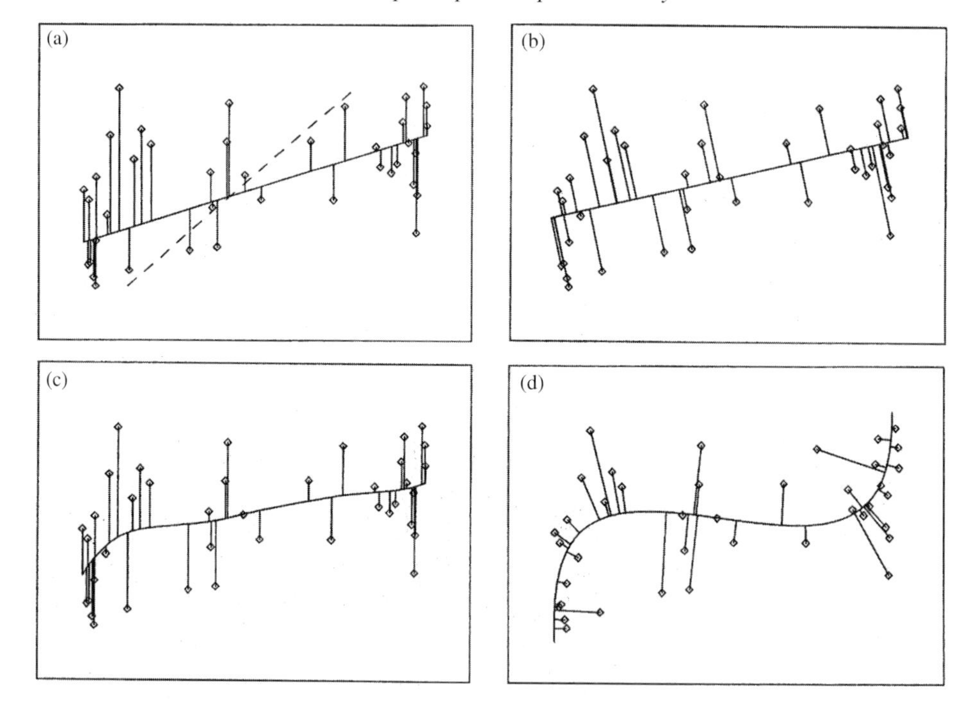

Fig. 10.1 Comparison of analysis methods. (a) The linear regression line minimizes the mean squared error (MSE) in the response variable (i.e. the predictand). The dashed line illustrates the dramatically different result when the role of the predictor variable and the response variable are reversed. (b) Principal component analysis (PCA) minimizes the MSE in all variables. (c) Nonlinear regression methods produce a curve minimizing the MSE in the response variable. (d) Nonlinear PCA methods use a curve which minimizes the MSE of all variables. In both (c) and (d), the smoothness of the curve can be varied by the method. (Reprinted from Hastie and Stuetzle (1989) with permission from the Journal of the American Statistical Association. Copyright 1989 by the American Statistical Association. All rights reserved.)

10.1 Auto-associative NN for nonlinear PCA

10.1.1 Open curves

Kramer (1991) proposed a neural-network-based nonlinear PCA (NLPCA) model where the straight line solution in PCA is replaced by a continuous open curve for approximating the data. The fundamental difference between NLPCA and PCA is that PCA allows only a linear mapping ($u = \mathbf{e} \cdot \mathbf{x}$) between $\mathbf{x}$ and the PC u, while NLPCA allows a nonlinear mapping. To perform NLPCA, the multi-layer perceptron (MLP) NN in Fig. 10.2(a) contains three hidden layers of neurons between the input and output layers of variables. The NLPCA is basically a standard MLP NN with four layers of activation functions mapping from the inputs to the outputs. One can view the NLPCA network as composed of two standard two layer MLP

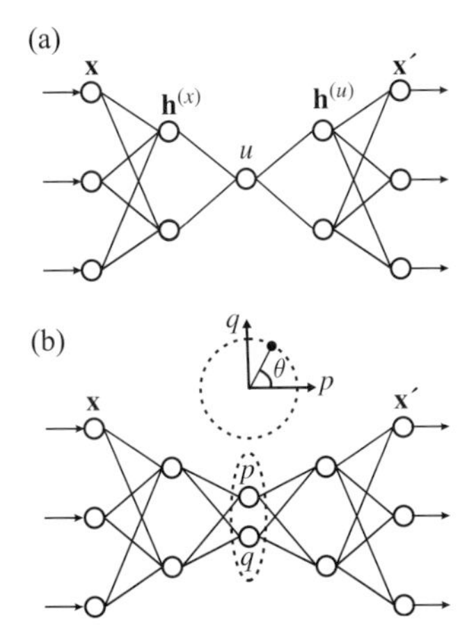

Fig. 10.2 (a) The NN model for calculating the NLPCA. There are three layers of hidden neurons sandwiched between the input layer $\mathbf{x}$ on the left and the output layer $\mathbf{x}'$ on the right. Next to the input layer is the encoding layer, followed by the 'bottleneck' layer (with a single neuron u), which is then followed by the decoding layer. A nonlinear function maps from the higher dimension input space to the 1-dimension bottleneck space, followed by an inverse transform mapping from the bottleneck space back to the original space represented by the outputs, which are to be as close to the inputs as possible by minimizing the objective function $J = \langle \|\mathbf{x} - \mathbf{x}'\|^2 \rangle$. Data compression is achieved by the bottleneck, with the bottleneck neuron giving u, the nonlinear principal component (NLPC). (b) The NN model for calculating the NLPCA with a circular node at the bottleneck (NLPCA(cir)). Instead of having one bottleneck neuron u, there are now two neurons p and q constrained to lie on a unit circle in the p-q plane, so there is only one free angular variable θ, the NLPC. This network is suited for extracting a closed curve solution. (Reproduced from Hsieh (2001b), with permission from Blackwell.)

NNs placed one after the other. The first two layer network maps from the inputs $\mathbf{x}$ through a hidden layer to the *bottleneck layer* with only one neuron u, i.e. a nonlinear mapping $u = f(\mathbf{x})$. The next two layer MLP NN inversely maps from the nonlinear PC (NLPC) u back to the original higher dimensional $\mathbf{x}$-space, with the objective that the outputs $\mathbf{x}' = \mathbf{g}(u)$ be as close as possible to the inputs $\mathbf{x}$, where $\mathbf{g}(u)$ nonlinearly generates a curve in the $\mathbf{x}$-space, hence a 1-dimensional approximation of the original data. Because the target data for the output neurons $\mathbf{x}'$ are simply the input data $\mathbf{x}$, such networks are called *auto-associative* NNs. To minimize the MSE (mean squared error) of this approximation, the objective function $J = \langle \|\mathbf{x} - \mathbf{x}'\|^2 \rangle$ (where $\langle \ldots \rangle$ denotes taking the average over all the data points) is minimized to solve for the weight and offset parameters of the NN. Squeezing the input information through a bottleneck layer with only one neuron accomplishes the dimensional reduction.

In Fig. 10.2(a), the activation function f_1 maps from $\mathbf{x}$, the input vector of length l, to the first hidden layer (the *encoding layer*), represented by $\mathbf{h}^{(x)}$, a vector of length m, with elements

$$h_k^{(x)} = f_1((\mathbf{W}^{(x)}\mathbf{x} + \mathbf{b}^{(x)})_k), \tag{10.1}$$

where $\mathbf{W}^{(x)}$ is an $m \times l$ weight matrix, $\mathbf{b}^{(x)}$, a vector of length m containing the offset parameters, and $k = 1, \ldots, m$. Similarly, a second activation function f_2 maps from the encoding layer to the bottleneck layer containing a single neuron, which represents the nonlinear principal component u,

$$u = f_2(\mathbf{w}^{(x)} \cdot \mathbf{h}^{(x)} + \overline{b}^{(x)}). \tag{10.2}$$

The activation function f_1 is generally nonlinear (usually the hyperbolic tangent or the sigmoidal function, though the exact form is not critical), while f_2 is usually taken to be the identity function.

Next, an activation function f_3 maps from u to the final hidden layer (the *decoding layer*) $\mathbf{h}^{(u)}$,

$$h_k^{(u)} = f_3((\mathbf{w}^{(u)}u + \mathbf{b}^{(u)})_k), \tag{10.3}$$

$(k = 1, \ldots, m)$; followed by f_4 mapping from $\mathbf{h}^{(u)}$ to $\mathbf{x}'$, the output vector of length l, with

$$x_i' = f_4((\mathbf{W}^{(u)}\mathbf{h}^{(u)} + \overline{\mathbf{b}}^{(u)})_i). \tag{10.4}$$

The objective function $J = \langle \|\mathbf{x} - \mathbf{x}'\|^2 \rangle$ is minimized by finding the optimal values of $\mathbf{W}^{(x)}$, $\mathbf{b}^{(x)}$, $\mathbf{w}^{(x)}$, $\overline{b}^{(x)}$, $\mathbf{w}^{(u)}$, $\mathbf{b}^{(u)}$, $\mathbf{W}^{(u)}$ and $\overline{\mathbf{b}}^{(u)}$. The MSE between the NN output $\mathbf{x}'$ and the original data $\mathbf{x}$ is thus minimized. The NLPCA was implemented using the hyperbolic tangent function for f_1 and f_3, and the identity function for f_2 and f_4, so that

$$u = \mathbf{w}^{(x)} \cdot \mathbf{h}^{(x)} + \overline{b}^{(x)}, \tag{10.5}$$

$$x_i' = (\mathbf{W}^{(u)}\mathbf{h}^{(u)} + \overline{\mathbf{b}}^{(u)})_i. \tag{10.6}$$

Furthermore, we adopt the normalization conditions that $\langle u \rangle = 0$ and $\langle u^2 \rangle = 1$. These conditions are approximately satisfied by modifying the objective function to

$$J = \langle \|\mathbf{x} - \mathbf{x}'\|^2 \rangle + \langle u \rangle^2 + (\langle u^2 \rangle - 1)^2. \tag{10.7}$$

The total number of (weight and offset) parameters used by the NLPCA is $2lm + 4m + l + 1$, though the number of effectively free parameters is two less due to the constraints on $\langle u \rangle$ and $\langle u^2 \rangle$.

The choice of m, the number of hidden neurons in both the encoding and decoding layers, follows a general principle of parsimony. A larger m increases the nonlinear modelling capability of the network, but could also lead to overfitted solutions (i.e. wiggly solutions which fit to the noise in the data). If f_4 is the identity function, and $m = 1$, then (10.6) implies that all x'_i are linearly related to a single hidden neuron, hence there can only be a linear relation between the x'_i variables. Thus, for nonlinear solutions, we need to look at $m \geq 2$. Actually, one can use different numbers of neurons in the encoding layer and in the decoding layer; however, keeping them both at m neurons gives roughly the same number of parameters in the forward mapping from $\mathbf{x}$ to u and in the inverse mapping from u to $\mathbf{x}'$. It is also possible to have more than one neuron at the bottleneck layer. For instance, with two bottleneck neurons, the mode extracted will span a 2-D surface instead of a 1-D curve.

Because of local minima in the objective function, there is no guarantee that the optimization algorithm reaches the global minimum. Hence a number of runs with random initial weights and offset parameters are needed. Also, a portion (e.g. 15%) of the data is randomly selected as validation data and withheld from training of the NNs. Runs where the MSE is larger for the validation dataset than for the training dataset are rejected to avoid overfitted solutions. Then the run with the smallest MSE is selected as the solution.

In general, the presence of local minima in the objective function is a major problem for NLPCA. Optimizations started from different initial parameters often converge to different minima, rendering the solution unstable or non-unique. Regularization of the objective function by adding weight penalty terms is an answer.

The purpose of the weight penalty terms is to limit the nonlinear power of the NLPCA, which came from the nonlinear activation functions in the network. The activation function tanh has the property that given x in the interval $[-L, L]$, one can find a small enough weight w, so that $\tanh(wx) \approx wx$, i.e. the activation function is almost linear. Similarly, one can choose a large enough w, so that tanh approaches a step function, thus yielding Z-shaped solutions. If we can penalize the use of excessive weights, we can limit the degree of nonlinearity in the NLPCA solution. This is achieved with a modified objective function

$$J = \langle \|\mathbf{x} - \mathbf{x}'\|^2 \rangle + \langle u \rangle^2 + (\langle u^2 \rangle - 1)^2 + P \sum_{ki} (W_{ki}^{(x)})^2, \tag{10.8}$$

where P is the weight penalty parameter. A large P increases the concavity of the objective function, and forces the weights in $\mathbf{W}^{(x)}$ to be small in magnitude, thereby yielding smoother and less nonlinear solutions than when P is small or zero. Hence, increasing P also reduces the number of effectively free parameters of the model. We have not penalized other weights in the network. In principle, $\mathbf{w}^{(u)}$ also controls the nonlinearity in the inverse mapping from u to $\mathbf{x}'$. However, if the nonlinearity in the forward mapping from $\mathbf{x}$ to u is already being limited by penalizing $\mathbf{W}^{(x)}$, then there is no need to limit further the weights in the inverse mapping.

The percentage of the variance explained by the NLPCA mode is given by

$$100\% \times \left(1 - \frac{\langle \|\mathbf{x} - \mathbf{x}'\|^2 \rangle}{\langle \|\mathbf{x} - \overline{\mathbf{x}}\|^2 \rangle}\right), \tag{10.9}$$

with $\overline{\mathbf{x}}$ being the mean of $\mathbf{x}$.

In effect, the linear relation ($u = \mathbf{e} \cdot \mathbf{x}$) in PCA is now generalized to $u = f(\mathbf{x})$, where f can be any nonlinear continuous function representable by an MLP NN mapping from the input layer to the bottleneck layer; and $\langle \|\mathbf{x} - \mathbf{g}(u)\|^2 \rangle$ is minimized. Limitations in the mapping properties of the NLPCA are discussed by Newbigging *et al.* (2003). The residual, $\mathbf{x} - \mathbf{g}(u)$, can be input into the same network to extract the second NLPCA mode, and so on for the higher modes.

That the classical PCA is indeed a linear version of this NLPCA can be readily seen by replacing all the activation functions with the identity function, thereby removing the nonlinear modelling capability of the NLPCA. Then the forward map to u involves only a linear combination of the original variables as in the PCA.

In the classical linear approach, there is a well-known dichotomy between PCA and rotated PCA (RPCA) (Section 2.2). In PCA, the linear mode which accounts for the most variance of the dataset is sought. However, as illustrated in Preisendorfer (1988, Fig. 7.3), the resulting eigenvectors may not align close to local data clusters, so the eigenvectors may not represent actual physical states well. One application of RPCA methods is to rotate the PCA eigenvectors, so they point closer to the local clusters of data points (Preisendorfer, 1988). Thus the rotated eigenvectors may bear greater resemblance to actual physical states (though they account for less variance) than the unrotated eigenvectors, hence RPCA is also widely used (Richman, 1986; von Storch and Zwiers, 1999). As there are many possible criteria for rotation, there are many RPCA schemes, among which the varimax (Kaiser, 1958) scheme is perhaps the most popular.

10.1.2 Application

The NLPCA has been applied to the Lorenz (1963) three component chaotic system (Monahan, 2000; Hsieh, 2001b). For the tropical Pacific climate variability, the NLPCA has been used to study the sea surface temperature (SST) field (Monahan, 2001; Hsieh, 2001b) and the sea level pressure (SLP) field (Monahan, 2001). In remote sensing, Del Frate and Schiavon (1999) applied NLPCA to the inversion of radiometric data to retrieve atmospheric profiles of temperature and water vapour. More examples of NLPCA applications are given in Chapter 12.

The tropical Pacific climate system contains the famous interannual variability known as the El Niño–Southern Oscillation (ENSO), a coupled atmosphere–ocean interaction involving the oceanic phenomenon El Niño and the associated atmospheric phenomenon, the Southern Oscillation. The coupled interaction results in anomalously warm SST in the eastern equatorial Pacific during El Niño episodes, and cool SST in the central equatorial Pacific during La Niña episodes (Philander, 1990; Diaz and Markgraf, 2000). The ENSO is an irregular oscillation, but spectral analysis does reveal a broad spectral peak at the 4-5 year period. Hsieh (2001b) used the tropical Pacific SST data (1950-1999) to make a three-way comparison between NLPCA, RPCA and PCA. The tropical Pacific SST anomaly (SSTA) data (i.e. the SST data with the climatological seasonal cycle removed) were pre-filtered by PCA, with only the three leading modes retained. The PCA modes 1, 2 and 3 accounted for 51.4%, 10.1% and 7.2%, respectively, of the variance in the SSTA data. Due to the large number of spatially gridded variables, NLPCA could not be applied directly to the SSTA time series, as this would lead to a huge NN with the number of model parameters vastly exceeding the number of observations. Instead, the first three PCs (PC1, PC2 and PC3) were used as the input $\mathbf{x}$ for the NLPCA network.

The data are shown as dots in a scatter plot in the PC1-PC2 plane in Fig. 10.3, where the cool La Niña states lie in the upper left corner, and the warm El Niño states in the upper right corner. The NLPCA solution is a U-shaped curve linking the La Niña states at one end (low u) to the El Niño states at the other end (high u), similar to that found originally by Monahan (2001). In contrast, the first PCA eigenvector lies along the horizontal line, and the second PCA, along the vertical line (Fig. 10.3). It is easy to see that the first PCA eigenvector describes a somewhat unphysical oscillation, as there are no dots (data) close to either end of the horizontal line. For the second PCA eigenvector, there are dots close to the bottom of the vertical line, but no dots near the top end of the line, i.e. one phase of the mode 2 oscillation is realistic, but the opposite phase is not. Thus if the underlying data have a nonlinear structure but we are restricted to finding linear solutions using PCA, the energy of the nonlinear oscillation is scattered into multiple PCA modes, many of which represent unphysical linear oscillations.

For comparison, a varimax rotation (Kaiser, 1958; Preisendorfer, 1988), was applied to the first three PCA eigenvectors, as mentioned in Section 2.2. The varimax criterion can be applied to either the loadings or the PCs depending on one's objectives; here it is applied to the PCs. The resulting first RPCA eigenvector, shown as a dashed line in Fig. 10.3, spears through the cluster of El Niño states in the upper right corner, thereby yielding a more accurate description of the El

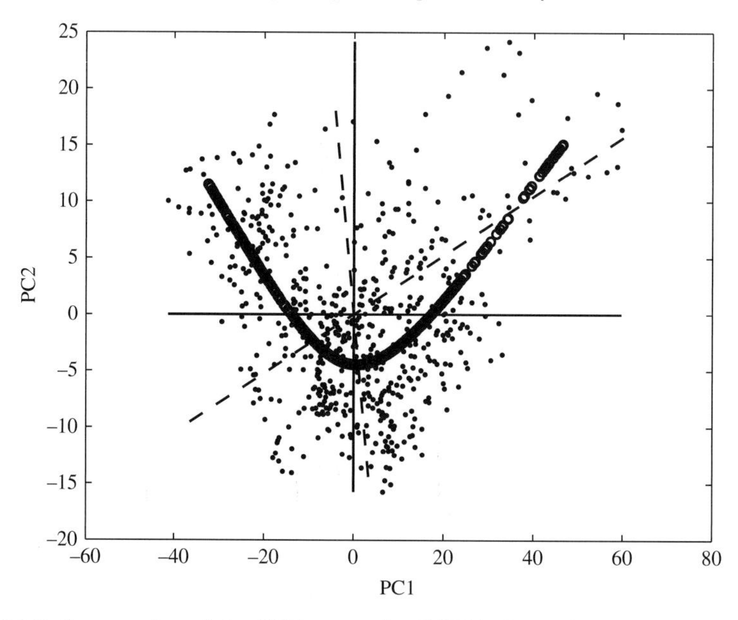

Fig. 10.3 Scatter plot of the SST anomaly (SSTA) data (shown as dots) in the PC1-PC2 plane, with the El Niño states lying in the upper right corner, and the La Niña states in the upper left corner. The PC2 axis is stretched relative to the PC1 axis for better visualization. The first mode NLPCA approximation to the data is shown by the (overlapping) small circles, which trace out a U-shaped curve. The first PCA eigenvector is oriented along the horizontal line, and the second PCA along the vertical line. The varimax method rotates the two PCA eigenvectors in a counterclockwise direction, as the rotated PCA (RPCA) eigenvectors are oriented along the dashed lines. (As the varimax method generates an orthogonal rotation, the angle between the two RPCA eigenvectors is 90° in the 3-dimensional PC1-PC2-PC3 space). (Reproduced from Hsieh (2001b), with permission from Blackwell.)

Niño anomalies (Fig. 10.4(c)) than the first PCA mode (Fig. 10.4(a)), which did not fully represent the intense warming of Peruvian waters – for better legibility, compare Fig. 2.13(a) with Fig. 2.5(a) instead. The second RPCA eigenvector, also shown as a dashed line in Fig. 10.3, did not improve much on the second PCA mode, with the PCA spatial pattern shown in Fig. 10.4(b), and the RPCA pattern in Fig. 10.4(d). In terms of variance explained, the first NLPCA mode explained 56.6% of the variance, versus 51.4% by the first PCA mode, and 47.2% by the first RPCA mode.

With the NLPCA, for a given value of the NLPC u, one can map from u to the three PCs. This is done by assigning the value u to the bottleneck neuron and mapping forward using the second half of the network in Fig. 10.2(a). Each of the three PCs can be multiplied by its associated PCA (spatial) eigenvector, and the three added together to yield the spatial pattern for that particular value of u. Unlike

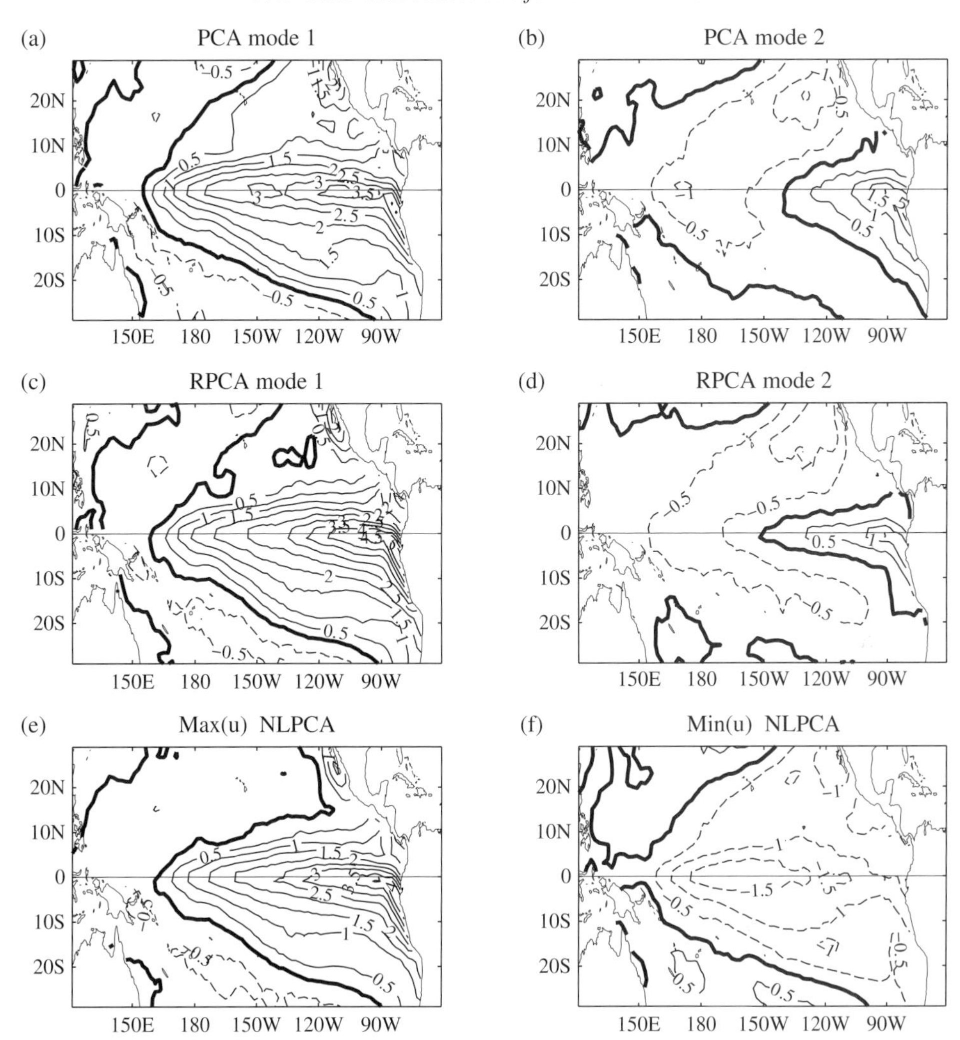

Fig. 10.4 The SSTA patterns (in °C) of the PCA, RPCA and the NLPCA. The first and second PCA spatial modes are shown in (a) and (b) respectively, (both with their corresponding PCs at maximum value). The first and second varimax RPCA spatial modes are shown in (c) and (d) respectively, (both with their corresponding RPCs at maximum value). The anomaly pattern as the NLPC u of the first NLPCA mode varies from (e) maximum (strong El Niño) to (f) its minimum (strong La Niña). With a contour interval of 0.5 °C , the positive contours are shown as solid curves, negative contours as dashed curves, and the zero contour as a thick curve. (Reproduced from Hsieh (2004) with permission from the American Geophysical Union.)

PCA which gives the same spatial anomaly pattern except for changes in the amplitude as the PC varies, the NLPCA spatial pattern generally varies continuously as the NLPC changes. Figure 10.4(e) and (f) show respectively the spatial anomaly patterns when u has its maximum value (corresponding to the strongest El Niño) and when u has its minimum value (strongest La Niña). Clearly the asymmetry between El Niño and La Niña, i.e. the cool anomalies during La Niña episodes (Fig. 10.4(f)) are observed to centre much further west of the warm anomalies during El Niño (Fig. 10.4(e)) (Hoerling *et al.*, 1997), is well captured by the first NLPCA mode – in contrast, the PCA mode 1 gives a La Niña which is simply the mirror image of the El Niño (Fig. 10.4(a)). While El Niño has been known by Peruvian fishermen for many centuries due to its strong SSTA off the coast of Peru and its devastation of the Peruvian fishery, the La Niña, with its weak manifestation in the Peruvian waters, was not appreciated until the last two decades of the 20th century.

In summary, PCA is used for two main purposes: (i) to reduce the dimensionality of the dataset, and (ii) to extract features or recognize patterns from the dataset. It is purpose (ii) where PCA can be improved upon. Both RPCA and NLPCA take the PCs from PCA as input. However, instead of multiplying the PCs by a fixed orthonormal rotational matrix, as performed in the varimax RPCA approach, NLPCA performs a nonlinear mapping of the PCs. RPCA sacrifices on the amount of variance explained, but by rotating the PCA eigenvectors, RPCA eigenvectors tend to point more towards local data clusters and are therefore more representative of physical states than the PCA eigenvectors.

With a linear approach, it is generally impossible to have a solution simultaneously (a) explaining maximum global variance of the dataset and (b) approaching local data clusters, hence the dichotomy between PCA and RPCA, with PCA aiming for (a) and RPCA for (b). With the more flexible NLPCA method, both objectives (a) and (b) may be attained together, thus the nonlinearity in NLPCA unifies the PCA and RPCA approaches (Hsieh, 2001b). It is easy to see why the dichotomy between PCA and RPCA in the linear approach automatically vanishes in the nonlinear approach. By increasing m, the number of hidden neurons in the encoding layer (and the decoding layer), the solution is capable of going through all local data clusters while maximizing the global variance explained. (In fact, for large enough m, NLPCA can pass through all data points, though this will in general give an undesirable, overfitted solution.)

The tropical Pacific SST example illustrates that with a complicated oscillation like the El Niño-La Niña phenomenon, using a linear method such as PCA results in the nonlinear mode being scattered into several linear modes (in fact, all three leading PCA modes are related to this phenomenon). In the study of climate variability, the wide use of PCA methods has created perhaps a slightly misleading view that our climate is dominated by a number of spatially fixed oscillatory patterns, which is in fact due to the limitations of the linear method. Applying NLPCA to the tropical Pacific SSTA, we found no spatially fixed oscillatory patterns, but an oscillation evolving in space as well as in time.

10.1.3 Overfitting

When using nonlinear machine learning methods, the presence of noise in the data can lead to overfitting. When plentiful data are available (i.e. far more observations than model parameters), overfitting is not a problem when performing nonlinear regression on noisy data (Section 6.2). Unfortunately, even with plentiful data, overfitting is a problem when applying NLPCA to noisy data (Hsieh, 2001b; Christiansen, 2005; Hsieh, 2007). As illustrated in Fig. 10.5, overfitting in NLPCA can arise from the geometry of the problem, rather than from the scarcity of data. Here for a Gaussian-distributed data cloud, a nonlinear model with enough flexibility will find the zigzag solution of Fig. 10.5(b) as having a smaller MSE than the linear solution in Fig. 10.5(a). Since the distance between the point A and a, its

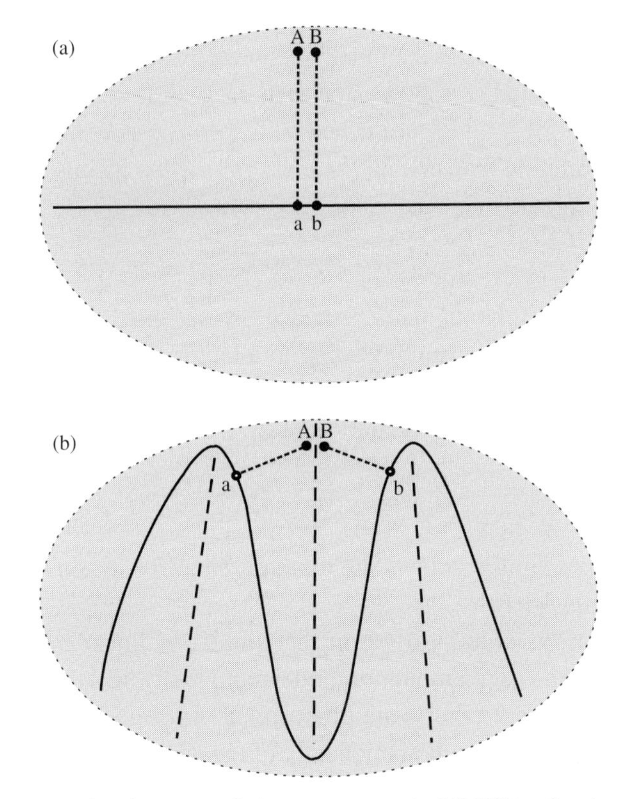

Fig. 10.5 Illustrating how overfitting can occur in NLPCA of noisy data (even in the limit of infinite sample size). (a) PCA solution for a Gaussian data cloud (shaded in grey), with two neighbouring points A and B shown projecting to the points a and b on the PCA straight line solution. (b) A zigzag NLPCA solution found by a flexible enough nonlinear model. Dashed lines illustrate 'ambiguity' lines where neighbouring points (e.g. A and B) on opposite sides of these lines are projected to a and b, far apart on the NLPCA curve. (Reproduced from Hsieh (2007) with permission of Elsevier.)

projection on the NLPCA curve, is smaller in Fig. 10.5(b) than the corresponding distance in Fig. 10.5(a), it is easy to see that the more zigzags there are in the curve, the smaller is the MSE. However, the two neighbouring points A and B, on opposite sides of an ambiguity line, are projected far apart on the NLPCA curve in Fig. 10.5(b). Thus simply searching for the solution which gives the smallest MSE does not guarantee that NLPCA will find a good solution in a highly noisy dataset.

Hsieh (2001b) added weight penalty to the Kramer (1991) NLPCA model to smooth out excessively wiggly solutions, but did not provide an objective way to select the optimal weight penalty parameter P. With NLPCA, if the overfitting arises from the data geometry (as in Fig. 10.5(b)) and not from data scarcity, using independent data to validate the MSE from the various models is not a viable method for choosing the appropriate P. Instead, Hsieh (2007) proposed an *'inconsistency' index* for detecting the projection of neighbouring points to distant parts of the NLPCA curve, and used the index to choose the appropriate P.

The index is calculated as follows: for each data point $\mathbf{x}$, find its nearest neighbour $\tilde{\mathbf{x}}$. The NLPC for $\mathbf{x}$ and $\tilde{\mathbf{x}}$ are u and $\tilde{u}$, respectively. With $C(u, \tilde{u})$ denoting the (Pearson) correlation between all the pairs $(u, \tilde{u})$, the inconsistency index I is defined as

$$I = 1 - C(u, \tilde{u}). \tag{10.10}$$

If for some nearest neighbour pairs, u and $\tilde{u}$ are assigned very different values, $C(u, \tilde{u})$ would have a lower value, leading to a larger I, indicating greater inconsistency in the NLPC mapping. With u and $\tilde{u}$ standardized to having zero mean and unit standard deviation, (10.10) is equivalent to

$$I = \frac{1}{2}\langle (u - \tilde{u})^2 \rangle. \tag{10.11}$$

The I index plays a similar role as the topographic error in self-organizing maps (SOM) (see Section 10.3).

In statistics, various criteria, often in the context of linear models, have been developed to select the right amount of model complexity so neither overfitting nor underfitting occurs. These criteria are often called *'information criteria'* (IC) (von Storch and Zwiers, 1999). An IC is typically of the form

$$\text{IC} = \text{MSE} + \text{complexity term}, \tag{10.12}$$

where MSE is evaluated over the training data and the complexity term is larger when a model has more free parameters. The IC is evaluated over a number of models with different numbers of free parameters, and the model with the minimum IC is selected as the best. As the presence of the complexity term in the IC penalizes models which use an excessive number of free parameters to attain low

MSE, choosing the model with the minimum IC would rule out complex models with overfitted solutions.

In Hsieh (2007), the data were randomly divided into a training data set and a validation set (containing 85% and 15% of the original data, respectively), and for every given value of P and m (the number of neurons in the first hidden layer in Fig. 10.2(a)), the model was trained a number of times from random initial weights, and model runs where the MSE evaluated over the validation data was larger than the MSE over the training data were discarded. To choose among the model runs which had passed the validation test, a new holistic IC to deal with the type of overfitting arising from the broad data geometry (Fig. 10.5(b)) was introduced as

$$H = \mathrm{MSE} + \text{inconsistency term}, \tag{10.13}$$

$$= \mathrm{MSE} - C(u, \tilde{u}) \times \mathrm{MSE} \;=\; \mathrm{MSE} \times I, \tag{10.14}$$

where MSE and C were evaluated over all (training and validation) data, inconsistency was penalized, and the model run with the smallest H value was selected as the best. As the inconsistency term only prevents overfitting arising from the broad data geometry, validation data are still needed to prevent 'local' overfitting from excessive number of model parameters, since H, unlike (10.12), does not contain a complexity term.

Consider the test problem in Hsieh (2007): For a random number t uniformly distributed in the interval $(-1, 1)$, the signal $\mathbf{x}^{(s)}$ was generated by using a quadratic relation

$$x_1^{(s)} = t, \qquad x_2^{(s)} = \frac{1}{2}t^2. \tag{10.15}$$

Isotropic Gaussian noise was added to the signal $\mathbf{x}^{(s)}$ to generate a noisy dataset of $\mathbf{x}$ containing 500 'observations'. NLPCA was performed on the data using the network in Fig. 10.2(a) with $m = 4$ and with the weight penalty P at various values $(10, 1, 10^{-1}, 10^{-2}, 10^{-3}, 10^{-4}, 10^{-5}, 0)$. For each value of P, the model training was done 30 times starting from random initial weights, and model runs where the MSE evaluated over the validation data was larger than the MSE over the training data were deemed ineligible. In the traditional approach, among the eligible runs over the range of P values, the one with the lowest MSE over all (training and validation) data was selected as the best. Figure 10.6(a) shows this solution where the zigzag curve retrieved by NLPCA is very different from the theoretical parabolic signal (10.15), demonstrating the pitfall of selecting the lowest MSE run.

In contrast, in Fig. 10.6(b), among the eligible runs over the range of P values, the one with the lowest information criterion H was selected. This solution, which has a much larger weight penalty ($P = 0.1$) than that in Fig. 10.6(a) ($P = 10^{-4}$),

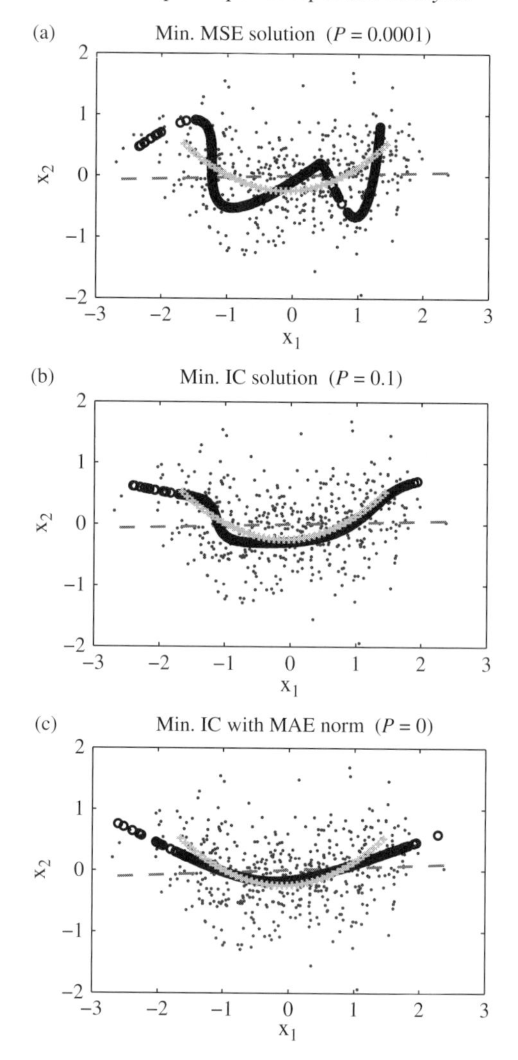

Fig. 10.6 The NLPCA solution (shown as densely overlapping black circles) for the synthetic dataset (dots), with the parabolic signal curve indicated by '+' and the linear PCA solution by the dashed line. The solution was selected from the multiple runs over a range of P values based on (a) minimum MSE, ($P = 0.0001$) (b) minimum information criterion (IC) $H(P = 0.1)$, and (c) minimum IC together with the MAE norm ($P = 0$). (Adapted from Hsieh (2007) with permission of Elsevier.)

shows less wiggly behaviour and better agreement with the theoretical parabolic signal.

Even less wiggly solutions can be obtained by changing the error norm used in the objective function from the mean squared error to the mean absolute error (MAE), i.e. replacing $\langle \|\mathbf{x} - \mathbf{x}'\|^2 \rangle$ by $\langle \sum_j |x_j - x'_j| \rangle$ in (10.8). The MAE norm is known to be robust to outliers in the data (Section 6.2). Figure 10.6(c) is the solution selected based on minimum H with the MAE norm used. While wiggles are eliminated, the solution underestimates the curvature in the parabolic signal. The rest of this section uses the MSE norm.

In summary, with noisy data, not having plentiful observations could cause a flexible nonlinear model to overfit. In the limit of infinite number of observations, overfitting cannot occur in nonlinear regression, but can still occur in NLPCA due to the geometric shape of the data distribution. The inconsistency index I for detecting the projection of neighbouring points to distant parts of the NLPCA curve has been introduced, and incorporated into a holistic IC H to select the model with the appropriate weight penalty parameter and the appropriate number of hidden neurons (Hsieh, 2007). An alternative approach for model selection was proposed by Webb (1999), who applied a constraint on the Jacobian in the objective function.

10.1.4 Closed curves

While the NLPCA is capable of finding a continuous open curve solution, there are many phenomena involving waves or quasi-periodic fluctuations, which call for a continuous closed curve solution. Kirby and Miranda (1996) introduced an NLPCA with a circular node at the network bottleneck (henceforth referred to as the NLPCA(cir)), so that the nonlinear principal component (NLPC) as represented by the circular node is an angular variable θ, and the NLPCA(cir) is capable of approximating the data by a closed continuous curve. Figure 10.2(b) shows the NLPCA(cir) network, which is almost identical to the NLPCA of Fig. 10.2(a), except at the bottleneck, where there are now two neurons p and q constrained to lie on a unit circle in the p-q plane, so there is only one free angular variable θ, the NLPC.

At the bottleneck in Fig. 10.2(b), analogous to u in (10.5), we calculate the pre-states p_o and q_o by

$$p_o = \mathbf{w}^{(x)} \cdot \mathbf{h}^{(x)} + \overline{b}^{(x)}, \quad \text{and} \quad q_o = \tilde{\mathbf{w}}^{(x)} \cdot \mathbf{h}^{(x)} + \tilde{b}^{(x)}, \tag{10.16}$$

where $\mathbf{w}^{(x)}$, $\tilde{\mathbf{w}}^{(x)}$ are weight parameter vectors, and $\overline{b}^{(x)}$ and $\tilde{b}^{(x)}$ are offset parameters. Let

$$r = (p_o^2 + q_o^2)^{1/2}, \tag{10.17}$$

then the circular node is defined with

$$p = p_o/r, \quad \text{and} \quad q = q_o/r, \tag{10.18}$$

satisfying the unit circle equation $p^2 + q^2 = 1$. Thus, even though there are two variables p and q at the bottleneck, there is only one angular degree of freedom from θ (Fig. 10.2(b)), due to the circle constraint. The mapping from the bottleneck to the output proceeds as before, with (10.3) replaced by

$$h_k^{(u)} = \tanh((\mathbf{w}^{(u)} p + \tilde{\mathbf{w}}^{(u)} q + \mathbf{b}^{(u)})_k). \tag{10.19}$$

When implementing NLPCA(cir), Hsieh (2001b) found that there are actually two possible configurations: (i) a restricted configuration where the constraints $\langle p \rangle = 0 = \langle q \rangle$ are applied; and (ii) a general configuration without the constraints. With (i), the constraints can be satisfied approximately by adding the extra terms $\langle p \rangle^2$ and $\langle q \rangle^2$ to the objective function. If a closed curve solution is sought, then (i) is better than (ii) as it has effectively two fewer parameters. However, (ii), being more general than (i), can more readily model open curve solutions like a regular NLPCA. The reason is that if the input data mapped onto the p-q plane cover only a segment of the unit circle instead of the whole circle, then the inverse mapping from the p-q space to the output space will yield a solution resembling an open curve. Hence, given a dataset, (ii) may yield either a closed curve or an open curve solution. It uses $2lm + 6m + l + 2$ parameters.

Hsieh (2007) found that the information criterion (IC) H (Section 10.1.3) not only alleviates overfitting in open curve solution, but also chooses between open and closed curve solutions when using NLPCA(cir) in configuration (ii). The inconsistency index I and the IC are now obtained from

$$I = 1 - \frac{1}{2}\left[C(p, \tilde{p}) + C(q, \tilde{q})\right], \quad \text{and} \quad H = \text{MSE} \times I, \tag{10.20}$$

where p and q are from the bottleneck (Fig. 10.2(b)), and $\tilde{p}$ and $\tilde{q}$ are the corresponding nearest neighbour values.

For a test problem, consider a Gaussian data cloud (with 500 observations) in 2-dimensional space, where the standard deviation along the x_1 axis was double that along the x_2 axis. The dataset was analyzed by the NLPCA(cir) model with $m = 2, \ldots, 5$ and $P = 10, 1, 10^{-1}, 10^{-2}, 10^{-3}, 10^{-4}, 10^{-5}, 0$. From all the runs, the solution selected based on the minimum MSE has $m = 5$ (and $P = 10^{-5}$) (Fig. 10.7(a)), while that selected based on minimum H has $m = 3$ (and $P = 10^{-5}$) (Fig. 10.7(b)). The minimum MSE solution has (normalized) MSE = 0.370, $I = 9.50$ and $H = 3.52$, whereas the minimum H solution has the corresponding values of 0.994, 0.839 and 0.833, respectively, where for easy comparison with the linear mode, these values for the nonlinear solutions have been normalized upon division by the corresponding values from the linear PCA mode 1. Thus the IC correctly selected a nonlinear solution (Fig. 10.7(b)) which is similar to the linear solution. It also rejected the closed curve solution of Fig. 10.7(a) in favour of the open curve solution of Fig. 10.7(b), despite its much larger MSE.

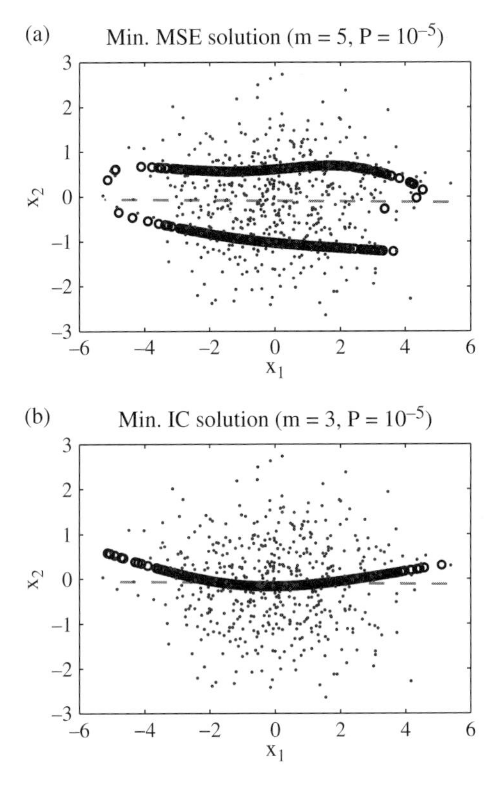

Fig. 10.7 The NLPCA(cir) mode 1 for a Gaussian dataset, with the solution selected based on (a) minimum MSE ($m = 5$, $P = 10^{-5}$), and (b) minimum IC ($m = 3$, $P = 10^{-5}$). The PCA mode 1 solution is shown as a dashed line. (Reproduced from Hsieh (2007) with permission from Elsevier.)

For an application of NLPCA(cir) on real data, consider the Quasi-Biennial Oscillation (QBO), which dominates over the annual cycle or other variations in the equatorial stratosphere, with the period of oscillation varying roughly between 22 and 32 months. Average zonal (i.e. the westerly component) winds at 70, 50, 40, 30, 20, 15 and 10 hPa (i.e. from about 20 to 30 km altitude) during 1956–2006 were studied. After the 51 year means were removed, the zonal wind anomalies U at seven vertical levels in the stratosphere became the seven inputs to the NLPCA(cir) network (Hamilton and Hsieh, 2002; Hsieh, 2007). Since the data were not very noisy (Fig. 10.8), a rather complex model was used, with m ranging from 5 to 9, and $P = 10^{-1}, 10^{-2}, 10^{-3}, 10^{-4}, 10^{-5}, 0$. The smallest H occurred when $m = 8$ and $P = 10^{-5}$, with the closed curve solution shown in Fig. 10.8. Thus in this example, by choosing a rather large m and a small P, the H IC justified having considerable model complexity, including the wiggly behaviour seen in the 70 hPa wind (Fig. 10.8(c)). The wiggly behaviour can be understood by viewing the

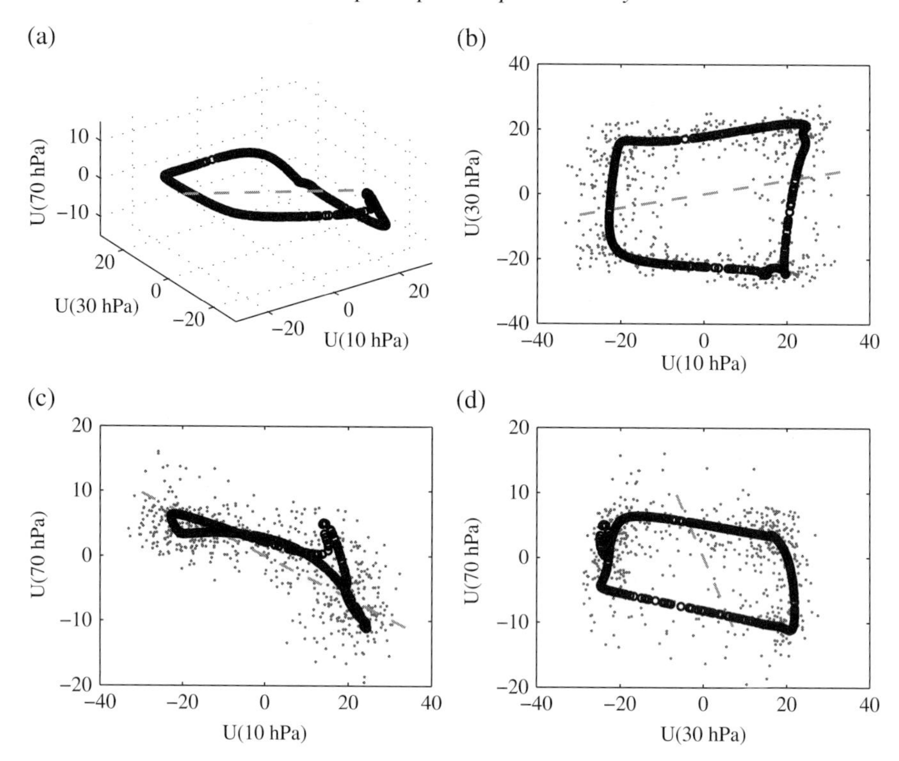

Fig. 10.8 The NLPCA(cir) mode 1 solution for the equatorial stratospheric zonal wind anomalies. For comparison, the PCA mode 1 solution is shown by the dashed line. Only three out of seven dimensions are shown, namely the zonal velocity anomaly U at the top, middle and bottom levels (10, 30 and 70 hPa). (a) A 3-D view. (b)–(d) 2-D views. (Reproduced from Hsieh (2007), with permission from Elsevier.)

phase–pressure contour plot of the zonal wind anomalies (Fig. 10.9): As the easterly wind anomaly descends with time (i.e. as phase increases), wavy behaviour is seen in the 40, 50 and 70 hPa levels at θ_{weighted} around 0.4–0.5. This example demonstrates the benefit of having an IC to decide objectively on how smooth or wiggly the fitted curve should be.

The observed strong asymmetries between the easterly and westerly phases of the QBO (Hamilton, 1998; Baldwin *et al.*, 2001) are captured by this NLPCA(cir) mode – e.g. the much more rapid transition from easterlies to westerlies than the reverse transition, and the much deeper descent of the easterlies than the westerlies (Fig. 10.9). For comparison, Hamilton and Hsieh (2002) constructed a *linear* model of θ, which was unable to capture the observed strong asymmetry between easterlies and westerlies. See Section 12.3.2 for more discussion of the QBO.

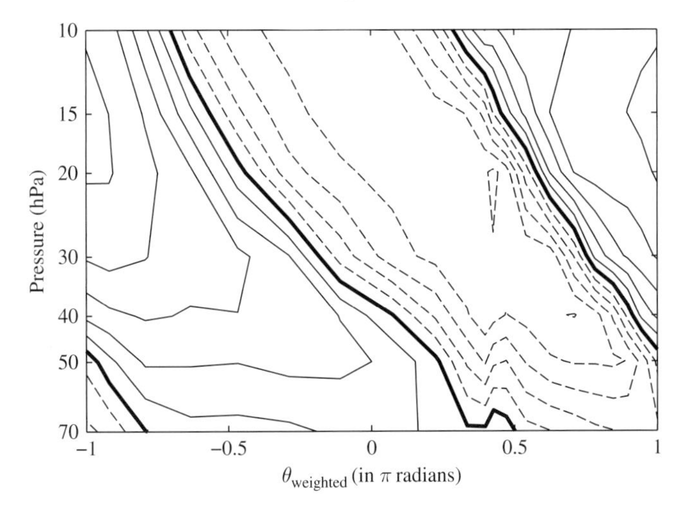

Fig. 10.9 Contour plot of the NLPCA(cir) mode 1 zonal wind anomalies as a function of pressure and phase θ_{weighted}, where θ_{weighted} is θ weighted by the histogram distribution of θ (see Hamilton and Hsieh, 2002). Thus θ_{weighted} is more representative of actual time during a cycle than θ. Contour interval is 5 ms^{-1}, with westerly winds indicated by solid lines, easterlies by dashed lines, and zero contours by thick lines. (Reproduced from Hsieh (2007), with permission from Elsevier.)

The NLPCA(cir) approach has also been used to study the tropical Pacific climate variability (Hsieh, 2001b; An *et al.*, 2005, 2006), the non-sinusoidal propagation of underwater sandbars off beaches in the Netherlands and Japan (Ruessink *et al.*, 2004), and the tidal cycle off the German North Sea coast (Herman, 2007). A nonlinear singular spectrum analysis method has also been developed based on the NLPCA(cir) model (Hsieh and Wu, 2002) (see Section 10.6).

10.2 Principal curves

Having originated from the statistics community, the principal curve method (Hastie and Stuetzle, 1989; Hastie *et al.*, 2001) offers another approach to nonlinearly generalize the PCA method. Figure 10.10 illustrates how to proceed from the PCA first mode to the principal curve. The mean of the data points within a bin (as confined between the two dashed lines) is computed (as indicated by the cross in Fig. 10.10). By moving the bin along the PCA line, the means for the various bins can be connected to form a curve C_1, a first approximation for the principal curve. For the next iteration, instead of projecting data to the PCA straight line, the data points are projected normally to the curve C_1. Those falling within a bin are used to obtain a mean value, and again by moving the bin along C_1, a new curve C_2

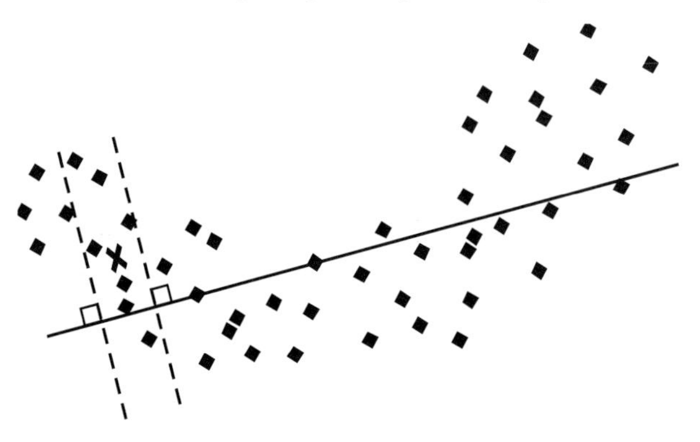

Fig. 10.10 Illustrating how to proceed towards finding the principal curve starting from the linear PCA solution of a dataset. First, the data points are projected normally to the PCA line, then all points falling within a bin (bounded by the two dashed lines) are used to compute a mean (indicated by the cross). The bin is shifted along the PCA line, and the means of the bins are connected to give a curve C_1, a first approximation to the principal curve solution. For the next step, data points are projected normally to the curve C_1, again the means within the bins are calculated to give the curve C_2. This procedure is iterated until convergence to the principal curve solution.

is given by the mean values. The process is repeated until the curves C_j converge to the principal curve. Thus, the procedure involves two basic steps: (i) given a curve C_j, data are projected normally to the curve; (ii) for all the data points projected to within a bin (i.e. interval) along the curve, the mean is calculated, which gives a new estimate for the curve C_{j+1}. After sufficient iterations of the two basic steps, a curve is found which passes through the middle of the data. It is not essential to start with the PCA straight line solution, though this is widely used as the first guess.

Next, we proceed to define the principal curve for random variables $\hat{\mathbf{x}}$ in $\mathbb{R}^l$. Let $\mathbf{g}(u)$ be a parameterized smooth curve in $\mathbb{R}^l$. It is convenient (but not essential) to choose the parameter u to be the arc-length along the curve from some fixed origin. Finding the normal projection of a data point $\mathbf{x}$ to the curve is equivalent to finding the value of u which minimizes the distance between the curve $\mathbf{g}(\tilde{u})$ and the data point $\mathbf{x}$, i.e.

$$u = \arg\min_{\tilde{u}} \|\mathbf{g}(\tilde{u}) - \mathbf{x}\|^2 = f(\mathbf{x}), \tag{10.21}$$

where we consider $u = f(\mathbf{x})$ to be the forward or encoding mapping, and u is referred to as the *projection index*. Note that if there are two or more points on the curve with the shortest distance to the point $\mathbf{x}$, then we choose u to be the largest value among the possible candidates. The curve $\mathbf{g}(u)$ is called a *principal curve* if

$$\mathbf{g}(u) = \mathrm{E}(\hat{\mathbf{x}}|f(\hat{\mathbf{x}}) = u). \tag{10.22}$$

This means the principal curve $\mathbf{g}(u)$ is the average of all the data points projected to u, a property known as *self-consistency*. With a finite dataset, the projection is to a bin or interval containing u. The 1-D principal curve can be generalized to a *principal surface* of two or higher dimensions (Hastie *et al.*, 2001).

A major disadvantage of the principal curve method is that when a new data point $\mathbf{x}_{\text{new}}$ arrives, unlike the case of NLPCA by auto-associative NN, there are no mapping functions to provide the NLPC u_{new} and the projection $\mathbf{x}'_{\text{new}}$ on the principal curve.

10.3 Self-organizing maps (SOM)

The goal of *clustering* or cluster analysis is to group the data into a number of subsets or 'clusters', such that the data within a cluster are more closely related to each other than data from other clusters. By projecting all data belonging to a cluster to the cluster centre, data compression can be achieved. In Section 1.7, the simple and widely used *K-means clustering* method was introduced, whereas in Section 8.6, we have seen how an NN model under unsupervised learning can perform clustering.

The *self-organizing map* (SOM) method, introduced by Kohonen (1982, 2001), approximates a dataset in multi-dimensional space by a flexible grid (typically of one or two dimensions) of cluster centres. Widely used for clustering, SOM can also be regarded as a discrete version of a principal curve/surface or an NLPCA (Cherkassky and Mulier, 1998).

As with many neural network models, self-organizing maps have a biological background. In neurobiology, it is known that many structures in the cortex of the brain are 2-D or 1-D. In contrast, even the perception of colour involves three types of light receptor. Besides colour, human vision also processes information about the shape, size, texture, position and movement of an object. So the question naturally arises as to how 2-D networks of neurons can process higher dimensional signals in the brain.

Among various possible grids, rectangular and hexagonal grids are most commonly used by SOM. For a 2-dimensional rectangular grid, the grid points or units $\mathbf{i}_j = (l, m)$, where l and m take on integer values, i.e. $l = 1, \ldots, L$, $m = 1, \ldots, M$, and $j = 1, \ldots, LM$. (If a 1-dimensional grid is desired, simply set $M = 1$.)

To initialize the training process, PCA is usually performed on the dataset, and the grid $\mathbf{i}_j$ is mapped to $\mathbf{z}_j(0)$ lying on the plane spanned by the two leading PCA eigenvectors. As training proceeds, the initial flat 2D surface of $\mathbf{z}_j(0)$ is bent to fit the data. The original SOM was trained in a flow-through manner (i.e. observations are presented one at a time during training), though algorithms for batch training are now also available. In flow-through training, there are two steps to be iterated, starting with $n = 1$:

Step (i): At the nth iteration, select an observation $\mathbf{x}(n)$ from the data space, and find among the points $\mathbf{z}_j(n-1)$, the one with the shortest (Euclidean) distance to $\mathbf{x}(n)$. Call this closest neighbour $\mathbf{z}_k$, with the corresponding unit $\mathbf{i}_k$ called the *best matching unit* (BMU).

Step (ii): Let

$$\mathbf{z}_j(n) = \mathbf{z}_j(n-1) + \eta\, h(\|\mathbf{i}_j - \mathbf{i}_k\|^2)\,[\mathbf{x}(n) - \mathbf{z}_j(n-1)], \tag{10.23}$$

where η is the learning rate parameter and h is a neighbourhood or kernel function. The neighbourhood function gives more weight to the grid points $\mathbf{i}_j$ near $\mathbf{i}_k$, than those far away, an example being a Gaussian drop-off with distance. Note that the distances between neighbours are computed for the fixed grid points ($\mathbf{i}_j = (l, m)$), not for their corresponding positions $\mathbf{z}_j$ in the data space. Typically, as n increases, the learning rate η is decreased gradually from the initial value of 1 towards 0, while the width of the neighbourhood function is also gradually narrowed (Cherkassky and Mulier, 1998).

While SOM has been commonly used as a clustering tool, it should be pointed out that it may underform simpler techniques such as K-means clustering. Balakrishnan *et al.* (1994) found that K-means clustering had fewer points misclassified compared with SOM, and the classification accuracy of SOM worsened as the number of clusters in the data increased. Mingoti and Lima (2006) tested SOM against K-means and other clustering methods over a large number of datasets, and found that SOM did not perform well in almost all cases. Hence the value of SOM lies in its role as discrete nonlinear PCA, rather than as a clustering algorithm.

As an example, consider the famous Lorenz 'butterfly'-shaped attractor from chaos theory (Lorenz, 1963). Describing idealized atmospheric convection, the Lorenz system is governed by three (non-dimensionalized) differential equations:

$$\dot{x} = -ax + ay, \quad \dot{y} = -xz + bx - y, \quad \dot{z} = xy - cz, \tag{10.24}$$

where the overhead dot denotes a time derivative, and a, b and c are three parameters. A chaotic system is generated by choosing $a = 10$, $b = 28$, and $c = 8/3$. The Lorenz data are fitted by a 2-dimensional SOM (from the *Matlab* neural network toolbox) in Fig. 10.11, and by a 1-dimensional SOM in Fig. 10.12. The 1-dimensional fit resembles a discrete version of the NLPCA solution found using auto-associative neural networks (Monahan, 2000).

A propagating wave can also be represented in a 2-dimensional SOM. Liu *et al.* (2006) illustrated a sinusoidal wave travelling to the right by a 3×4 SOM (Fig. 10.13). As time progresses, the best matching unit (BMU) rotates counterclockwise around the SOM (i.e. the patterns of the 3×4 SOM are manifested

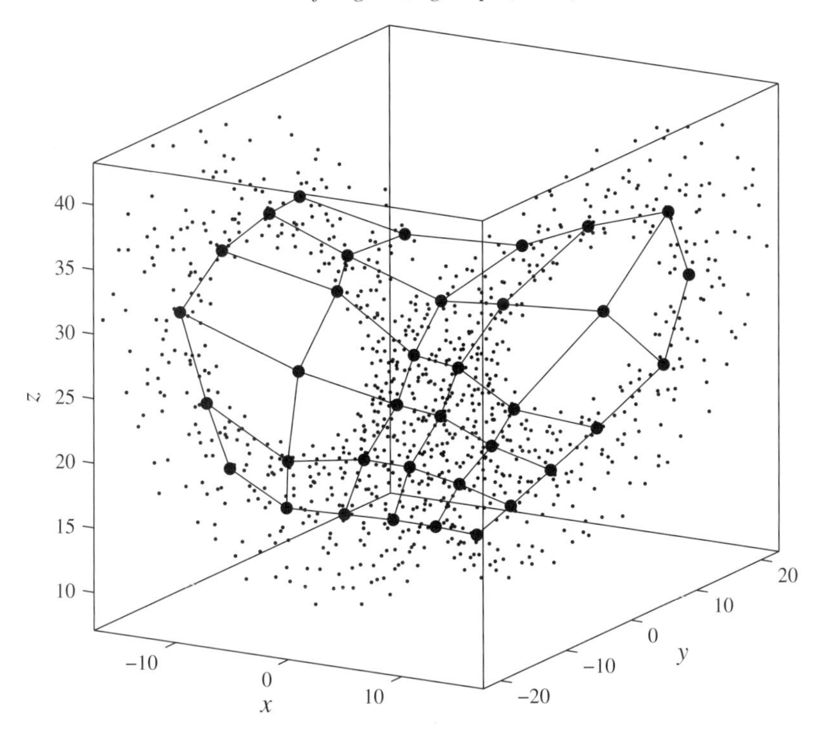

Fig. 10.11 A 2-dimensional self-organizing map (SOM) where a 6 × 6 mesh is fitted to the Lorenz (1963) attractor data.

sequentially as time progresses), producing a travelling wave pattern. The counter-clockwise movement around the SOM means that the two nodes at the centre of the SOM are not excited, hence their associated patterns (patterns (5) and (8) in the figure) are spurious.

How many grid points or units should one use in the SOM? Again the problem of underfitting with too few units and overfitting with too many units presents itself. Two quantitative measures of mapping quality are commonly used: average *quantization error* (QE) and *topographic error* (TE) (Kohonen, 2001). The measure QE is the average distance between each data point $\mathbf{x}$ and $\mathbf{z}_k$ of its BMU. The TE value gives the fraction of data points for which the first BMU and the second BMU are not neighbouring units. Smaller QE and TE values indicate better mapping quality. By increasing the number of units, QE can be further decreased; however, TE will eventually rise, indicating that one is using an excessive number of units. Thus QE and TE allow one to choose the appropriate number of units in the SOM. The TE plays a similar role as the inconsistency index (10.10) in NLPCA by MLP NN.

The SOM method has been widely used to classify satellite data, including ocean colour (Yacoub *et al.*, 2001), sea surface temperature (Richardson *et al.*, 2003), sea

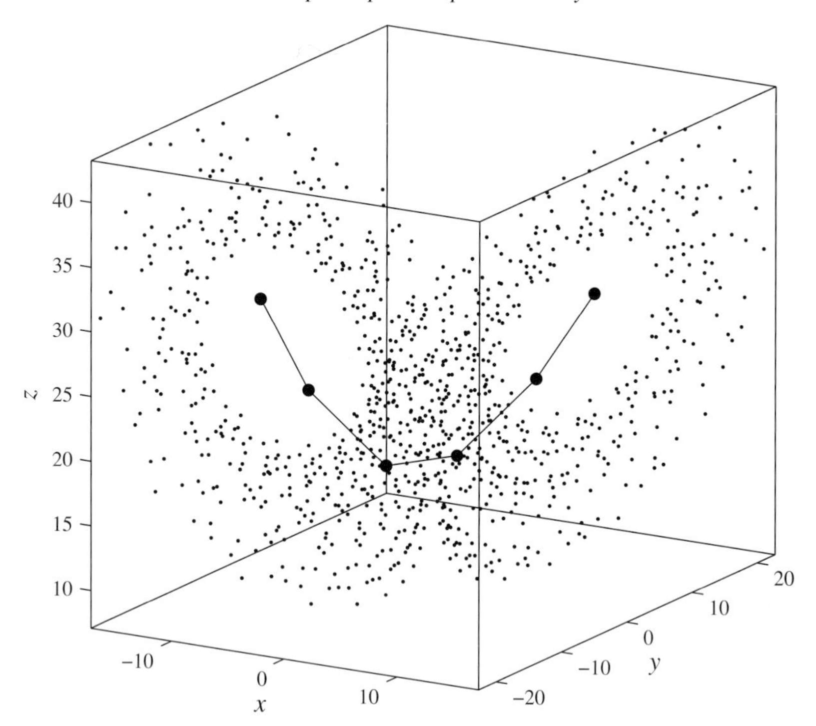

Fig. 10.12 A 1-dimensional self-organizing map (SOM) where a curve with six units is fitted to the Lorenz (1963) attractor data.

level height (Hardman-Mountford *et al.*, 2003), scatterometer winds (Richardson *et al.*, 2003), aerosol type and optical thickness (Niang *et al.*, 2006) and ocean currents (Liu and Weisberg, 2005; Liu *et al.*, 2007). Villmann *et al.* (2003) applied SOM not only to clustering low-dimensional spectral data from the LANDSAT thematic mapper, but also to high-dimensional hyperspectral AVIRIS (Airborne Visible-Near Infrared Imaging Spectrometer) data where there are about 200 frequency bands. A 2-D SOM with a mesh of 40×40 was applied to AVIRIS data to classify the geology of the land surface.

Cavazos (1999) applied a 2×2 SOM to cluster the winter daily precipitation over 20 grid points in northeastern Mexico and southeastern Texas. From the wettest and driest clusters, composites of the 500 hPa geopotential heights and sea level pressure were generated, yielding the large scale meteorological conditions associated with the wettest and driest clusters. Hewitson and Crane (2002) applied SOM to identify the January SLP anomaly patterns in northeastern USA. For seismic data, SOM has been used to identify and classify multiple events (Essenreiter *et al.*, 2001), and in well log calibration (Taner *et al.*, 2001). More applications of SOM are given in Chapter 12.

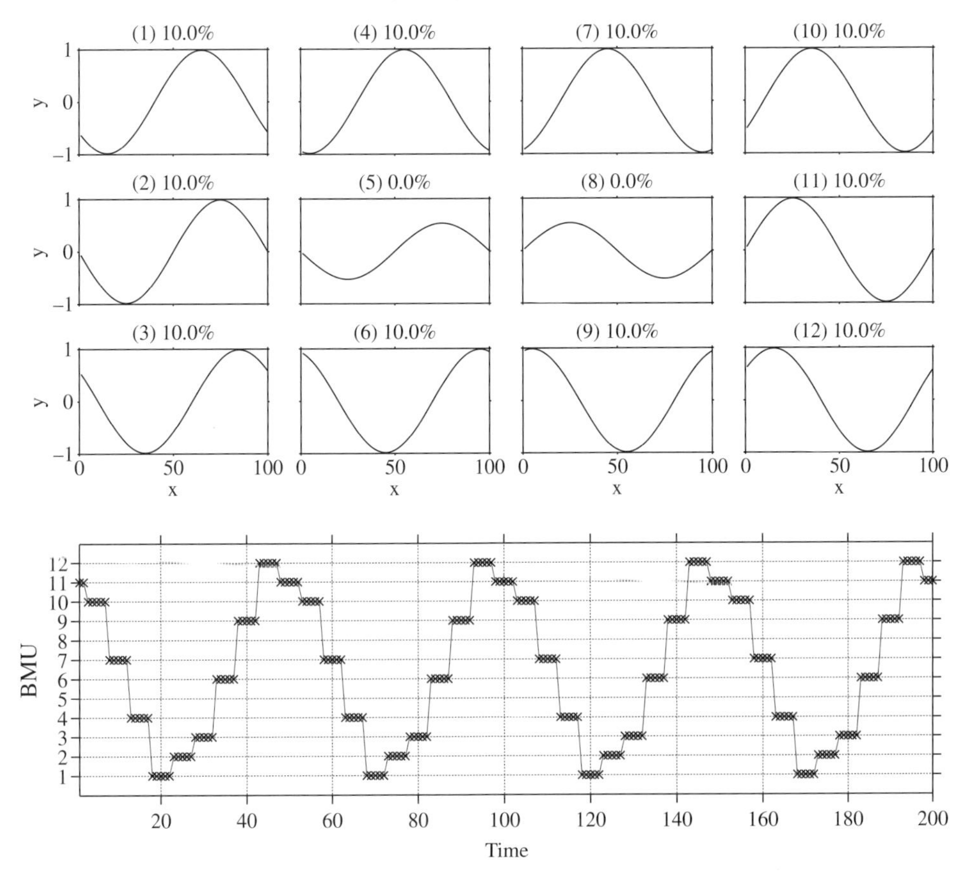

Fig. 10.13 A wave propagating to the right is analyzed by a 3×4 SOM. The frequency of occurrence of each SOM pattern is given on top of each panel. As time progresses, the best matching unit (BMU) rotates counterclockwise around the 3×4 SOM patterns, where the SOM patterns (5) and (8) are bypassed (as indicated by their frequency of occurrence of 0.0%). (Reproduced from Liu *et al.* (2006) with permission of the American Geophysical Union.)

10.4 Kernel principal component analysis

The kernel method of Chapter 7 allows another approach for nonlinear PCA. We recall that in the kernel method, points in the input space of dimension m are mapped to a 'feature' space by a nonlinear mapping function $\boldsymbol{\phi}$. The feature space is of dimension M, which is usually much larger than m and can be infinite. Essentially, (linear) PCA is performed in the high-dimensional feature space, which corresponds to finding nonlinear modes in the original data space (i.e. the input space).

Note that PCA is meaningful only when applied to centred data, i.e. data with zero mean. How to centre data in the feature space is an issue which we will deal with later, but for now assume that data in the feature space have been centred, i.e.

$$\frac{1}{n}\sum_{i=1}^{n}\boldsymbol{\phi}(\mathbf{x}_i) = 0, \tag{10.25}$$

where $\mathbf{x}_i$ $(i = 1, \ldots, n)$ is the ith observation in the input space. The covariance matrix is

$$\mathbf{C} = \frac{1}{n}\sum_{i=1}^{n}\boldsymbol{\phi}(\mathbf{x}_i)\boldsymbol{\phi}^{\mathrm{T}}(\mathbf{x}_i). \tag{10.26}$$

The PCA method involves finding the eigenvalues λ and eigenvectors $\mathbf{v}$ satisfying

$$\mathbf{C}\mathbf{v} = \lambda\mathbf{v}. \tag{10.27}$$

The eigenvectors can be expressed as a linear combination of the data points in the feature space, i.e.

$$\mathbf{v} = \sum_{j=1}^{n}\alpha_j\boldsymbol{\phi}(\mathbf{x}_j). \tag{10.28}$$

Substituting (10.26) and (10.28) into (10.27), we get

$$\sum_{i=1}^{n}\sum_{j=1}^{n}\alpha_j\boldsymbol{\phi}(\mathbf{x}_i)K(\mathbf{x}_i, \mathbf{x}_j) = n\lambda\sum_{j=1}^{n}\alpha_j\phi(\mathbf{x}_j), \tag{10.29}$$

where $K(\mathbf{x}_i, \mathbf{x}_j)$ is an inner-product *kernel* defined by

$$K(\mathbf{x}_i, \mathbf{x}_j) = \boldsymbol{\phi}^{\mathrm{T}}(\mathbf{x}_i)\boldsymbol{\phi}(\mathbf{x}_j). \tag{10.30}$$

As with other kernel methods, we aim to eliminate $\boldsymbol{\phi}$ (which can be prohibitively costly to compute) by clever use of the kernel trick. Upon multiplying both sides of the equation by $\boldsymbol{\phi}^{\mathrm{T}}(\mathbf{x}_k)$, we obtain

$$\sum_{i=1}^{n}\sum_{j=1}^{n}\alpha_j K(\mathbf{x}_k, \mathbf{x}_i)K(\mathbf{x}_i, \mathbf{x}_j) = n\lambda\sum_{j=1}^{n}\alpha_j K(\mathbf{x}_k, \mathbf{x}_j), \quad k = 1, \ldots, n. \tag{10.31}$$

In matrix notation, this is simply

$$\mathbf{K}^2\boldsymbol{\alpha} = n\lambda\mathbf{K}\boldsymbol{\alpha}, \tag{10.32}$$

where $\mathbf{K}$ is the $n \times n$ *kernel matrix*, with $K(\mathbf{x}_i, \mathbf{x}_j)$ as its ijth element, and $\boldsymbol{\alpha}$ is an $n \times 1$ vector, with α_j as its jth element. All the solutions of interest for this eigenvalue problem are also found by solving the simpler eigenvalue problem (Schölkopf and Smola, 2002)

$$\mathbf{K}\boldsymbol{\alpha} = n\lambda\boldsymbol{\alpha}. \tag{10.33}$$

This eigenvalue equation can be rewritten in the more familiar form

$$\mathbf{K}\boldsymbol{\alpha} = \lambda' \boldsymbol{\alpha}, \tag{10.34}$$

where $\lambda' = n\lambda$. Let $\lambda_1 \geq \lambda_2 \geq \ldots \geq \lambda_n$ denote the solution for λ' in this eigenvalue problem. Suppose λ^p is the smallest non-zero eigenvalue. The eigenvectors $\mathbf{v}^{(1)}, \ldots, \mathbf{v}^{(p)}$ are all normalized to unit length, i.e.

$$\mathbf{v}^{(k)\mathrm{T}}\mathbf{v}^{(k)} = 1, \quad k = 1, \ldots, p. \tag{10.35}$$

This normalization of the eigenvectors translates into a normalization condition for $\alpha^{(k)}$ $(k = 1, \ldots, p)$ upon invoking (10.28), (10.30) and (10.34):

$$\begin{aligned} 1 &= \sum_{i=1}^{n}\sum_{j=1}^{n} \alpha_i^{(k)} \alpha_j^{(k)} \boldsymbol{\phi}(\mathbf{x}_i)^{\mathrm{T}} \boldsymbol{\phi}(\mathbf{x}_i) = \sum_{i=1}^{n}\sum_{j=1}^{n} \alpha_i^{(k)} \alpha_j^{(k)} K_{ij} \\ &= \boldsymbol{\alpha}^{(k)\mathrm{T}}\, \mathbf{K} \boldsymbol{\alpha}^{(k)} = \lambda_k\, \boldsymbol{\alpha}^{(k)\mathrm{T}} \boldsymbol{\alpha}^{(k)}. \end{aligned} \tag{10.36}$$

Let us return to the problem of centering data in the feature space. In the kernel evaluations, we actually need to work with

$$\tilde{K}(\mathbf{x}_i, \mathbf{x}_j) = (\boldsymbol{\phi}(\mathbf{x}_i) - \bar{\boldsymbol{\phi}})^{\mathrm{T}} (\boldsymbol{\phi}(\mathbf{x}_j) - \bar{\boldsymbol{\phi}}), \tag{10.37}$$

where the mean

$$\bar{\boldsymbol{\phi}} = \frac{1}{n} \sum_{l=1}^{n} \boldsymbol{\phi}(\mathbf{x}_l). \tag{10.38}$$

Hence

$$\begin{aligned} \tilde{K}(\mathbf{x}_i, \mathbf{x}_j) &= \left(\boldsymbol{\phi}(\mathbf{x}_i) - \frac{1}{n}\sum_{l=1}^{n} \boldsymbol{\phi}(\mathbf{x}_l) \right)^{\mathrm{T}} \left(\boldsymbol{\phi}(\mathbf{x}_j) - \frac{1}{n}\sum_{l=1}^{n} \boldsymbol{\phi}(\mathbf{x}_l) \right) \\ &= \boldsymbol{\phi}(\mathbf{x}_i)^{\mathrm{T}} \boldsymbol{\phi}(\mathbf{x}_j) - \boldsymbol{\phi}(\mathbf{x}_i)^{\mathrm{T}} \frac{1}{n}\sum_{l=1}^{n} \boldsymbol{\phi}(\mathbf{x}_l) - \boldsymbol{\phi}(\mathbf{x}_j)^{\mathrm{T}} \frac{1}{n}\sum_{l=1}^{n} \boldsymbol{\phi}(\mathbf{x}_l) \\ &\quad + \frac{1}{n^2} \sum_{l=1}^{n}\sum_{l'=1}^{n} \boldsymbol{\phi}(\mathbf{x}_l)^{\mathrm{T}} \boldsymbol{\phi}(\mathbf{x}_{l'}), \end{aligned} \tag{10.39}$$

yielding

$$\begin{aligned} \tilde{K}(\mathbf{x}_i, \mathbf{x}_j) &= K(\mathbf{x}_i, \mathbf{x}_j) - \frac{1}{n}\sum_{l=1}^{n} K(\mathbf{x}_i, \mathbf{x}_l) - \frac{1}{n}\sum_{l=1}^{n} K(\mathbf{x}_j, \mathbf{x}_l) \\ &\quad + \frac{1}{n^2} \sum_{l=1}^{n}\sum_{l'=1}^{n} K(\mathbf{x}_l, \mathbf{x}_{l'}). \end{aligned} \tag{10.40}$$

The eigenvalue problem is now solved with $\tilde{\mathbf{K}}$ replacing $\mathbf{K}$ in (10.34).

For any test point $\mathbf{x}$, with a corresponding point $\boldsymbol{\phi}(\mathbf{x})$ in the feature space, we project $\boldsymbol{\phi}(\mathbf{x}) - \bar{\boldsymbol{\phi}}$ onto the eigenvector $\mathbf{v}^{(k)}$ to obtain the kth principal component or *feature*:

$$\mathbf{v}^{(k)\mathrm{T}}(\boldsymbol{\phi}(\mathbf{x}) - \bar{\boldsymbol{\phi}}) = \sum_{j=1}^{n} \alpha_j^{(k)} (\boldsymbol{\phi}(\mathbf{x}_j) - \bar{\boldsymbol{\phi}})^{\mathrm{T}} (\boldsymbol{\phi}(\mathbf{x}) - \bar{\boldsymbol{\phi}})$$
$$= \sum_{j=1}^{n} \alpha_j^{(k)} \tilde{K}(\mathbf{x}_j, \mathbf{x}), \quad k = 1, \ldots, p. \qquad (10.41)$$

In summary, the basic steps of kernel PCA are as follows: first, having chosen a kernel function, we compute the kernel matrix $\tilde{K}(\mathbf{x}_i, \mathbf{x}_j)$ where $\mathbf{x}_i$ and $\mathbf{x}_j$ are among the data points in the input space. Next, the eigenvalue problem is solved with $\tilde{\mathbf{K}}$ in (10.34). The eigenvector expansion coefficients $\boldsymbol{\alpha}^{(k)}$ are then normalized by (10.36). Finally, the PCs are calculated from (10.41). As the kernel has an adjustable parameter (e.g. the standard deviation σ controlling the shape of the Gaussian in the case of the Gaussian kernel), one searches over various values of the kernel parameter for the optimal one – the one explaining the most variance in the feature space, as determined by the magnitude of the leading PC(s) in (10.41).

Suppose n, the number of observations, exceeds m, the dimension of the input space. With PCA, no more than m PCs can be extracted. In contrast, with kernel PCA, where PCA is performed in the feature space of dimension M (usually much larger than m and n), up to n PCs or features can be extracted, i.e. the number of features that can be extracted by kernel PCA is determined by the number of observations. The NLPCA method by auto-associative neural networks involves nonlinear optimization, hence local minima problems, whereas kernel PCA, which only performs linear PCA in the feature space, does not involve nonlinear optimization, hence no local minima problem. On the other hand, with NLPCA, the inverse mapping from the nonlinear PCs to the original data space (input space) is entirely straightforward, while for kernel PCA, there is no obvious inverse mapping from the feature space to the input space. This is the pre-image problem common to kernel methods, and only an approximate solution can be found, as discussed in Section 7.6.

10.5 Nonlinear complex PCA

Complex principal component analysis (CPCA) is PCA applied to complex variables. In the first type of application, a 2-dimensional vector field such as the wind (u, v) can be analyzed by applying CPCA to $w = u + iv$ (Section 2.3). In the second type of application, a real time-varying field can be complexified by the

Hilbert transform and analyzed by CPCA, often called Hilbert PCA (Section 3.7) to distinguish from the first type of application.

Earlier in this chapter, we have examined the auto-associative multi-layer perceptron NN approach of Kramer (1991) for performing nonlinear PCA. Here we will discuss how the same approach can be applied to complex variables, giving rise to nonlinear complex PCA (NLCPCA).

In the real domain, a common nonlinear activation function is the hyperbolic tangent function $\tanh(x)$, bounded between -1 and $+1$ and analytic everywhere. For a complex activation function to be bounded and analytic everywhere, it has to be a constant function (Clarke, 1990), as Liouville's theorem states that entire functions (i.e. functions that are analytic on the whole complex plane) which are bounded are always constants. The function $\tanh(z)$ in the complex domain has an infinite number of singularities located at $(\frac{1}{2} + l)\pi \mathrm{i}$, $l \in \mathbb{N}$ and $\mathrm{i}^2 = -1$. Using functions like $\tanh(z)$ (without any constraint) leads to non-convergent solutions (Nitta, 1997).

Traditionally, the complex activation functions used focused mainly on overcoming the unbounded nature of the analytic functions in the complex domain. Some complex activation functions basically scaled the magnitude (amplitude) of the complex signals but preserved their arguments (phases) (Georgiou and Koutsougeras, 1992; Hirose, 1992), hence they are less effective in learning nonlinear variations in the argument. A more traditional approach has been to use a 'split' complex nonlinear activation function (Nitta, 1997), where the real and imaginary components are used as separate real inputs for the activation function. This approach avoids the unbounded nature of the nonlinear complex function but results in a nowhere analytic complex function, as the Cauchy-Riemann equations are not satisfied (Saff and Snider, 2003).

Recently, a set of elementary activation functions has been proposed by Kim and Adali (2002) with the property of being *almost everywhere* (*a.e.*) bounded and analytic in the complex domain. The complex hyperbolic tangent, $\tanh(z)$, is among them, provided that the complex optimization is performed with certain constraints on z. If the magnitude of z is within a circle of radius $\pi/2$, then the singularities do not pose any problem, and the boundedness property is also satisfied. In reality, the dot product of the input and weight vectors may be $\geq \pi/2$. Thus a restriction on the magnitudes of the input and weights is needed.

The NLCPCA model proposed by Rattan and Hsieh (2004, 2005) uses basically the same architecture (Fig. 10.2(a)) as the NLPCA model by auto-associative NN (with three layers of hidden neurons where the middle layer is the bottleneck layer), except that all the input variables, and the weight and offset parameters are now complex-valued. The magnitudes of input data are scaled by dividing each element in the rth row of the $m \times n$ data matrix $\mathbf{Z}$ (with m the number of variables and

n the number of observations) by the maximum magnitude of an element in that row, so each element of $\mathbf{Z}$ has magnitude ≤ 1. The weights at the first hidden layer are randomly initialized with small magnitude, thus limiting the magnitude of the dot product between the input vector and weight vector to be about 0.1, and a weight penalty term is added to the objective function J to restrict the weights to small magnitude during optimization. The weights at subsequent layers are also randomly initialized with small magnitude and penalized during optimization by the objective function

$$J = \langle \|\mathbf{z} - \mathbf{z}'\|^2 \rangle + P \sum_j |w_j|^2, \tag{10.42}$$

where $\mathbf{z}$ is the model output, $\mathbf{z}'$, the target data, w_j, the individual weights (including the offset parameters) from hidden layers 1, 2 and 3, and P, the weight penalty parameter.

Since the objective function J is a real function with complex weights, optimization of J is equivalent to finding the vanishing gradient of J with respect to the real and the imaginary parts of the weights (Rattan and Hsieh, 2005). All the weights (and offsets) in the model are combined into a single weight vector $\mathbf{w}$. Hence the gradient of the objective function with respect to the complex weights can be split into (Georgiou and Koutsougeras, 1992):

$$\frac{\partial J}{\partial \mathbf{w}} = \frac{\partial J}{\partial \mathbf{w}^{\mathrm{R}}} + \mathrm{i}\frac{\partial J}{\partial \mathbf{w}^{\mathrm{I}}}, \tag{10.43}$$

where $\mathbf{w}^{\mathrm{R}}$ and $\mathbf{w}^{\mathrm{I}}$ are the real and the imaginary components of the weight vector. The two components can be put into a single real parameter vector during nonlinear optimization using an algorithm for real variables.

The tropical Pacific wind anomalies (expressed as $w = u + \mathrm{i}v$) have been analyzed by NLCPCA in Rattan and Hsieh (2004), where a comparison between the first mode of CPCA and that of NLCPCA revealed a large difference in the spatial anomaly patterns during strong El Niño episodes but a much smaller difference during strong La Niña episodes, indicating that stronger nonlinearity was manifested in the El Niño side than the La Niña side of the oscillation.

The second type of NLCPCA application is for nonlinear Hilbert PCA. In Rattan *et al.* (2005), evolution of the offshore bottom topography at three sandy beaches was studied. All three sites were characterized by sandbars with interannual quasi-periodic offshore propagation. A bar cycle comprises bar birth in the inner nearshore, followed by up to several years of net offshore migration and final disappearance in the outer nearshore zone. The CPCA method was applied to the complexified topographic anomaly data, and the five leading complex PCs were retained as inputs for the NLCPCA NN model. The

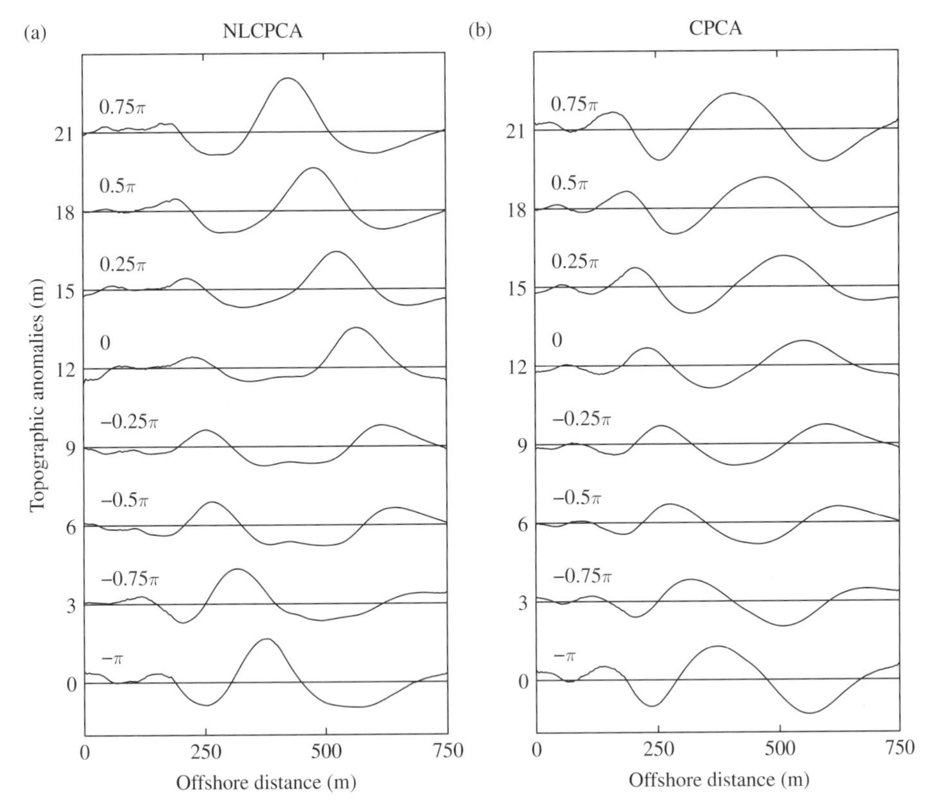

Fig. 10.14 Sequence of topographic anomalies as a function of the offshore distance at Egmond in $\pi/4$-wide θ classes centered around $\theta = -\pi$ to $\theta = 0.75\pi$ based on (a) NLCPCA mode 1 and (b) CPCA mode 1. The results for each phase class have been vertically shifted for better visualization. The phase generally decreases with time, so the anomalies gradually propagate offshore. (Modified from Rattan *et al.* (2005) with permission of the European Geosciences Union.)

first NLCPCA mode and the first CPCA mode of the topographic anomalies at Egmond aan Zee (The Netherlands) were compared. The topographic anomalies reconstructed from the nonlinear and linear mode were divided in eight θ classes, each $\pi/4$ in width, where θ is the phase of the (nonlinear or linear) complex PC. Figure 10.14 shows how the shape of the topographic anomalies changes with phase. The CPCA shows sinusoidal-shaped topographic anomalies propagating offshore, while the NLCPCA shows non-sinusoidal anomalies – relatively steep sandbars and shallow, broad troughs. The percentage variance explained by the first NLCPCA mode was 81.4% versus 66.4% by the first CPCA mode. Thus, using the NLCPCA as nonlinear Hilbert PCA successfully captures the non-sinusoidal wave properties which were missed by the linear method.

10.6 Nonlinear singular spectrum analysis

In Sections 3.4 and 3.5, we have learned that by incorporating time lagged versions of the dataset, the PCA method can be extended to the singular spectrum analysis (SSA) method. We have also learned earlier in this chapter, that the leading PCs of a dataset can be nonlinearly combined by an auto-associative NN to produce nonlinear PCA. We will see in this section that the leading PCs from an SSA can also be nonlinearly combined to produce nonlinear SSA (NLSSA).

The NLSSA procedure proposed in Hsieh and Wu (2002) is as follows: first SSA is applied to the dataset as a prefilter, and after discarding the higher modes, we retain the leading SSA PCs, $\mathbf{x}(t) = [x_1, \ldots, x_l]$, where each variable x_i, $(i = 1, \ldots, l)$, is a time series of length n. The variables $\mathbf{x}$ are the inputs to the NLPCA(cir) model, i.e. the auto-associative NN model for nonlinear principal component analysis with a circular node at the bottleneck (Section 10.1.4). The NLPCA(cir), with its ability to extract closed curve solutions due to its circular bottleneck node, is ideal for extracting periodic or wave modes in the data. In SSA, it is common to encounter periodic modes, each of which has to be split into a pair of SSA modes (Sections 3.4 and 3.5), as the underlying PCA technique is not capable of modelling a periodic mode (a closed curve) by a single mode (a straight line). Thus, two (or more) SSA modes can easily be combined by NLPCA(cir) into one NLSSA mode, taking the shape of a closed curve. When implementing NLPCA(cir), Hsieh (2001b) found that there were two possible configurations, a restricted configuration and a general configuration (see Section 10.1.4). Here we will use the general configuration (which if needed is also able to model an open curve). After the first NLSSA mode has been extracted, it can be subtracted from $\mathbf{x}$ to get the residual, which can be input again into the same NN to extract the second NLSSA mode, and so forth for the higher modes.

To illustrate the difference between the NLSSA and the SSA, Hsieh (2004) considered a test problem with a non-sinusoidal wave of the form

$$f(t) = \begin{cases} 3 & \text{for } t = 1, ..., 7 \\ -1 & \text{for } t = 8, ..., 28 \\ & \text{periodic thereafter.} \end{cases} \tag{10.44}$$

This is a square wave with the peak stretched to be 3 times as tall but only 1/3 as broad as the trough, and has a period of 28. Gaussian noise with twice the standard deviation as this signal was added, and the time series was normalized to unit standard deviation (Fig. 10.15). This time series y has 600 data points.

The SSA method with window $L = 50$ was applied to this y time series. The first eight SSA modes individually accounted for 6.3, 5.6, 4.6, 4.3, 3.3, 3.3, 3.2, and 3.1 % of the variance of the augmented data. The leading pair of modes displays

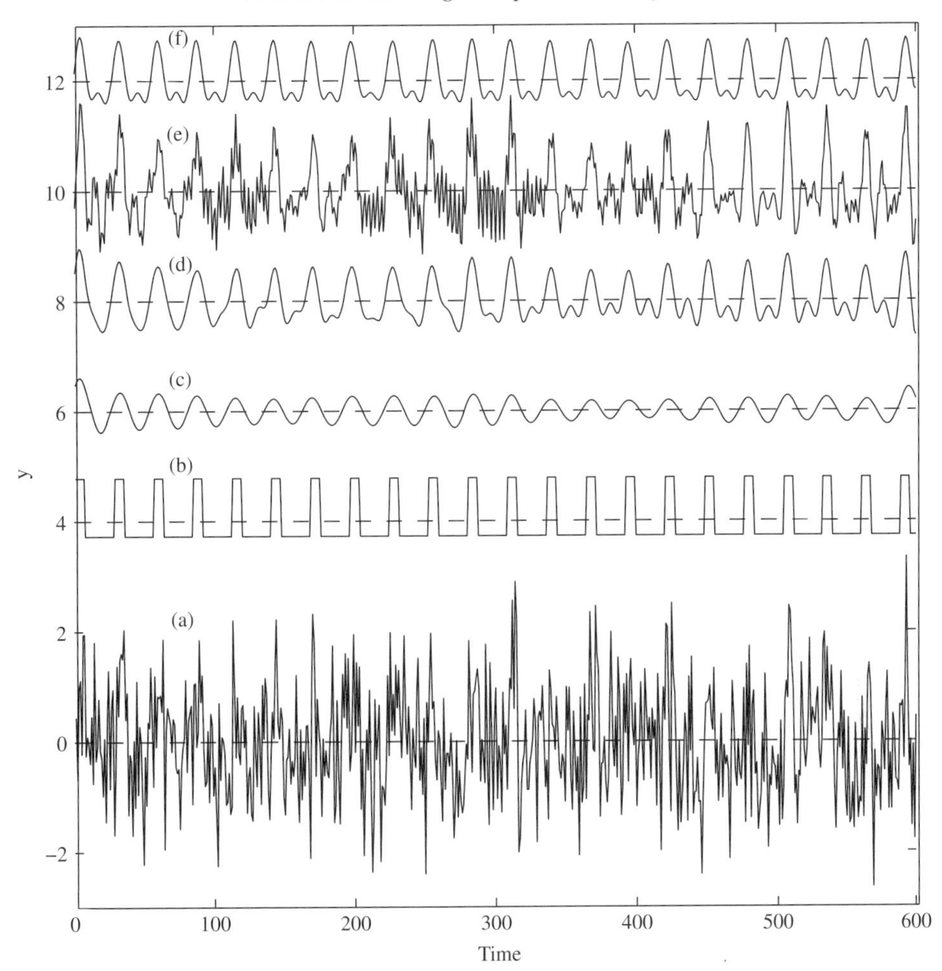

Fig. 10.15 The difference between NLSSA and SSA. (a) The noisy time series y containing a stretched square wave signal. (b) The stretched square wave signal, which we will try to extract from the noisy time series. Curves (c), (d) and (e) are the reconstructed components (RC) from SSA leading modes, using 1, 3 and 8 modes, respectively. Curve (f) is the NLSSA mode 1 RC (NLRC1). The dashed lines show the means of the various curves, which have been vertically shifted for better visualization. (Reproduced from Hsieh (2004) with permission from the American Geophyscial Union.)

oscillations of period 28, while the next pair manifests oscillations at a period of 14, i.e. the first harmonic (Hsieh, 2004). The first eight SSA PC time series x_i ($i = 1, \ldots, 8$) were served as inputs to the NLPCA(cir) network (Fig. 10.2(b)), where the network with $m = 5$ was selected (m being the number of neurons in the first hidden layer).

The NLSSA *reconstructed component* 1 (NLRC1) is the approximation of the original time series y by the NLSSA mode 1. The neural network output $\mathbf{x}'$ are the NLSSA mode 1 approximation for the eight leading PCs. Multiplying these approximated PCs by their corresponding SSA eigenvectors, and summing over the eight modes allows reconstruction of the time series from the NLSSA mode 1. As each eigenvector contains the loading over a range of lags, each value in the reconstructed time series at time t_j also involves averaging over the contributions at t_j from the various lags (similarly to computing the RC in Section 3.4). Thus, with a window $L = 50$, the NLRC1 at any time involves averaging over 50 values, (except towards the end of the record where one has fewer than 50 values to average).

In Fig. 10.15, NLRC1 (curve (f)) from NLSSA is to be compared with the reconstructed component (RC) from SSA mode 1 (RC1) (curve (c)). The non-sinusoidal nature of the oscillations is not revealed by the RC1, but is clearly manifested in the NLRC1, where each strong narrow peak is followed by a weak broad trough, similar to the original stretched square wave. Also, the wave amplitude is more steady in the NLRC1 than in the RC1. Using contributions from the first two SSA modes, RC1-2 (not shown) is rather similar to RC1 in appearance, except for a larger amplitude.

In Fig. 10.15, curves (d) and (e) show the RC from SSA using the first three modes, and the first eight modes, respectively. These curves, referred to as RC1-3 and RC1-8, respectively, show increasing noise as more modes are used. Among the RCs, with respect to the stretched square wave time series (curve (b)), RC1-3 attains the most favourable correlation (0.849) and root mean squared error (RMSE) (0.245), but remains behind the NLRC1, with correlation (0.875) and RMSE (0.225).

The stretched square wave signal accounted for only 22.6% of the variance in the noisy data. For comparison, NLRC1 accounted for 17.9%, RC1, 9.4%, and RC1-2, 14.1% of the variance. With more modes, the RCs account for increasingly more variance, but beyond RC1-3, the increased variance is only from fitting to the noise in the data.

When classical Fourier spectral analysis was performed, the most energetic bands were the sine and cosine at a period of 14, the two together accounting for 7.0% of the variance. In this case, the strong scattering of energy to higher harmonics by the Fourier technique has actually assigned 38% more energy to the first harmonic (at period 14) than to the fundamental period of 28. Next, the data record was slightly shortened from 600 to 588 points, so that the data record is exactly 21 times the fundamental period of our known signal – this is to avoid violating the periodicity assumption of Fourier analysis and the resulting spurious energy scatter into higher spectral bands. The most energetic Fourier bands were the sine and cosine at the fundamental period of 28, the two together accounting for 9.8% of the variance, compared with 14.1% of the variance accounted for by the first two SSA modes. Thus even with great care, the Fourier method scatters the spectral energy considerably more than the SSA method.

Let us illustrate the use of NLSSA on a multivariate dataset by applying the method to study the tropical Pacific climate variability (Hsieh and Wu, 2002). We need to choose a lag window wide enough to resolve the ENSO (El Niño-Southern Oscillation) variability. With a lag interval of 3 months, the original plus 24 lagged copies of the sea surface temperature anomaly (SSTA) data in the tropical Pacific were chosen to form the augmented SSTA dataset – with a lag window of ($1 + 24 \times 3 =$) 73 months. The first eight SSA modes respectively explain 12.4%, 11.7%, 7.1%, 6.7%, 5.4%, 4.4%, 3.5% and 2.8% of the total variance of the augmented dataset. In Fig. 3.9, we have seen that the first two modes have space–time eigenvectors (i.e. loading patterns) displaying an oscillatory time scale of about 48 months, comparable to the ENSO time scale, with the mode 1 anomaly pattern occurring about 12 months before a very similar mode 2 pattern, i.e. the two patterns are in quadrature. The PC time series also show similar time scales for modes 1 and 2. Modes 3 and 5 show longer time scale fluctuations, while modes 4 and 6 show shorter time scale fluctuations – around the 30 month time scale.

With the eight PCs as input $x_1, \ldots, x_8$ to the NLPCA(cir) network, the resulting NLSSA mode 1 is a closed curve in the 8-D PC space, and is plotted in the x_1-x_2-x_3 space in Fig. 10.16. The NLSSA mode 1 is basically a large loop aligned parallel to the x_1-x_2 plane, thereby combining the first two SSA modes. The solution also shows some modest variations in the x_3 direction. This NLSSA mode 1 explains 24.0% of the total variance of the augmented dataset, essentially that explained by the first two SSA modes together. The linear PCA solution is shown as a straight line in the figure, which is of course simply the SSA mode 1. Of interest is r, the ratio of the mean squared error (MSE) of the nonlinear solution to the MSE of the linear solution. Here $r = 0.71$, reflecting the strong departure of the nonlinear solution from the linear solution.

In the linear case of PCA or SSA, as the PC varies, the loading pattern is unchanged except for scaling by the PC. In the nonlinear case, as the NLPC θ varies, the loading pattern changes as it does not generally lie along a fixed eigenvector. The NLSSA mode 1 space-time loading pattern for a given value of θ, can be obtained by mapping from θ to the outputs $\mathbf{x}'$, which are the eight PC values corresponding to the given θ. Multiplying each PC value by its corresponding SSA eigenvector and summing over the eight modes, we obtain the NLSSA mode 1 pattern corresponding to the given θ.

The NLSSA mode 1 loading pattens for various θ values are shown in Fig. 10.17. As the NLPC θ was found generally to increase with time, the patterns in Fig. 10.17

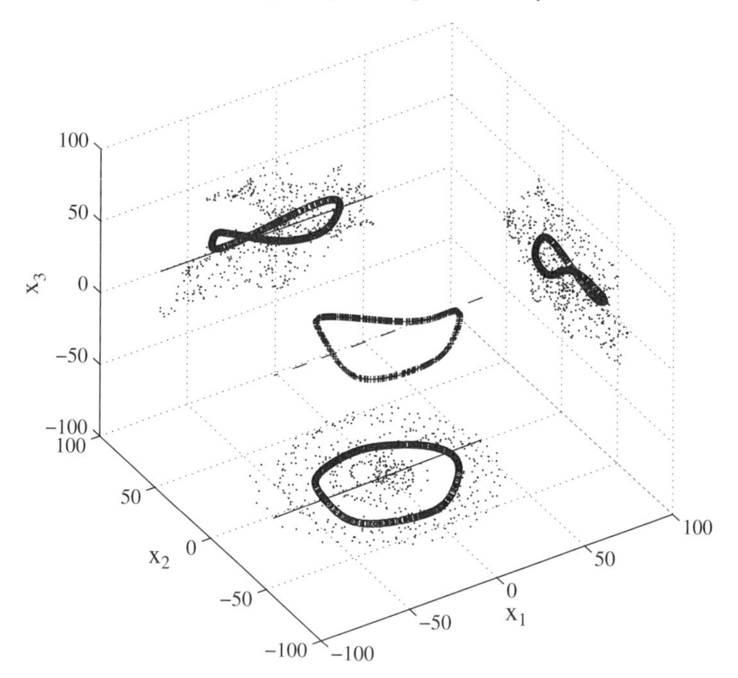

Fig. 10.16 The NLSSA mode 1 for the tropical Pacific SSTA. The PCs of SSA modes 1 to 8 were used as inputs $x_1, \ldots, x_8$ to the NLPCA(cir) network, with the resulting NLSSA mode 1 shown as (densely overlapping) crosses in the x_1-x_2-x_3 3-D space. The projections of this mode onto the x_1-x_2, x_1-x_3 and x_2-x_3 planes are denoted by the (densely overlapping) circles, and the projected data by dots. For comparison, the linear SSA mode 1 is shown by the dashed line in the 3-D space, and by the projected solid lines on the 2-D planes. (Reproduced from Hsieh and Wu (2002) with permission of the American Geophysical Union.)

generally evolve from (a) to (f) with time. Comparing the patterns in this figure with the patterns from the first two SSA modes in Fig. 3.9, we find three notable differences.

(1) The presence of warm anomalies for 24 months followed by cool anomalies for 24 months in the first two SSA modes, is replaced by warm anomalies for 18 months followed by cool anomalies for about 33 months in the NLSSA mode 1. Although the cool anomalies can be quite mild for long periods, they can develop into full La Niña cool episodes (Fig. 10.17(c)).
(2) The El Niño warm episodes are strongest near the eastern boundary, while the La Niña episodes are strongest near the central equatorial Pacific in the NLSSA mode 1, an asymmetry not found in the individual SSA modes.
(3) The magnitude of the peak positive anomalies is significantly larger than that of the peak negative anomalies in the NLSSA mode 1 (Fig. 10.17(c)), again an asymmetry not found in the individual SSA modes.

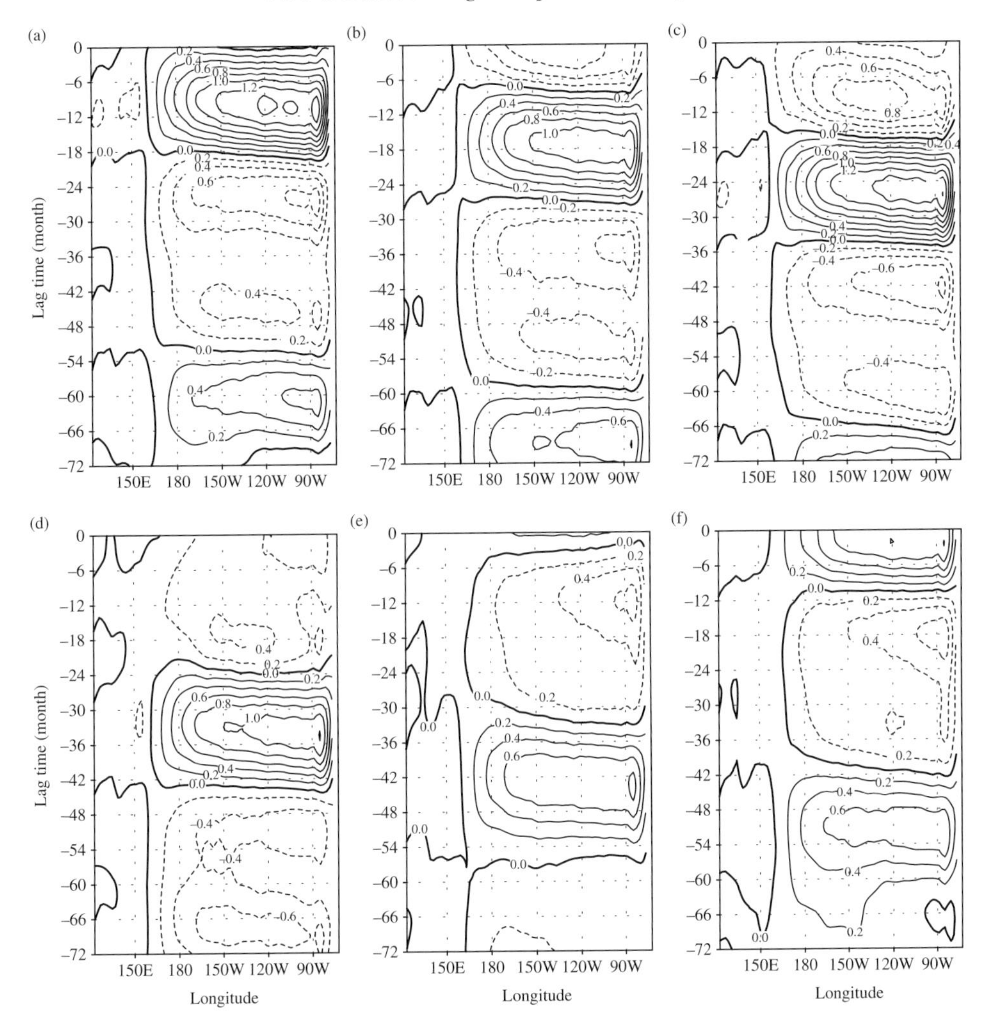

Fig. 10.17 The SSTA NLSSA mode 1 space-time loading patterns for (a) $\theta = 0°$, (b) $\theta = 60°$, (c) $\theta = 120°$, (d) $\theta = 180°$, (e) $\theta = 240°$ and (f) $\theta = 300°$. The contour plots display the SSTA anomalies along the equator as a function of the lag time. Contour interval is 0.2 °C. (Reproduced from Hsieh and Wu (2002) with permission of the American Geophysical Union.)

All three differences indicate that the NLSSA mode 1 is much closer to the observed ENSO properties than the first two SSA modes are.

Next, we try to reconstruct the SSTA field from the NLSSA mode 1 for the whole record. The PC at a given time gives not only the SSTA field at that time, but a total of 25 SSTA fields (spread 3 months apart) covering the 73 months in the lag window. Thus (except for the final 72 months of the record, where one has fewer points to perform averaging), the SSTA field at any time is taken to be

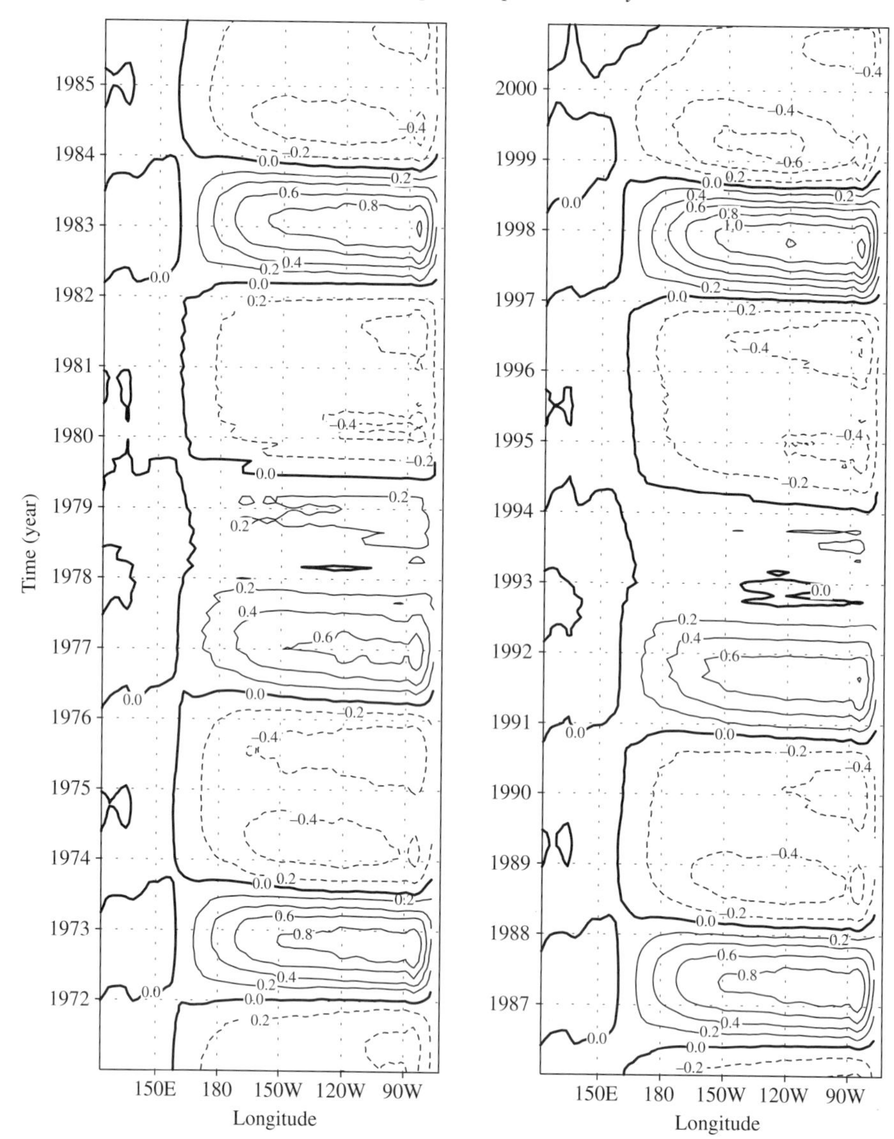

Fig. 10.18 The reconstructed SSTA of the NLSSA mode 1, from January 1971 to December 2000. The contour plots display the SSTA along the equator as a function of time, with a contour interval of 0.2 °C. (Reproduced from Hsieh and Wu (2002) with permission of the American Geophysical Union.)

the average of the 25 SSTA fields produced by the PC of the current month and the preceding 72 months. The NLRC1 during 1971–2000 is shown in Fig. 10.18, which compares well with the observations, except for its weaker amplitude.

To extract the NLSSA mode 2, the NLSSA mode 1 $\mathbf{x}'$ was removed from the data $\mathbf{x}$, and the residual $(\mathbf{x}-\mathbf{x}')$ was input into the same NLPCA(cir) network (Hsieh and

Wu, 2002). Similarly, NLSSA was applied to study the tropical Pacific monthly sea level pressure anomaly (SLPA) data during 1950–2000, with the first mode revealing the Southern Oscillation (Hsieh, 2004). The NLSSA technique has also been used to study the stratospheric equatorial winds for the QBO phenomenon (Hsieh and Hamilton, 2003).

In summary, the NLSSA mode has been developed by nonlinearly combining the leading SSA modes. In general, NLSSA has several advantages over SSA.

(1) The PCs from different SSA modes are linearly uncorrelated; however, they may have relationships that can be detected by the NLSSA.
(2) Although the SSA modes are not restricted to sinusoidal oscillations in time like the Fourier spectral components, in practice they are inefficient in modelling non-sinusoidal periodic signals, scattering the signal energy into many SSA modes, similar to the way Fourier spectral analysis scatters the energy of a non-sinusoidal wave to its higher harmonics. The NLSSA recombines the SSA modes to extract the non-sinusoidal signal, alleviating spurious transfer of energy to higher frequencies.
(3) As different SSA modes are associated with different time scales (e.g. time scales of the ENSO, QBO and decadal oscillations), the relations found by the NLSSA reveal the time scales between which there are interactions, thereby disclosing potential relations between seemingly separate phenomena.

Exercises

(10.1) Let x_1 be 60 random numbers uniformly distributed in $(-1.5\pi, 1.5\pi)$, and let $x_2 = \cos(x_1)$. Add Gaussian noise with standard deviation of 0.1 to both x_1 and x_2. Compute nonlinear principal component analysis (NLPCA) on this dataset. If you are using an auto-associate MLP neural network, vary the number of hidden neurons.

(10.2) Consider a rectangle with one side twice as long as the other in the (x_1, x_2) plane. Select 60 points on the boundary of this rectangle and add Gaussian noise to these points. Use the closed curve NLPCA (Section 10.1.4) to retrieve the signal (i.e. the rectangle) from this dataset. Vary the number of hidden neurons and use the information criterion to select the best solution. Vary the amount of Gaussian noise.

(10.3) Fit a 1-dimensional self-organizing map (SOM) to $y = \sin(x)$ $(0 \le x \le 2\pi)$. Use 50 data points for x and add Gaussian noise with standard deviation of 0.3 to both x and y. Vary the number of nodes from 5 to 40. From the quantization error and the topographic error, determine the best choice for the number of nodes.

11

Nonlinear canonical correlation analysis

In Section 2.4 the canonical correlation analysis (CCA) was presented as the generalization of correlation to the multivariate case. CCA is widely used but is limited by being a linear model. A number of different approaches has been proposed to generalize CCA nonlinearly (Lai and Fyfe, 1999, 2000; Hsieh, 2000; Suykens *et al.*, 2002; Hardoon *et al.*, 2004; Shawe-Taylor and Cristianini, 2004).

To perform nonlinear CCA (NLCCA), the simplest way is to use three multi-layer perceptron (MLP) NNs, where the linear mappings in the CCA are replaced by nonlinear mapping functions via 1-hidden-layer NNs. Alternatively, with the kernel approach, CCA can be performed in the feature space to yield a nonlinear CCA method. Nonlinear principal predictor analysis, a somewhat related method for nonlinearly linking predictors to predictands, has also been developed using MLP NNs (Cannon, 2006).

In Section 11.1, the MLP approach to NLCCA is formulated, then illustrated using the tropical Pacific climate variability and its connection to the mid-latitude climate variability. In Section 11.2, NLCCA is made more robust to outliers.

11.1 MLP-based NLCCA model

Consider two vector variables $\mathbf{x}$ and $\mathbf{y}$, each with n observations. CCA looks for linear combinations

$$u = \mathbf{f}^{\mathrm{T}}\mathbf{x}, \quad \text{and} \quad v = \mathbf{g}^{\mathrm{T}}\mathbf{y}, \tag{11.1}$$

where the canonical variates u and v have maximum correlation, i.e. the weight vectors $\mathbf{f}$ and $\mathbf{g}$ are chosen such that $\mathrm{cor}(u, v)$, the Pearson correlation coefficient between u and v, is maximized (see Section 2.4.1). For NLCCA, the linear maps $\mathbf{f}$ and $\mathbf{g}$, and their inverse maps, are replaced below by nonlinear mapping functions using MLP NNs (Hsieh, 2000, 2001a).

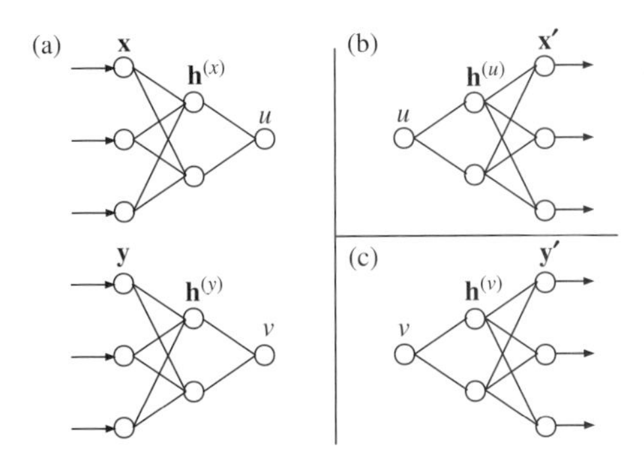

Fig. 11.1 The three MLP NNs used to perform NLCCA. (a) The double-barrelled NN maps from the inputs $\mathbf{x}$ and $\mathbf{y}$ to the canonical variates u and v. The objective function J forces the correlation between u and v to be maximized. (b) The NN maps from u to the output layer $\mathbf{x}'$. The objective function J_1 basically minimizes the MSE of $\mathbf{x}'$ relative to $\mathbf{x}$. (c) The NN maps from v to the output layer $\mathbf{y}'$. The objective function J_2 basically minimizes the MSE of $\mathbf{y}'$ relative to $\mathbf{y}$. (Reproduced from Hsieh (2001a) with permission of the American Meteorological Society.)

The mappings from $\mathbf{x}$ to u and $\mathbf{y}$ to v are represented by the double-barrelled NN in Fig. 11.1(a). By minimizing the objective function $J = -\text{cor}(u, v)$, one finds the parameters which maximize the correlation $\text{cor}(u, v)$. After the forward mapping with the double-barrelled NN has been solved, inverse mappings from the canonical variates u and v to the original variables, as represented by the two standard MLP NNs in Fig. 11.1(b) and (c), are to be solved, where the mean squared error (MSE) of their outputs $\mathbf{x}'$ and $\mathbf{y}'$ are minimized with respect to $\mathbf{x}$ and $\mathbf{y}$, respectively.

In Fig. 11.1, the inputs $\mathbf{x}$ and $\mathbf{y}$ are mapped to the neurons in the hidden layer:

$$h_k^{(x)} = \tanh((\mathbf{W}^{(x)}\mathbf{x} + \mathbf{b}^{(x)})_k), \qquad h_n^{(y)} = \tanh((\mathbf{W}^{(y)}\mathbf{y} + \mathbf{b}^{(y)})_n), \tag{11.2}$$

where $\mathbf{W}^{(x)}$ and $\mathbf{W}^{(y)}$ are weight matrices, and $\mathbf{b}^{(x)}$ and $\mathbf{b}^{(y)}$, the offset or bias parameter vectors. The dimensions of $\mathbf{x}$, $\mathbf{y}$, $\mathbf{h}^{(x)}$ and $\mathbf{h}^{(y)}$ are l_1, m_1, l_2 and m_2 respectively.

The canonical variate neurons u and v are calculated from a linear combination of the hidden neurons $\mathbf{h}^{(x)}$ and $\mathbf{h}^{(y)}$, respectively, with

$$u = \mathbf{w}^{(x)} \cdot \mathbf{h}^{(x)} + \overline{b}^{(x)}, \qquad v = \mathbf{w}^{(y)} \cdot \mathbf{h}^{(y)} + \overline{b}^{(y)}. \tag{11.3}$$

These mappings are standard MLP NNs, and are capable of representing any continuous functions mapping from $\mathbf{x}$ to u and from $\mathbf{y}$ to v to any given accuracy, provided large enough l_2 and m_2 are used.

To maximize $\text{cor}(u, v)$, the objective function $J = -\text{cor}(u, v)$ is minimized by finding the optimal values of $\mathbf{W}^{(x)}$, $\mathbf{W}^{(y)}$, $\mathbf{b}^{(x)}$, $\mathbf{b}^{(y)}$, $\mathbf{w}^{(x)}$, $\mathbf{w}^{(y)}$, $\overline{b}^{(x)}$ and $\overline{b}^{(y)}$. The constraints $\langle u\rangle = 0 = \langle v\rangle$, and $\langle u^2\rangle = 1 = \langle v^2\rangle$ are also used, where $\langle \ldots \rangle$ denotes calculating the average over all the data points. The constraints are approximately satisfied by modifying the objective function to

$$J = -\mathrm{cor}(u, v) + \langle u \rangle^2 + \langle v \rangle^2 + (\langle u^2 \rangle^{1/2} - 1)^2 + (\langle v^2 \rangle^{1/2} - 1)^2. \tag{11.4}$$

A number of runs minimizing J starting from different random initial parameters is needed to deal with the presence of local minima in the objective function. The correlation is calculated over independent validation data and over training data. Runs where the correlation over validation data is worse than over training data are rejected to avoid overfitting, and among those remaining, the run attaining the highest $\mathrm{cor}(u, v)$ is selected as the solution.

We need to be able to map inversely from the canonical variates u and v back to the original data space. In Fig. 11.1(b), the NN (a standard MLP) maps from u to $\mathbf{x}'$ in two steps:

$$h_k^{(u)} = \tanh((\mathbf{w}^{(u)} u + \mathbf{b}^{(u)})_k), \quad \text{and} \quad \mathbf{x}' = \mathbf{W}^{(u)} \mathbf{h}^{(u)} + \overline{\mathbf{b}}^{(u)}. \tag{11.5}$$

The objective function $J_1 = \langle \|\mathbf{x}' - \mathbf{x}\|^2 \rangle$ is minimized by finding the optimal values of $\mathbf{w}^{(u)}$, $\mathbf{b}^{(u)}$, $\mathbf{W}^{(u)}$ and $\overline{\mathbf{b}}^{(u)}$. Again multiple runs are needed due to local minima, and validation data are used to reject overfitted runs. The MSE between the NN output $\mathbf{x}'$ and the original data $\mathbf{x}$ is thus minimized.

Similarly, the NN in Fig. 11.1(c) maps from v to $\mathbf{y}'$:

$$h_n^{(v)} = \tanh((\mathbf{w}^{(v)} v + \mathbf{b}^{(v)})_n), \quad \text{and} \quad \mathbf{y}' = \mathbf{W}^{(v)} \mathbf{h}^{(v)} + \overline{\mathbf{b}}^{(v)}, \tag{11.6}$$

with the objective function $J_2 = \langle \|\mathbf{y}' - \mathbf{y}\|^2 \rangle$ minimized.

The total number of parameters used by the NLCCA is $2(l_1 l_2 + m_1 m_2) + 4(l_2 + m_2) + l_1 + m_1 + 2$, though the number of effectively free parameters is four fewer due to the constraints on $\langle u \rangle$, $\langle v \rangle$, $\langle u^2 \rangle$ and $\langle v^2 \rangle$. After the first NLCCA mode has been retrieved from the data, the method can be applied again to the residual (i.e. $\mathbf{x} - \mathbf{x}'$ and $\mathbf{y} - \mathbf{y}'$) to extract the second mode, and so forth.

That the CCA is indeed a linear version of this NLCCA can be readily seen by replacing the hyperbolic tangent activation functions in (11.2), (11.5) and (11.6) with the identity function, thereby removing the nonlinear modelling capability of the NLCCA. Then the forward maps to u and v involve only a linear combination of the original variables $\mathbf{x}$ and $\mathbf{y}$, as in the CCA.

With three NNs in NLCCA, overfitting can occur in any of the three networks. With noisy data, the three objective functions are modified to:

$$J = -\mathrm{cor}(u, v) + \langle u \rangle^2 + \langle v \rangle^2 + (\langle u^2 \rangle^{1/2} - 1)^2 + (\langle v^2 \rangle^{1/2} - 1)^2$$
$$+ P \left[\sum_{ki} (W_{ki}^{(x)})^2 + \sum_{nj} (W_{nj}^{(y)})^2 \right], \tag{11.7}$$

$$J_1 = \langle \|\mathbf{x}' - \mathbf{x}\|^2 \rangle + P_1 \sum_k (w_k^{(u)})^2, \tag{11.8}$$

$$J_2 = \langle \|\mathbf{y}' - \mathbf{y}\|^2 \rangle + P_2 \sum_n (w_n^{(v)})^2, \tag{11.9}$$

where P, P_1 and P_2 are non-negative weight penalty parameters. Since the nonlinearity of a network is controlled by the weights in the hyperbolic tangent activation function, only those weights are penalized.

In prediction problems where one wants to obtain the values of a multivariate predictand from a multivariate predictor, i.e. $\tilde{\mathbf{y}} = f(\mathbf{x})$, values of the canonical variate $\tilde{v}$ must be predicted from values of the canonical variate u. The procedure is as follows: first the NLCCA model is trained from some training data and $\mathrm{cor}(u, v)$ is obtained, then from the predictor data $\mathbf{x}$ (which can either be from the training dataset or new data), corresponding u values are computed by the model. For canonical variates normalized to unit variance and zero mean, the linear least-squares regression solution for estimating $\tilde{v}$ from u is given by

$$\tilde{v} = u \,\mathrm{cor}(u, v), \tag{11.10}$$

(von Storch and Zwiers, 1999, p. 325). From $\tilde{v}$, one obtains a value for $\tilde{\mathbf{y}}$ via the inverse mapping NN.

To illustrate NLCCA, consider the following test problem (Hsieh, 2000): let

$$X_1 = t - 0.3t^2, \quad X_2 = t + 0.3t^3, \quad X_3 = t^2, \tag{11.11}$$

$$Y_1 = \tilde{t}^3, \quad Y_2 = -\tilde{t} + 0.3\tilde{t}^3, \quad Y_3 = \tilde{t} + 0.3\tilde{t}^2, \tag{11.12}$$

where t and $\tilde{t}$ are uniformly distributed random numbers in $[-1, 1]$. Also let

$$X_1' = -s - 0.3s^2, \quad X_2' = s - 0.3s^3, \quad X_3' = -s^4, \tag{11.13}$$

$$Y_1' = \mathrm{sech}(4s), \quad Y_2' = s + 0.3s^3, \quad Y_3' = s - 0.3s^2, \tag{11.14}$$

where s is a uniformly distributed random number in $[-1, 1]$. The shapes described by the $\mathbf{X}$ and $\mathbf{X}'$ vector functions are displayed in Fig. 11.2, and those by $\mathbf{Y}$ and $\mathbf{Y}'$ in Fig. 11.3. To lowest order, (11.11) for $\mathbf{X}$ describes a quadratic curve, and (11.13) for $\mathbf{X}'$, a quartic. Similarly, to lowest order, $\mathbf{Y}$ is a cubic, and $\mathbf{Y}'$ a hyperbolic secant. The signal in the test data was produced by adding the second mode ($\mathbf{X}'$, $\mathbf{Y}'$) to the first mode ($\mathbf{X}$, $\mathbf{Y}$), with the variance of the second mode being 1/3 that of the first mode. A small amount of Gaussian random noise, with standard deviation equal to

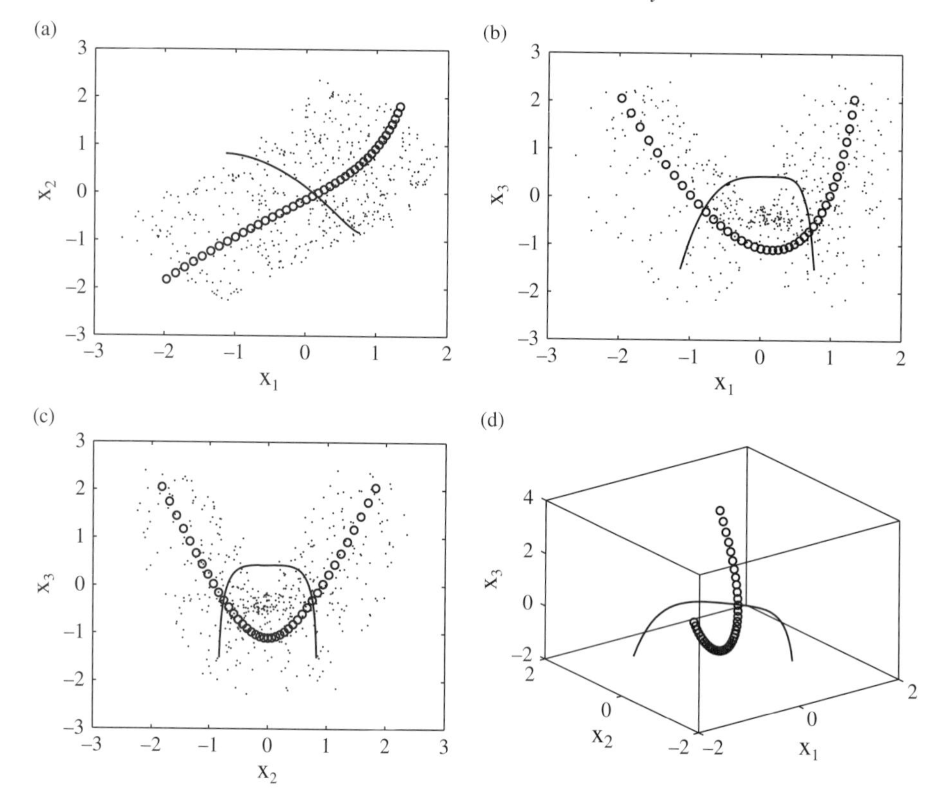

Fig. 11.2 The curve made up of small circles shows the first theoretical mode **X** generated from (11.11), and the solid curve, the second mode **X**′, from (11.13). (a) A projection in the x_1-x_2 plane. (b) A projection in the x_1-x_3 plane. (c) A projection in the x_2-x_3 plane. (d) A 3-dimensional plot. The actual data set of 500 points (shown by dots) was generated by adding mode 2 to mode 1 (with mode 2 having 1/3 the variance of mode 1) and adding a small amount of Gaussian noise. (Follows Hsieh (2000) with permission from Elsevier.)

10% of the signal standard deviation, was also added to the dataset. The dataset of $n = 500$ points was then standardized (i.e. each variable with mean removed was normalized by the standard deviation). Note that different sequences of random numbers t_i and $\tilde{t}_i$ ($i = 1, \ldots, n$) were used to generate the first modes **X** and **Y**, respectively. Hence these two dominant modes in the **x**-space and the **y**-space are unrelated. In contrast, as **X**′ and **Y**′ were generated from the same sequence of random numbers s_i, they are strongly related. The NLCCA was applied to the data, and the first NLCCA mode retrieved (Fig. 11.4 and 11.5) resembles the expected theoretical mode (**X**′, **Y**′). This is quite remarkable considering that **X**′ and **Y**′ have only 1/3 the variance of **X** and **Y**, i.e. the NLCCA ignores the large variance of **X** and **Y**, and succeeded in detecting the nonlinear correlated mode (**X**′, **Y**′). In

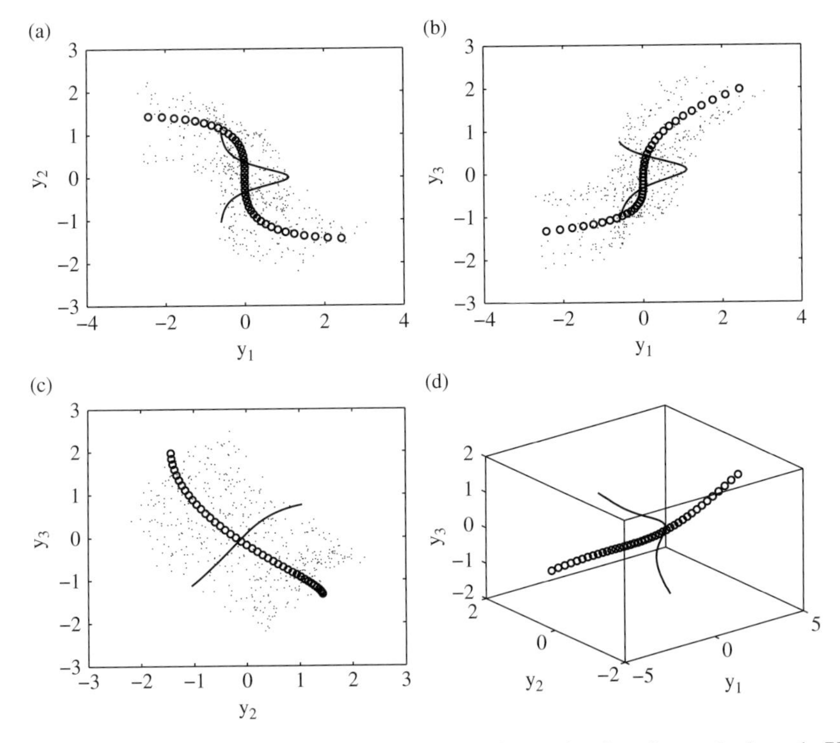

Fig. 11.3 The curve made up of small circles shows the first theoretical mode **Y** generated from (11.12), and the solid curve, the second mode **Y**′, from (11.14). (a) A projection in the y_1-y_2 plane. (b) A projection in the y_1-y_3 plane. (c) A projection in the y_2-y_3 plane. (d) A 3-dimensional plot. The data set of 500 points was generated by adding mode 2 to mode 1 (with mode 2 having 1/3 the variance of mode 1) and adding a small amount of Gaussian noise. (Follows Hsieh (2000) with permission from Elsevier.)

contrast, if nonlinear principal component analysis (NLPCA) is applied to **x** and **y** separately, then the first NLPCA mode retrieved from **x** will be **X**, and the first mode from **y** will be **Y**. This illustrates the essential difference between NLPCA and NLCCA. In the next two subsections, NLCCA is illustrated using real world problems.

11.1.1 Tropical Pacific climate variability

The NLCCA has been applied to analyze the tropical Pacific sea level pressure anomaly (SLPA) and sea surface temperature anomaly (SSTA) fields (where 'anomaly' means the deviation from the climatological monthly mean value). The six leading principal components (PC) of the SLPA and the six PCs of the SSTA during 1950–2000 were inputs to an NLCCA model (Hsieh, 2001a). The first

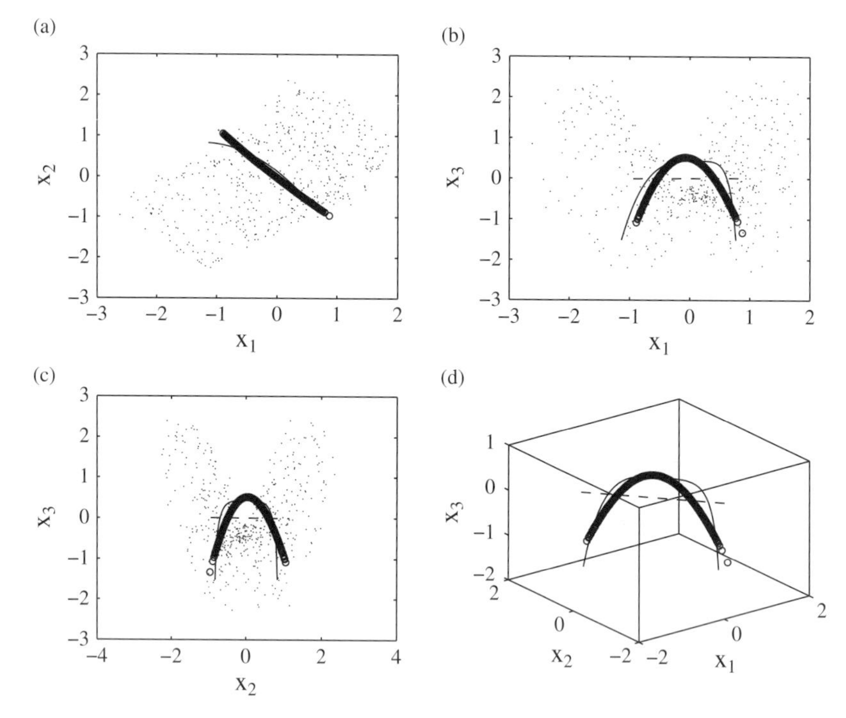

Fig. 11.4 The NLCCA mode 1 in **x**-space shown as a string of (densely overlapping) small circles. The theoretical mode $\mathbf{X}'$ is shown as a thin solid curve and the linear (CCA) mode is shown as a thin dashed line. The dots display the 500 data points. The number of hidden neurons used is $l_2 = m_2 = 3$. (Follows Hsieh (2000) with permission from Elsevier.)

NLCCA mode is plotted in the PC-spaces of the SLPA and the SSTA (Fig. 11.6), where only the three leading PCs are shown. For the SLPA (Fig. 11.6(a)), in the PC1-PC2 plane, the cool La Niña states are on the far left side (corresponding to low u values), while the warm El Niño states are in the upper right corner (high u values). The CCA solutions are shown as thin straight lines. For the SSTA (Fig. 11.6(b)), in the PC1-PC2 plane, the first NLCCA mode is a U-shaped curve linking the La Niña states in the upper left corner (low v values) to the El Niño states in the upper right corner (high v values). In general, the nonlinearity is greater in the SSTA than in the SLPA, as the difference between the CCA mode and the NLCCA mode is greater in Fig. 11.6(b) than in Fig. 11.6(a).

The MSE of the NLCCA divided by the MSE of the CCA is a useful measure of how different the nonlinear solution is relative to the linear solution – a smaller ratio means greater nonlinearity, while a ratio of 1 means that the NLCCA can find only a linear solution. This ratio is 0.951 for the SLPA and 0.935 for the SSTA,

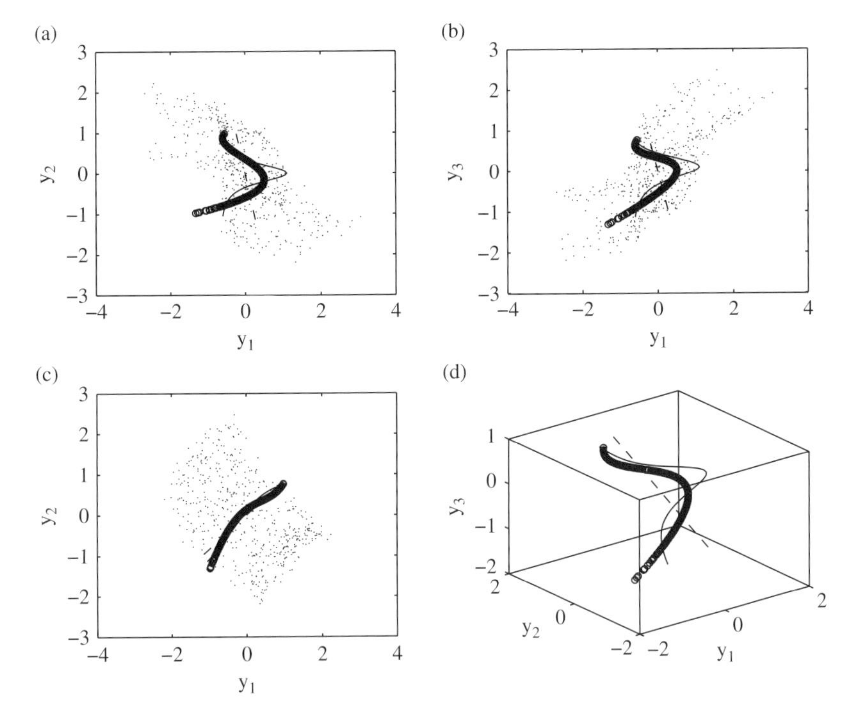

Fig. 11.5 The NLCCA mode 1 in **y**-space shown as a string of overlapping small circles. The thin solid curve is the theoretical mode $\mathbf{Y}'$, and the thin dashed line, the CCA mode. (Follows Hsieh (2000) with permission from Elsevier.)

confirming that the mapping for the SSTA was more nonlinear than that for the SLPA. When the data record was divided into two halves (1950–1975, and 1976–1999) to be separatedly analyzed by the NLCCA, Hsieh (2001a) found that this ratio decreased for thc second half, implying an increase in the nonlinearity of ENSO during the more recent period.

For the NLCCA mode 1, as u varies from its minimum value to its maximum value, the SLPA field varies from the strong La Niña phase to the strong El Niño phase (Fig. 11.7). The zero contour is further west during La Niña (Fig. 11.7(a)) than during strong El Niño (Fig. 11.7(c)). Meanwhile, as v varies from its minimum to its maximum, the SSTA field varies from strong La Niña to strong El Niño, where the SST anomalies during La Niña (Fig. 11.7(b)) are centred further west of the anomalies during El Niño (Fig. 11.7(d)).

Relation between the tropical Pacific wind stress anomaly (WSA) and SSTA fields have also been studied using the NLCCA (Wu and Hsieh, 2002, 2003), where interdecadal changes of ENSO behaviour before and after the mid 1970s climate regime shift were found, with greater nonlinearity found during 1981–99 than during 1961–75.

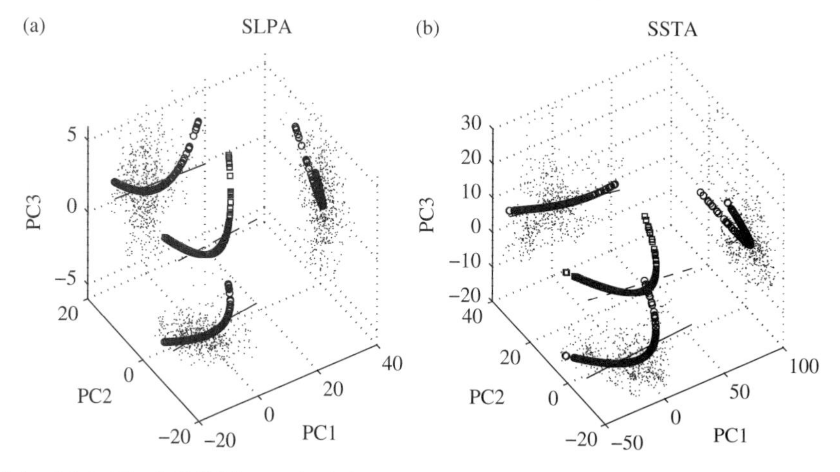

Fig. 11.6 The NLCCA mode 1 between the tropical Pacific (a) SLPA and (b) SSTA, plotted as (overlapping) squares in the PC_1-PC_2-PC_3 3-D space. The linear (CCA) mode is shown as a dashed line. The NLCCA mode and the CCA mode are also projected onto the PC_1-PC_2 plane, the PC_1-PC_3 plane, and the PC_2-PC_3 plane, where the projected NLCCA is indicated by (overlapping) circles, and the CCA by thin solid lines, and the projected data points (during 1950–2000) by the scattered dots. There is no time lag between the SLPA and the corresponding SSTA data. The NLCCA solution was obtained with the number of hidden neurons $l_2 = m_2 = 2$ (if $l_2 = m_2 = 1$, only a linear solution can be found). (Reproduced from Hsieh (2004) with permission from the American Geophysical Union.)

11.1.2 Atmospheric teleconnection

Relations between the tropical Pacific sea surface temperature anomalies (SSTA) and the Northern Hemisphere mid latitude winter atmospheric variability simulated in an atmospheric general circulation model (GCM) have also been explored using the NLCCA, which shows the value of NLCCA as a nonlinear diagnostic tool for GCMs (Wu *et al.*, 2003). Monthly 500 hPa geopotential height (Z500) and surface air temperature (SAT) data were produced by the CCCMA (Canadian Centre for Climate Modelling and Analysis) GCM2. An ensemble of six 47-year runs of GCM2 were carried out, in which each integration started from minor different initial conditions and was forced by the observed SST.

The five leading SST principal components (PC), and the five leading Z500 (or SAT) PCs (from January 1950 to November 1994) were the inputs to the NLCCA model. Here, minimum u and maximum u are chosen to represent the La Niña states and the El Niño states, respectively. For the Z500 field, instead of showing the spatial anomaly patterns during minimum v and maximum v, patterns are shown for the values of v when minimum u and maximum u occurred. As u takes its minimum value, the SSTA field presents a La Niña with negative anomalies (about

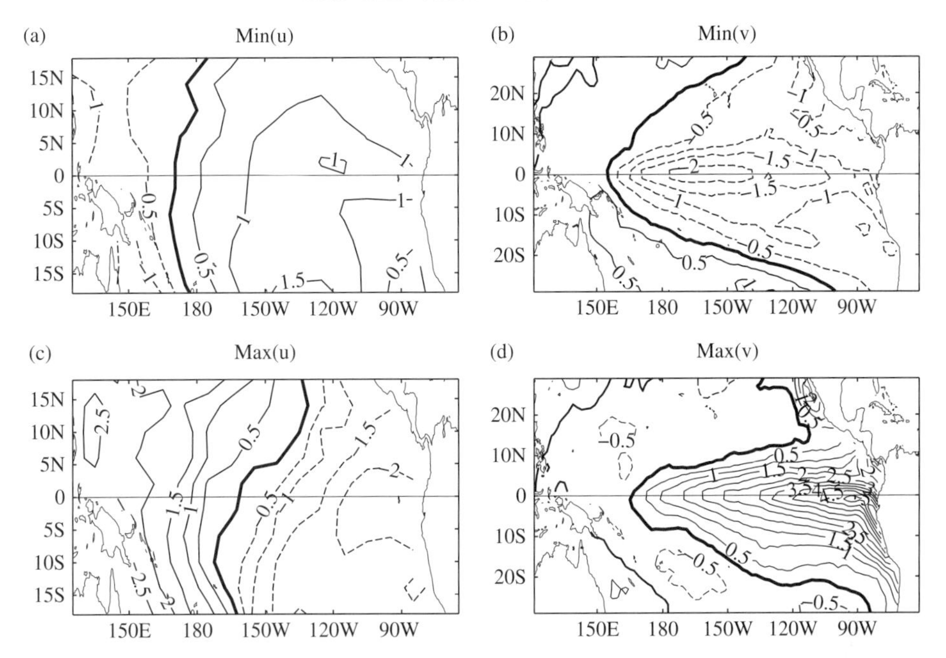

Fig. 11.7 The SLPA field when the canonical variate u of the NLCCA mode 1 is at (a) its minimum (strong La Niña), and (c) its maximum (strong El Niño); and the SSTA field when the canonical variate v is at (b) its minimum (strong La Niña), and (d) its maximum (strong El Niño). Contour interval is 0.5 hPa for the SLPA and 0.5 °C for the SSTA. Negative contours are shown as dashed lines, and the zero contour as a thick line. (Reproduced from Hsieh (2001a), with permission of the American Meteorological Society.)

−2.0 °C) over the central-western equatorial Pacific (Fig. 11.8(a)). *Atmospheric teleconnection* patterns are preferred states of large-scale flow in the atmosphere. The response field of Z500 is a negative *Pacific-North American (PNA) teleconnection* pattern (Horel and Wallace, 1981) with a positive anomaly centred over the North Pacific, a negative anomaly centred over western Canada and a positive anomaly centred over eastern USA.

As u takes its maximum value, the SSTA field presents a fairly strong El Niño with positive anomalies (about 2.5-3.0 °C) over the central-eastern Pacific (Fig. 11.8(b)). The SSTA warming centre shifts eastward by 30–40° longitude relative to the cooling centre in Fig. 11.8(a). The warming in Fig. 11.8(b) does not display peak warming off Peru, in contrast to the NLPCA mode 1 of the SSTA (Fig. 10.4(e)), where the El Niño peak warming occurred just off Peru. This difference between the NLPCA and the NLCCA mode implies that warming confined solely to the eastern equatorial Pacific waters does not have a corresponding strong mid-latitude atmospheric response, in agreement with Hamilton (1988).

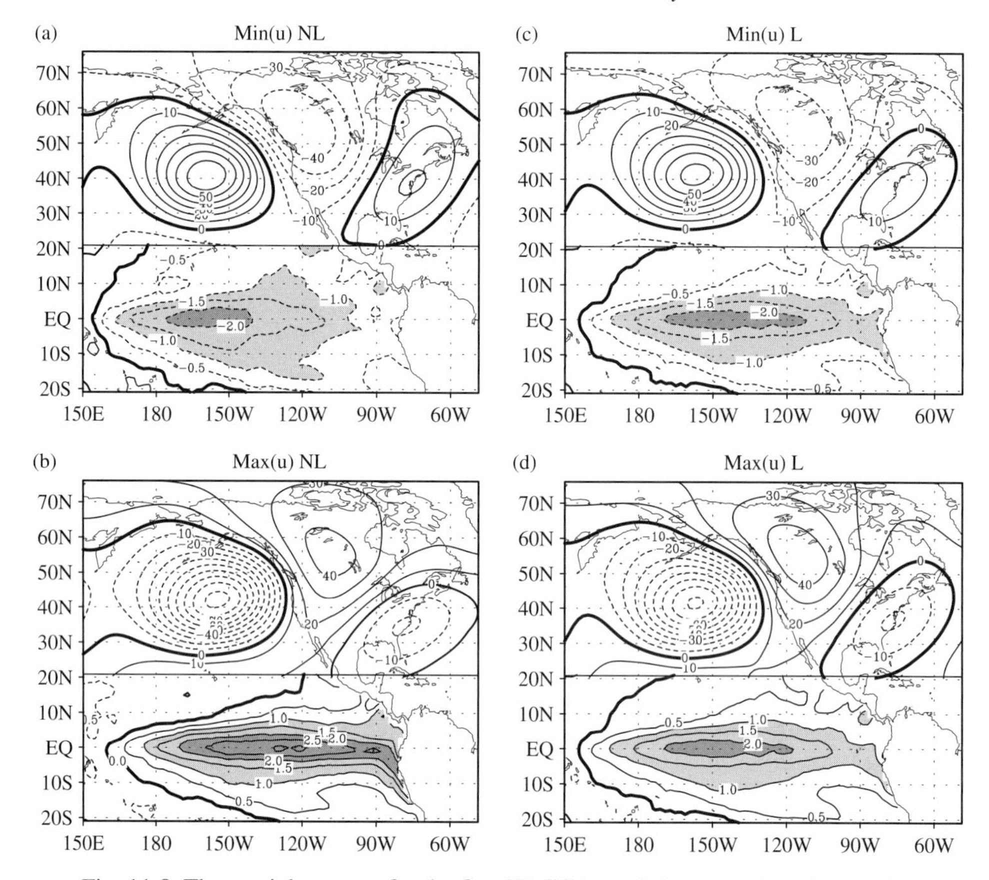

Fig. 11.8 The spatial patterns for the first NLCCA mode between the winter Z500 anomalies and the tropical Pacific SST anomalies as the canonical variate u takes its (a) minimum value and (b) maximum value. The Z500 anomalies with contour intervals of 10 m are shown north of 20°N. SST anomalies with contour intervals of 0.5 °C are displayed south of 20°N. The SST anomalies greater than +1 °C or less than −1 °C are shaded, and heavily shaded if greater than +2 °C or less than −2 °C. The linear CCA mode 1 is shown in panels (c) and (d) for comparison. (Reproduced from Wu *et al.* (2003), with permission of the American Meteorological Society).

The response field of Z500 is a PNA pattern (Fig. 11.8(b)), roughly opposite to that shown in Fig. 11.8(a), but with a notable eastward shift. The zero contour of the North Pacific anomaly is close to the western coastline of North America during the El Niño period (Fig. 11.8(b)), while it is about 10–15° further west during the La Niña period (Fig. 11.8(a)). The positive anomaly over eastern Canada and USA in Fig. 11.8(a) becomes a negative anomaly shifted southeastward in Fig. 11.8(b). The amplitude of the Z500 anomaly over the North Pacific is stronger during El Niño than La Niña, but the anomaly over western Canada and USA is weaker during El Niño than La Niña (Fig. 11.8(a) and (b)).

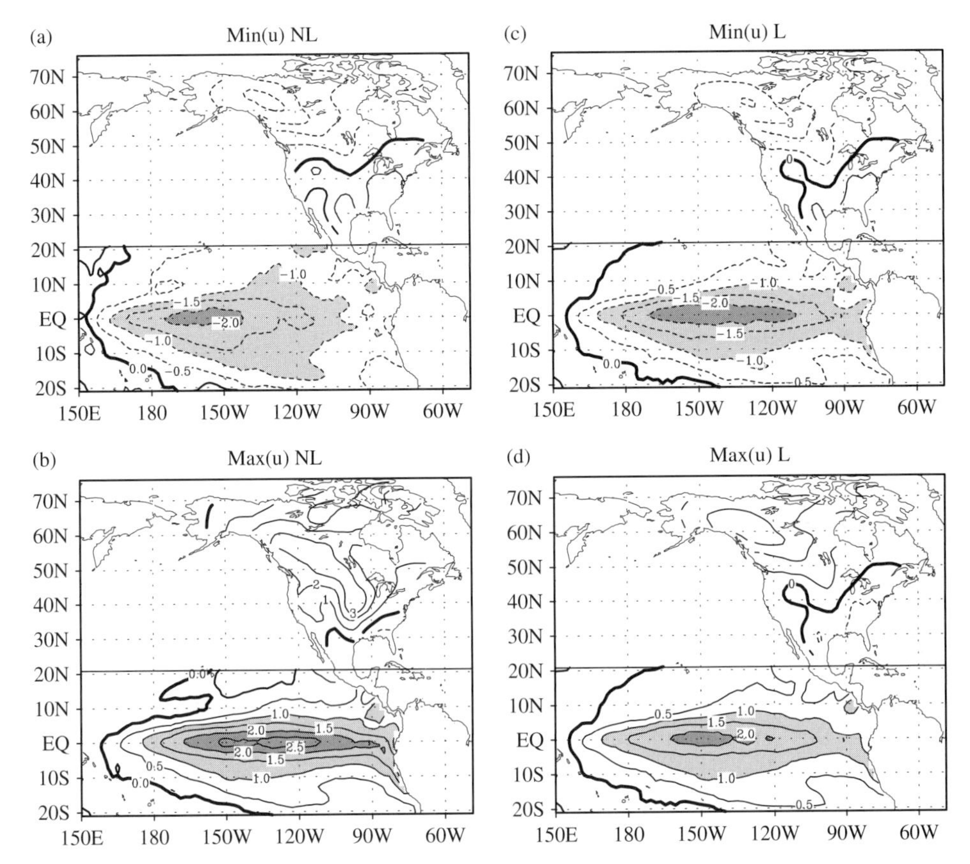

Fig. 11.9 Similar to Fig. 11.8, but for the NLCCA mode 1 between the surface air temperature (SAT) anomalies over North America and the tropical SST anomalies. The contour interval for the SAT anomalies is 1 °C. (Reproduced from Wu *et al.* (2003) with permission of the American Meteorological Society.)

For comparison, the spatial patterns extracted by the CCA mode 1 are shown in Fig. 11.8(c) and 11.8(d), where the atmospheric response to La Niña is exactly opposite to that for El Niño, though the El Niño pattens have somewhat stronger amplitudes. The SSTA patterns are also completely anti-symmetrical between the two extremes.

The NLCCA method was then applied to the SAT anomalies over North America and the tropical Pacific SSTA, with Fig. 11.9(a) and (b) showing the spatial anomaly patterns for both the SST and SAT associated with La Niña and El Niño, respectively. As u takes its minimum value, positive SAT anomalies (about 1 °C) appear over the southeastern USA, while much of Canada and northwestern USA are dominated by negative SAT anomalies. The maximum cooling centre (−4 °C) is located around northwestern Canada and Alaska (Fig. 11.9(a)). As u takes its maximum value (Fig. 11.9(b)), the warming centre (3 °C) is shifted to the southeast of the cooling centre in Fig. 11.9(a), with warming over almost the whole of North America except southeastern USA. The CCA mode 1 between the SAT and SST anomalies is also shown for reference (Fig. 11.9(c) and (d)).

11.2 Robust NLCCA

The Pearson correlation is not a robust measure of association between two variables, as its estimates can be affected by the presence of a single outlier (Section 1.3). To make the NLCCA model more robust to outliers, the Pearson correlation in the objective function J needs to be replaced by a more robust measure of correlation. The Spearman rank correlation (Section 1.3.1) is an obvious candidate, but since it is not a continuous function, it suffers from poor convergence when used in NLCCA (Cannon and Hsieh, 2008). Instead we turn to a different robust correlation function.

11.2.1 Biweight midcorrelation

The biweight midcorrelation (Mosteller and Tukey, 1977, pp. 203–209) is calculated in the same manner as the Pearson correlation coefficient, except that non-robust measures are replaced by robust measures (Wilcox, 2004). For instance, the mean and the deviation from the mean are replaced by the median and the deviation from the median, respectively.

To calculate the biweight midcorrelation function $\text{bicor}(x, y)$, first rescale x and y by

$$p = \frac{x - M_x}{9\,\text{MAD}_x}, \quad q = \frac{y - M_y}{9\,\text{MAD}_y}, \tag{11.15}$$

where M_x and M_y are the median values of x and y respectively and MAD_x and MAD_y (the median absolute deviations) are the median values of $|x - M_x|$ and $|y - M_y|$ respectively. Next, the sample *biweight midcovariance* is given by

$$\text{bicov}(x, y) = \frac{N \sum_i a_i b_i c_i^2 d_i^2 (x_i - M_x)(y_i - M_y)}{[\sum_i a_i c_i (1 - 5p_i^2)][\sum_i b_i d_i (1 - 5q_i^2)]}, \tag{11.16}$$

where there are $i = 1, \ldots, n$ observations, $a_i = 1$ if $-1 \le p_i \le 1$, otherwise $a_i = 0$; $b_i = 1$ if $-1 \le q_i \le 1$, otherwise $b_i = 0$; $c_i = (1 - p_i^2)$; and $d_i = (1 - q_i^2)$. Note that the outliers (with $|x - M_x| > 9\,\text{MAD}_x$ or $|y - M_y| > 9\,\text{MAD}_y$) are excluded from (11.16) by having either $a_i = 0$ or $b_i = 0$. The *biweight midcorrelation* is then given by

$$\text{bicor}(x, y) = \frac{\text{bicov}(x, y)}{\sqrt{\text{bicov}(x, x)\,\text{bicov}(y, y)}}. \tag{11.17}$$

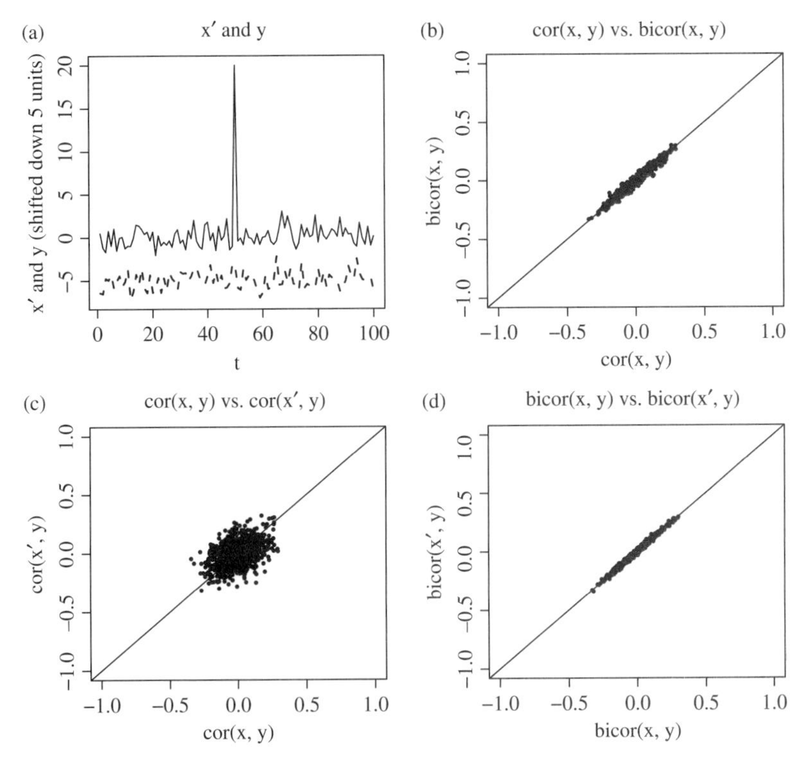

Fig. 11.10 Empirical comparison between the Pearson correlation (cor) and the biweight midcorrelation (bicor) on random variables x and y, each with samples drawn from a standard Gaussian distribution, and x' and y, where x' is the same as x but with one case replaced with an outlier. (a) Sample time series of x' (solid) and y (dashed). (b) Compares cor(x, y) and bicor(x, y), with the diagonal line indicating perfect agreement between the two. (c) Compares cor(x, y) and cor(x', y). (d) Compares bicor(x, y) and bicor(x', y). Plots are for 1000 randomly generated datasets. (Reproduced from Cannon and Hsieh (2008) with permission of the European Geosciences Union.)

The biweight midcorrelation, like the Pearson correlation, ranges from -1 (negative association) to $+1$ (positive association).

Figure 11.10 shows estimates of the Pearson correlation and the biweight midcorrelation between Gaussian random variables x and y (with unit standard deviation and zero mean) and between x' and y, where x' is the same as x but with one data point replaced by an outlier (Fig. 11.10(a)). On the outlier-free dataset, both cor(x, y) and bicor(x, y) give approximately equal estimates of the strength of association between the variables (Fig. 11.10(b)). Estimates of cor(x', y) are strongly affected by the outlier, showing almost no association between values calculated with and without the outlying data point (Fig. 11.10(c)), whereas estimates of bicor(x', y) are essentially unaffected by the outlier (Fig. 11.10(d)).

Note that NLCCA with the Pearson correlation objective function may fail when there are *common outliers*, i.e. outliers occurring simultaneously in both x and y. Consider two identical sinusoidal series, each with a common outlier

$$x_i = y_i = \sin(0.5i) + \delta(i), \tag{11.18}$$

where

$$\delta(i) = \begin{cases} 6 & \text{at } i = 150, \\ 0 & \text{elsewhere}, \end{cases} \tag{11.19}$$

with $i = 1, 2, ..., 300$. Next, create new series x' and y' by adding Gaussian noise (with standard deviation of 0.5) to x and y. The expected values of $\text{cor}(x', y')$ and $\text{bicor}(x', y')$ are found to be 0.71 and 0.68 respectively. Now, consider the effects of very nonlinear functions $u(x')$ and $v(y')$ on the values of $\text{cor}(u, v)$ and $\text{bicor}(u, v)$. Let us take $u = x'^p$, $v = y'^p$, where p is a positive odd integer. Increasing the value of p effectively increases the nonlinearity of u and v, as well as the separation between the outlier and the non-outliers (compare u when $p = 1$ in Fig. 11.11(a) with u when $p = 9$ in Fig. 11.11(b)). Values of $\text{cor}(u, v)$ and $\text{bicor}(u, v)$ for values of p from 1 to 9 are shown in Fig. 11.11(c). The Pearson correlation can be increased simply by increasing p, whereas the biweight midcorrelation decreases as p increases. This case illustrates how increasing the nonlinearity of the mapping functions u and v (by increasing p) can lead to very high Pearson correlation.

In the context of NLCCA, spuriously high values of $\text{cor}(u, v)$ can be found by the double-barrelled network (Fig. 11.1(a)) when the nonlinear NN mapping greatly magnifies a common outlier in both x and y. This artefact can be particularly dangerous when NLCCA is applied to datasets affected by strong, concurrent climate signals, for example those with large El Niño or La Niña anomalies (Hsieh, 2001a). The NLCCA method performed worse than CCA when weight penalty terms were not used to reduce the nonlinearity of the double-barrelled network. Based on results shown in Fig. 11.11, adopting bicor in the objective function should prevent this artefact.

11.2.2 Inverse mapping

To complete the robust NLCCA model, we need to consider the inverse mapping from u and v back to $\mathbf{x}'$ and $\mathbf{y}'$ (i.e. the two networks in Fig. 11.1(b) and (c)). For the objective functions of these two networks, we have used the MSE in (11.8) and (11.9). In Section 6.2, we showed that in the limit of an infinite number of observations and with a flexible enough model (e.g. an MLP NN with enough hidden neurons), the model converges to the conditional mean if the MSE is used and the conditional median if the mean absolute error (MAE) is used. As the median

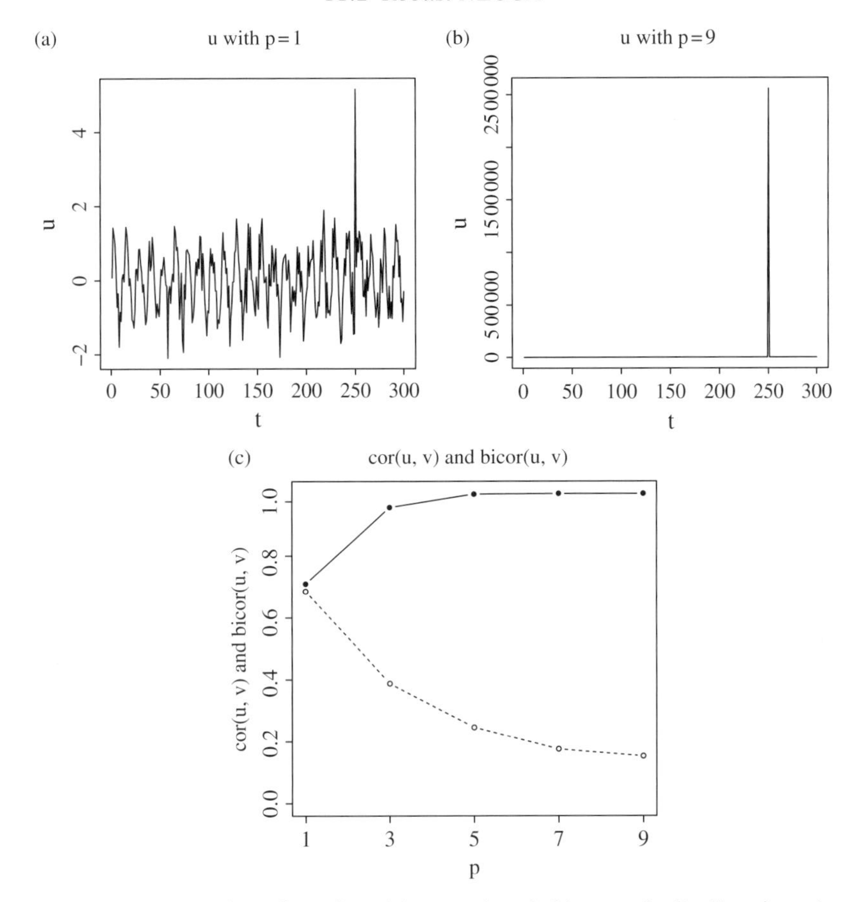

Fig. 11.11 Time series of u when (a) $p = 1$ and (b) $p = 9$. Outliers in v (not shown) occur at the same time as those in u. (c) The effect on cor(u, v) (solid line) and bicor(u, v) (dashed line) from increasing the separation between common outlier and non-outlier points by increasing p. (Reproduced from Cannon and Hsieh (2008) with permission of the European Geosciences Union.)

is robust to outliers whereas the mean is not, replacing the MSE by the MAE in (11.8) and (11.9) will make the inverse mappings more robust.

In summary, for the robust NLCCA, bicor(u, v) replaces cor(u, v) in the objective function J in (11.7), while the MAE replaces the MSE in J_1 and in J_2, in (11.8) and (11.9). There are two other variants of NLCCA: instead of using bicor and MAE, one variant uses bicor and MSE, while a second variant uses cor and MAE in the objective functions. Together with the original standard NLCCA (with cor and MSE), there is thus a total of four variants of NLCCA.

To test the performance of the four variants of the NLCCA models, 50 training and test datasets, each with 500 observations, were generated as follows in Cannon and Hsieh (2008). The first synthetic mode, consisting of **X** and **Y**, was generated as before from (11.11) and (11.12), except that the random number $\tilde{t}$ in (11.12) was replaced by t, i.e. $\mathbf{X}$ and $\mathbf{Y}$ were generated from the same sequence of random numbers t. Mode 2, consisting of $\mathbf{X}'$ and $\mathbf{Y}'$, was generated from (11.13) and (11.14). The signal in each dataset was produced by adding the second mode to the first mode, with the variance of the second equal to one third that of the first. Gaussian random noise with standard deviation equal to 50% of the signal standard deviation was added to the data. The variables were then standardized to zero mean and unit standard deviation.

The four variants of the NLCCA model were developed separately on the training datasets and applied to the test datasets. All MLP mappings used three hidden neurons in a single hidden layer and the NNs were trained without weight penalty terms. To avoid local minima, each network in Fig. 11.1 was trained 30 times, each time starting from different random initial weight and offset parameters, and the network with the lowest objective function value was then selected and applied to the test data.

Root MSE (RMSE) values between the first synthetic mode and the first mode extracted from the test data by the NLCCA variants are shown in Fig. 11.12 for the 50 test datasets. On average, all models performed approximately the same, although, for the leading NLCCA mode of the $\mathbf{x}$ dataset (Fig. 11.12(a)), NLCCA with bicor/MSE objective functions yielded the lowest median RMSE (0.44), followed by NLCCA with bicor/MAE (0.45) and NLCCA with cor/MSE (0.45). The NLCCA with cor/MAE performed worst with a median RMSE of 0.47. Median RMSE values and relative rankings of the models were the same for the leading NLCCA mode of the $\mathbf{y}$ dataset (Fig. 11.12(b)).

Of the four variants, NLCCA with the robust objective functions (bicor/MAE) was the most stable. No trial yielded an RMSE in excess of the series standard deviation of one (as indicated by the horizontal dashed line in Fig. 11.12). The other models had at least one trial with an RMSE value greater than one, indicating severe overfitting.

Overall, results for the synthetic dataset suggest that replacing the cor/MSE objective functions in NLCCA with bicor/MAE objective functions leads to a more stable model that is less susceptible to overfitting and poor test performance. All models were run without weight penalty terms in this comparison. In practice, the non-robust models will need weight penalty terms to reduce overfitting, as is done in the next example, where NLCCA models are applied to a real-world climate prediction problem.

11.2.3 Prediction

To use NLCCA for multivariate prediction, a regression model is needed to estimate v from u (or vice versa). For the standard NLCCA model, the linear least

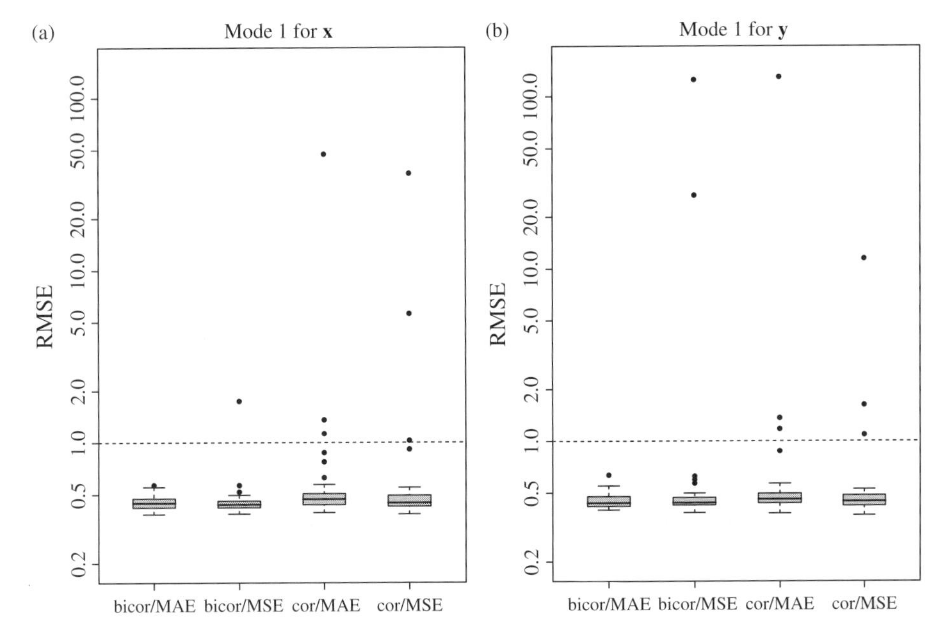

Fig. 11.12 Boxplots showing the distribution of RMSE between the first synthetic mode and the first mode extracted by NLCCA models for (a) **x** and (b) **y**, with different combinations of non-robust and robust objective functions over 50 trials. Boxes extend from the 25th to 75th percentiles, with the line indicating the median. Whiskers represent the most extreme data within ± 1.5 times the interquartile range (i.e. the box height); values outside this range are plotted as dots. The dashed line indicates an RMSE equal to one. The ordinate is log-scaled to accommodate the large range in RMSE. (Reproduced from Cannon and Hsieh (2008) with permission of the European Geosciences Union.)

squares estimate for the regression coefficient is given by (11.10). Similarly, the biweight midcorrelation is associated with a robust regression model which can be used to predict values of one canonical variate from the other. Following Lax (1985) and Hou and Koh (2004), the regression solution for estimating $\tilde{v}$ from u is given by

$$\tilde{v} = u \, \text{bicor}(u, v), \tag{11.20}$$

for canonical variates normalized to unit variance and zero mean. From $\tilde{v}$, one can then obtain a value for **y** via the inverse mapping NN.

Next test the prediction of tropical Pacific SST anomalies using tropical Pacific SLP anomalies as predictors. With data from 1948 to 2003, the climatological seasonal cycle was removed, data were linearly detrended, and a 3 month running mean filter was applied. The six leading SLP principal components (PC) and the six leading SST PCs are the **x** and **y** inputs to the NLCCA model. Only three variants are tested here, as the model with cor/MAE objective functions was dropped from consideration. The SLP state was used to forecast SST at lead times of 0, 3,

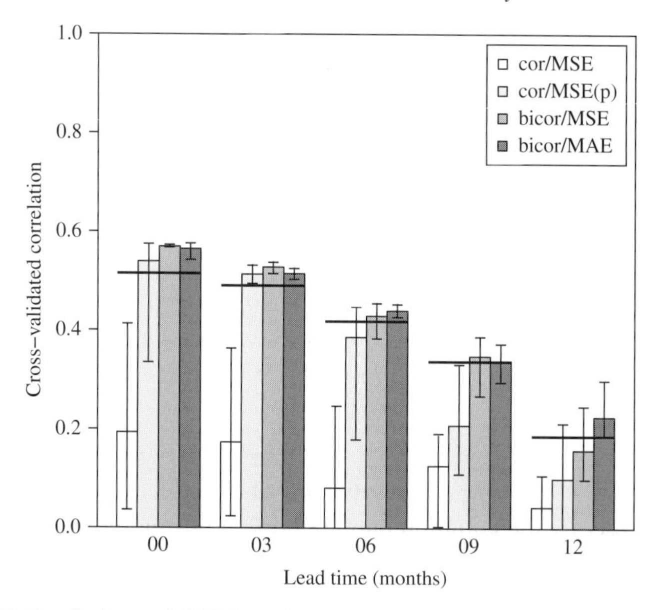

Fig. 11.13 Predictions of SST based on mode 1: cross-validated correlation skill for NLCCA models trained with cor/MSE, bicor/MSE, and bicor/MAE objective functions. Weight penalty was applied to the model denoted cor/MSE(p). Bars show the mean correlation over the spatial domain, averaged over ten trials. Vertical lines extend from the minimum to the maximum spatial mean correlation from the ten trials. Horizontal lines show correlation skill from the CCA model. The ordinate is limited to showing positive cross-validated skill. (Reproduced from Cannon and Hsieh (2008) with permission of the European Geosciences Union.)

6, 9, and 12 months. Lead times are defined as the number of months from the predictor observation to the predictand, e.g. a forecast at 3 month lead time from January would use the January SLP to forecast the April SST.

Forecast results from the NLCCA mode 1 are shown in Fig. 11.13. For reference, results from linear CCA models are also shown. Cross-validated Pearson correlation skill is averaged over the entire domain following reconstruction of the SST anomaly field from the six predicted SST PCs. Similar performance was found for the RMSE (not shown). Results with weight penalty are only given for the NLCCA model with cor/MSE objective functions as the addition of penalty terms to models with the bicor objective function did not generally lead to significant changes in skill in this example.

Without weight penalty, the NLCCA model with cor/MSE objective functions performed poorly, exhibiting mean skills worse than CCA at all lead times. The NLCCA method with bicor/MSE objective functions and bicor/MAE objective functions performed much better. In general, NLCCA models with bicor exhibited the least variability in skill between repeated trials. For NLCCA with cor/MSE objective functions, the large range in skill for the various trials indicates a very unstable model.

Little difference in skill was evident between bicor/MSE and bicor/MAE models, which suggests that the switch from cor to bicor in the double-barrelled network objective function was responsible for most of the increase in skill relative to the standard NLCCA model. Inspection of the canonical variates shows that this was due to the insensitivity of the bicor objective function to the common outlier artefact described in Section 11.2.1 and illustrated in Fig. 11.11.

Plots of the canonical variates u and v for the first mode of NLCCA models with cor/MSE and bicor/MSE objective functions at the 0 month lead time are shown in Fig. 11.14 along with the leading SST and SLP PCs. For these series, values of cor(u, v) and bicor(u, v) were 1.00 and 0.96 respectively. The high correlation between u and v for the NLCCA model with the cor objective function was driven almost exclusively by the common outliers present during the extremely strong El Niño of 1997–1998. With the 1997–1998 outliers removed, cor(u, v) dropped to 0.28. On the other hand, the high correlation between u and v for the NLCCA model with the bicor objective function was indicative of the strong relationship between SLP and SST, as evidenced by the Pearson correlation of 0.91 between the leading SST and SLP PCs (Fig. 11.14(a)).

Results discussed to this point have been for NLCCA models without weight penalty. Hsieh (2001a) found that the addition of weight penalty to the standard NLCCA model led to improvements in performance, due in part to the avoidance of the common outlier artefact. Addition of weight penalty to the standard NLCCA model resulted in improvements in mean correlation skill, although performance still lagged behind NLCCA with the bicor objective function at 9 and 12 month lead times (Fig. 11.13). At 0, 3, and 6 month lead times, maximum skill over the ten trials did, however, exceed the mean level of skill of the bicor-based models, which suggests that an appropriate amount of weight penalty can result in a well performing model. Inspection of the time series of u and v for the best performing runs suggests that improved performance was due to avoidance of the common outlier artefact. However, the wide range in performance over the ten trials (e.g. at 0 and 6 month lead times) reflects the instability of the training and the cross-validation procedure used to choose the weight penalty parameters. In practice, it may be difficult to reach the performance level of the robust model consistently by relying solely on weight penalty to control overfitting of the standard NLCCA model.

NLCCA models with the bicor/MSE and bicor/MAE objective functions tended to perform slightly better than CCA. For the bicor/MAE model, the small improvement in performance was significant (i.e. minimum skill over the ten trials exceeded the CCA skill) at 0, 3, 6, and 12 month lead times, while the same was true of the bicor/MSE model at 0 and 3 month lead times.

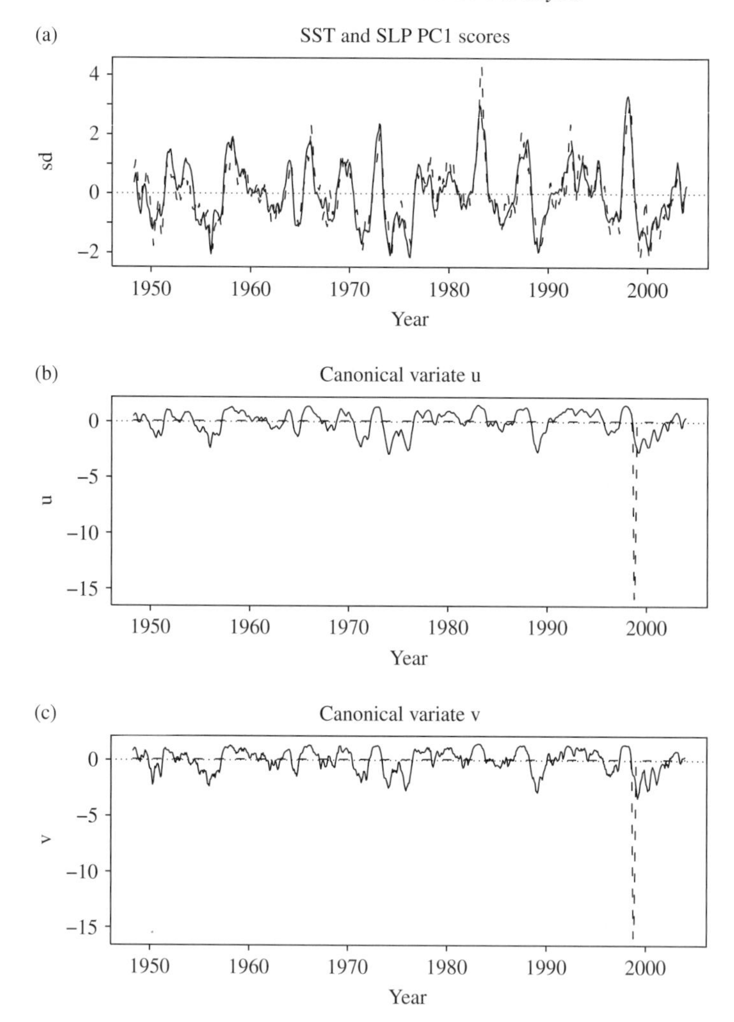

Fig. 11.14 (a) The first SST PC (solid line) and the first SLP PC (dashed line) in units of standard deviation. (b) The canonical variate u for the leading NLCCA mode using cor/MSE objective functions (dashed line) and using bicor/MSE objective functions (solid line). (c) Canonical variate v for the leading NLCCA mode using cor/MSE objective functions (dashed line) and bicor/MSE objective functions (solid line). (Reproduced from Cannon and Hsieh (2008) with permission of the European Geosciences Union.)

Concluding remarks

In this chapter, we have presented NLCCA via MLP NN models: NLCCA by kernel methods have also been developed (Lai and Fyfe, 2000; Suykens *et al.*, 2002; Hardoon *et al.*, 2004; Shawe-Taylor and Cristianini, 2004), though there do not seem to be any applications of kernel NLCCA methods to environmental problems.

These kernel approaches all used the Pearson correlation, which we have shown to be non-robust to outliers. In particular, common outliers occurring simultaneously in both $\mathbf{x}$ and $\mathbf{y}$ can lead to spuriously high Pearson correlation values. Use of biweight midcorrelation in NLCCA greatly improves the robustness of the NLCCA method. However, for climate variability outside the tropics, even robust NLCCA may not succeed in extracting a nonlinear signal, because time-averaging daily data to form seasonal (or longer) climate data strongly linearizes the data via the central limit theorem effect, as pointed out in Section 6.10. Nevertheless, it is encouraging that extra-tropical climate signals have been successfully extracted from GCM model data by NLCCA (Wu *et al.*, 2003). The advantage of using GCM results is that there were six runs in the ensemble, which provide far more data than the observed records, thereby allowing successful extraction of nonlinear atmospheric teleconnection patterns by NLCCA.

Exercises

(11.1) In NLCCA, the objective function J in (11.4) includes terms like $(\langle u^2 \rangle^{1/2} - 1)^2$ so that the canonical variate u approximately satisfies the normalization condition $\langle u^2 \rangle = 1$. If instead the simpler term $(\langle u^2 \rangle - 1)^2$ is used in J, then sometimes the converged solution unexpectedly has $\langle u^2 \rangle \approx 0$. With $\langle u^2 \rangle = (\sum_{i=1}^{n} u_i^2)/n$, where n is the number of observations, use $\partial f/\partial u_i$ to explain the different convergence properties at $\langle u^2 \rangle = 0$ for $f = (\langle u^2 \rangle - 1)^2$ and $f = (\langle u^2 \rangle^{1/2} - 1)^2$.

(11.2) Choose two time series x and y and compare the Pearson correlation with the biweight midcorrelation for the two time series. Repeat the comparison when a data point in x is replaced by an outlier (i.e. replace the data point by a very large number). Repeat the comparison when one particular pair of (x, y) values is replaced by a pair of outliers.

12

Applications in environmental sciences

In this final chapter, we survey the applications of machine learning methods in the various areas of environmental sciences – e.g. remote sensing, oceanography, atmospheric science, hydrology and ecology. More examples of applications are given in Haupt *et al.* (2009).

In the early applications of methods like NN to environmental problems, researchers often did not fully understand the problem of overfitting, and the need to validate using as much independent data as possible, by e.g. using cross-validation. For historical reasons, some of these early papers are still cited here, although the control of overfitting and/or the validation process may seem primitive by latter day standards.

In some of the papers, the authors used more than one type of machine learning method and compared the performance of the various methods. A word of caution is needed here. In our present peer-reviewed publication system, a new method can be published only if it is shown to be better in some way than established methods. Authors therefore tend to present their new methods in the best possible light to enhance their chance of passing the peer review. For instance, an author might test his new method A against the traditional method B on two different datasets and find method A and B each outperformed the other on one dataset. The author would then write up a journal paper describing his new method A and its application to the dataset which showed method A outperforming method B. Publications biased in favour of new methods can arise even without the authors doing 'cherry-picking'. Suppose author X applied method A to a dataset and found A performing better than the traditional method B, while author Y applied the same method A to a different dataset and unfortunately found method A worse than method B. The paper by author X was published but the paper by the unlucky author Y was rejected by the journal. Actually, a simple recommendation like 'method A is better than method B' is not meaningful for many reasons. For example: (i) method A outperforms B when the dataset has high signal-to-noise ratio, but B beats A when the ratio is low; (ii) A beats B when there are many independent observations but vice versa when there are few observations; (iii) A beats B when the noise is Gaussian but vice versa when the noise is non-Gaussian; (iv) A beats B when there are few predictor variables, but vice versa when there are many predictors, etc. In short, *caveat emptor*!

12.1 Remote sensing

Satellite remote sensing has greatly facilitated mankind's ability to monitor the environment. By deploying a radiometer (i.e. an instrument measuring radiation at certain frequency bands) on the satellite, remote sensing allows us to measure indirectly an astonishing number of variables in the air, land and water (Elachi and van Zyl, 2006), e.g. surface temperatures, winds, clouds, sea ice, sea level displacements, vegetation on land and chlorophyll concentration in the ocean, etc. There are two types of remote sensing – *passive* sensing where only a receiver is used, and *active* sensing where both a transmitter and a receiver are used. Active sensing, which operates similarly to radar, requires far more power to operate than passive sensing, as the transmitter has to emit a powerful enough signal beam to Earth below so that the return signal can be detected on the satellite. However, active sensing has good control of the signal to be detected since the signal is generated originally from the transmitter. One requirement of active remote sensing is that the transmitted radiation must be able to reflect strongly from the Earth's surface back to the receiver, hence only microwaves have been used in active sensing, as other radiation (e.g. visible and infrared) are too strongly absorbed at the Earth's surface. In contrast to active sensing, passive sensing simply receives whatever radiation is coming from the Earth. As the radiation passes through the atmosphere, it is scattered, absorbed and re-emitted by the clouds, aerosols and molecules in the atmosphere. Trying to use this radiation to infer variables at the Earth's surface is obviously a very difficult task, so it is amazing how scientists/engineers have managed to produce the great variety of good quality satellite data products.

Let $\mathbf{v}$ denote some variables of interest on Earth and $\mathbf{s}$, measurements made by the satellite. For instance, $\mathbf{s}$ can be the measurements made at several frequency bands by the radiometer, while $\mathbf{v}$ can be the chlorophyll and sediment concentrations in the surface water. The goal is to find a function $\mathbf{f}$ (a retrieval algorithm), where

$$\mathbf{v} = \mathbf{f}(\mathbf{s}) \tag{12.1}$$

allows the retrieval of $\mathbf{v}$ from the satellite measurements. This retrieval process is not always straightforward, since two different $\mathbf{v}$ can yield the same $\mathbf{s}$ (e.g. snow on the ground and clouds may both appear white as seen by the satellite), so the problem can be ill-posed. In fact, this is an inverse problem to a forward problem (which is well-posed):

$$\mathbf{s} = \mathbf{F}(\mathbf{v}), \tag{12.2}$$

where the variables on earth cause a certain **s** to be measured by the satellite's radiometer. One common way to resolve the ill-posedness is to use more frequency bands (e.g. snow and clouds are both white in the visible band, but in the thermal infrared (IR) band, they can be separated by their different temperatures). In recent years, NN methods for nonlinear regression and classification have been widely applied to solve the retrieval problem (12.1) for many types of satellite data (see e.g. the review by Krasnopolsky (2007)).

12.1.1 Visible light sensing

Ocean colour

Oceanographers have long known that tropical waters are blue due to conditions unfavourable to phytoplankton (i.e. photosynthetic plankton) growth, while in biologically productive regions, the waters are more greenish due to the presence of phytoplankton. By viewing the ocean from space in the visible light range, it is possible to extract the chlorophyll concentrations in the surface water, from which one can estimate the phytoplankton concentration, key to the ocean's biological productivity. There is atmospheric interference for the radiation travelling from the ocean's surface to the satellite, e.g. aerosol concentration could affect the ocean colour viewed from the satellite. Further complications arise when there are also suspended particulate materials (mainly sediments), and/or dissolved organic matter from decayed vegetation (known as 'gelbstoff' or simply 'yellow substance') in the water. As coastal water quality gradually degrades from increased pollution and human activities in many parts of the world, satellite data also offer a relatively inexpensive way to monitor water quality over wide areas of coastal waters.

Waters are divided into two classes: Case 1 waters have their optical properties dominated by phytoplankton and their degradation products only, whereas Case 2 waters have, in addition to phytoplankton, non-chlorophyll related sediments and/or yellow substance. Fortunately open ocean waters tend to be Case 1 waters, which are much simpler for retrieving the chlorophyll concentration than Case 2 waters, which tend to be found in coastal regions. Case 1 waters cover over 90% of the world's oceans.

In Keiner and Yan (1998), an MLP NN model was used to model the function **f** for retrieving the chlorophyll and suspended sediment concentrations from the satellite-measured radiances. The algorithm was developed for estuarine water (Case 2 water) using data from Delaware Bay. The MLP has three inputs (namely measurements from the three visible frequency channels of the Landsat Thematic

Mapper), a hidden layer with two hidden neurons, and one output (either the chlorophyll or the sediment concentration, i.e. two separate networks were trained for the two variables). *In situ* measurements of chlorophyll-a and sediment concentration in Delaware Bay were used as training data. The NN approach was compared with the more standard approach using multiple (linear and log-linear) regression analysis. The root mean squared (RMS) errors for the NN approach were $<10\%$, while the errors for regression analysis were $>25\%$.

In Schiller and Doerffer (1999), an MLP NN model was developed to derive the concentrations of phytoplankton pigment, suspended sediments and gelbstoff, and aerosol (expressed as the horizontal ground visibility) over turbid coastal waters from satellite data. The procedure was designed for the Medium Resolution Imaging Spectrometer (MERIS) flown onboard the satellite Envisat of the European Space Agency (ESA). This instrument has 15 frequency channels in the 390 nm to 1040 nm spectral range (from visible to near-infrared). Synthetic data from a radiative transfer model were used to train the NN model. The inputs of the NN model are from the multiple channels, and the four outputs are the three water constituent concentrations and the visibility. The NN has two hidden layers with 45 and 12 neurons respectively. The two hidden layer network performed better than a network with a single hidden layer of some hundreds of neurons, as well as the more traditional functional approximation using Chebychev polynomials. The NN model works well for both Case 1 and Case 2 waters. Although the training of the NN is time consuming, once trained, the NN model runs very fast, being eight times faster than the forward model (i.e. the radiative transfer model). Since the model inversion by a minimization method typically needs 10-20 calls of the forward model, the speedup by using the NN model instead is at least two orders of magnitude. Further improvements are given in Schiller and Doerffer (2005) and Schiller (2007).

In Gross *et al.* (2000), an MLP NN model was used to model the transfer function f mapping from the satellite-measured radiances to the phytoplankton pigment concentration (chlorophyll-a plus phaeophytin) for Case 1 water. The five inputs of the NN model are the five visible frequency channels from the Sea-viewing Wide Field-of-view Sensor (SeaWiFS), and the single output is the pigment concentration. There are two hidden layers with six and four hidden neurons respectively. The NN model was trained with synthetic data generated by the Morel (1988) model with Gaussian noise added. When applied to *in situ* California Cooperative Oceanic Fisheries Investigations data, the NN algorithm performed better than two standard reflectance ratio algorithms – relative RMS errors on pigment concentration were reduced from 61 and 62 to 38%, while absolute RMS errors were reduced from 4.43 and 3.52 to 0.83 mg m^{-3}. When applied to SeaWiFS derived imagery, there was statistical evidence showing the NN algorithm to filter residual atmospheric correction errors more efficiently than the standard SeaWiFS bio-optical algorithm. Note that NN methods have also been used to extract information about absorbing aerosols from SeaWiFS data (Jamet *et al.*, 2005; Brajard *et al.*, 2006).

Ocean colour data are available from the Moderate Resolution Imaging Spectroradiometer (MODIS) aboard the Terra satellite and from SeaWiFS. Data from these two sensors show apparent discrepancies originating from differences in sensor design, calibration, processing algorithms, and from the rate of change in the atmosphere and ocean between sensor imaging of the same regions on the ground. To eliminate incompatibilities between sensor data from different missions and produce merged daily global ocean colour coverage, Kwiatkowska and Fargion (2003) used support vector regression (SVR) to bring MODIS data to the SeaWiFS representation, with SeaWiFS data considered to exemplify a consistent ocean colour baseline. Clustering was first applied to the data, and an SVR trained for each of the 256 clusters. The method worked accurately in low chlorophyll waters and showed a potential to eliminate sensor problems, such as scan angle dependencies and seasonal and spatial trends in data.

Rojo-Alvarez *et al.* (2003) developed an SVR model using the ϵ-insensitive Huber error function (9.26). Using satellite data to retrieve ocean surface chorophyll concentration, Camps-Valls *et al.* (2006) found that SVR using the ϵ-insensitive Huber error function outperformed the standard SVR using an ϵ-insensitive error function, an SVR using an L_2 (i.e. MSE) error function, and MLP NN (Fig. 12.1), especially when there were not many data points in the training dataset.

Chen *et al.* (2004) used data from four visible light channels of the Landsat Thematic Mapper to classify coastal surface waters off Hong Kong into five types, which are characterized by different amounts of turbidity, suspended sediments, total volatile solid, chlorophyll-a and phaeo-pigment. Three classifiers, maximum likelihood, NN and support vector machine (SVM), were built. Over test data, the accuracies of the methods of maximum likelihood, NN and SVM were 78.3%, 82.6%, and 91.3%, respectively.

Land cover

In vegetative land cover applications, Benediktsson *et al.* (1990) used NN to classify Landsat satellite images into nine forest classes and water, using four channels from Landsat and three pieces of topographic information (elevation, slope and aspect) as predictors. Performance comparison between SVM, MLP NN and other classifiers in vegetation classification using satellite data were undertaken by Huang *et al.* (2002) and Foody and Mathur (2004). Dash *et al.* (2007) used SVM to classify the vegetative land cover from Envisat's Medium Resolution Imaging Spectrometer (MERIS) data.

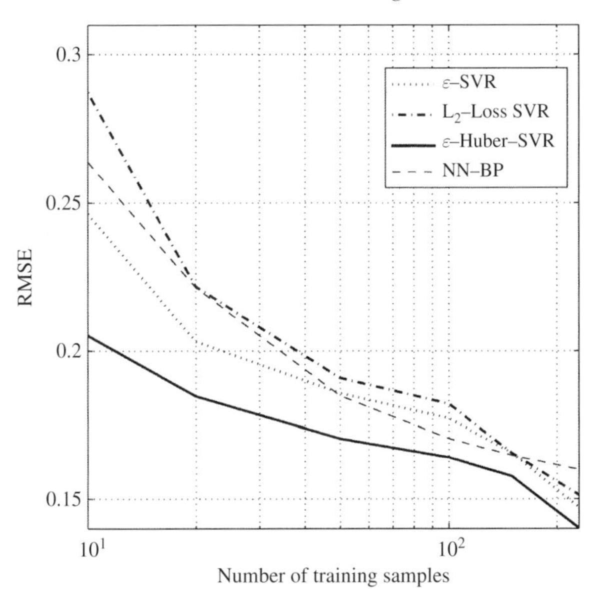

Fig. 12.1 Performance (RMSE) of four models on retrieving chlorophyll concentration from satellite data as the number of observations in the training set varied, with ϵ-SVR, L_2-loss-SVR and ϵ-Huber-SVR denoting, respectively, the SVR using the standard ϵ-insensitive error function, the L_2 function and the ϵ-Huber function, and NN-BP, the back-propagating MLP NN. (Reproduced from Camps-Valls *et al.* (2006) with permission from IEEE.)

Gopal and Woodcock (1996) used MLP NN to estimate the fraction of dead coniferous trees in an area. Forests are also burnt in forest fires, then regenerated. Trees in a regenerated area tend all to have the same age, so much of the boreal forest zone appears as a patchwork of areas with trees of the same vintage. Using data from four channels (two visible, one short-wave infrared and one near-infrared) of the SPOT satellite's vegetation sensor, Fraser and Li (2002) used MLP NN to estimate the postfire regeneration age of a forest area up to 30 years old, with an RMS error of 7 years.

In urban land cover applications, Del Frate *et al.* (2007) and Pacifici *et al.* (2007) applied MLP NN classifiers to very high-resolution (about 1m resolution) satellite images to monitor changes in urban land cover.

12.1.2 *Infrared sensing*

Clouds

The role of clouds is critical to our understanding of anthropogenic global warming. Since clouds reflect incoming solar radiation back to space and trap outgoing infrared radiation emitted by the Earth's surface, they exert both a cooling influence and a warming influence on the climate. Some cloud types exert a net cooling influence, others a net warming influence. Overall in the current climate, clouds exert a net cooling influence. Under increasing greenhouse gases in the atmosphere, the net cooling effect of clouds may be enhanced or reduced. The response of the global climate system to a given forcing is called *climate sensitivity*. Among the current global climate models, the primary source of inter-model differences in climate sensitivity is due to the large uncertainty in cloud radiative feedback (IPCC, 2007, Sect. 8.6).

Lee *et al.* (1990) used MLP NN models to classify cloud types (cumulus, stratocumulus and cirrus) from a single visible channel of Landsat satellite image. However, most studies of remote sensing of clouds tend to rely more on the infrared channels from instruments such as the Advanced Very High Resolution Radiometer (AVHRR).

Bankert (1994) used probabilistic NN to classify cloud types (cirrus, cirrocumulus, cirrostratus, altostratus, nimbostratus, stratocumulus, stratus, cumulus, cumulonimbus, and clear) in satellite images over maritime regions observed by the AVHRR instrument. The method has been extended to work over a much larger database of clouds over both land and ocean (Tag *et al.*, 2000). Lee *et al.* (2004) used SVM to classify cloud types.

In other remote sensing applications, undetected clouds are a source of contamination. For instance, when measuring sea surface temperature from satellite radiometers like the AVHRR, the presence of clouds can introduce large errors in the temperature. Using the thermal infrared and visible channels in the AVHRR, Yhann and Simpson (1995) successfully detected clouds (including subpixel clouds and cirrus clouds) with MLP NN classifiers. Miller and Emery (1997) also used an MLP NN classifier model to detect clouds in AVHRR images, where clouds were classified into eight types in addition to the cloudless condition. Simpson *et al.* (2005) used MLP NN to classify pixels into clouds, glint and clear, for the Along Track Scanning Radiometer-2 (ATSR-2) instrument. Their new method was able to avoid a pitfall of an existing cloud detection scheme, which was found consistently to misclassify cloud-free regions of cold ocean containing surface thermal gradient as clouds, with this overdetection of clouds leading to a warm bias in the sea surface temperature distributions. Srivastava *et al.* (2005) used MLP NN and SVM to improve the ability of the AVHRR/2 sensor to detect clouds over snow and ice.

Precipitation

From satellite images of clouds, one can estimate the precipitation rate. The PERSIANN (Precipitation Estimation from Remotely Sensed Information using Artificial Neural Networks) system is an automated system for retrieving rainfall from the Geostationary Operational Environmental Satellites (GOES) longwave infrared images (GOES-IR) at a resolution of 0.25° × 0.25° every half-hour (Hsu *et al.*, 1997; Sorooshian *et al.*, 2000). The PERSIANN algorithm uses the pixel brightness temperature (T_b, obtained from the infrared measurement) of cloud and its neighbouring temperature textures (in terms of means and standard deviations computed over a small region surrounding the pixel) as inputs to a self-organizing map (SOM). The SOM clusters the inputs into several cloud groups. For each cloud group, a multivariate linear function maps from the inputs to the rain rate (using gauge-corrected radar rain-rate data).

In Hong *et al.* (2004), the PERSIANN Cloud Classification System (CCS) was developed. Instead of direct pixel-to-pixel fitting of infrared cloud images to the rain rate, the PERSIANN CCS used segmentation and classification methods to process cloud images into a set of disjointed cloud-patch regions. Informative features were extracted from cloud-patches and input into a SOM. The SOM took in 23 inputs, and clustered them to a 20 × 20 grid of 400 cloud groups. For each cloud group, an exponential function was fitted between T_b and the rain rate R (Fig. 12.2). The spatial resolution was refined to 0.04° × 0.04°.

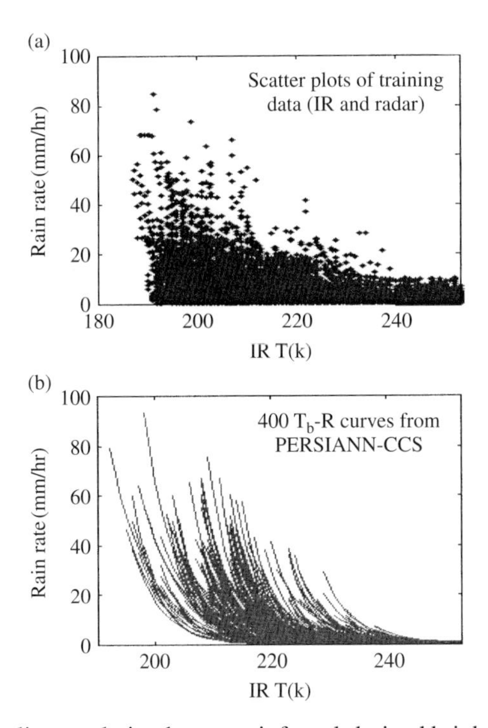

Fig. 12.2 The nonlinear relation between infrared-derived brightness temperature T_b (in degrees Kelvin) and rain rate. (a) The training data. (b) The 400 fitted exponential curves for the 400 cloud groups in PERSIANN-CCS. (Reproduced from Hong *et al.* (2004) with permission from the American Meteorological Society.)

Using Doppler weather radar data as input, MLP NN models were developed to estimate precipitation (Teschl *et al.*, 2007) and runoff (Teschl and Randeu, 2006).

Sea ice

Sea ice plays a major role in the global climate system: (a) it directly affects the planetary *albedo* (the fraction of incoming solar energy reflected by the Earth's surface), hence the net solar energy absorbed by the Earth; (b) its presence affects the exchange of heat and momentum between the atmosphere and the ocean; (c) it is crucial for deep water formation at high latitudes – when sea water freezes to form ice, salt is ejected into the surrounding water, thereby increasing the water's salinity and density, hence the sinking of surface waters to great depths. A major worry in the global warming scenario is the positive feedback associated with sea ice, i.e. when some sea ice melts, the albedo drops since water is not as reflective as ice, hence more solar radiation will be absorbed by the water leading to increased warming and more melting. Even more alarming is that the current observed melting rate of the Arctic sea ice is considerably greater than that predicted by the global climate models (Stroeve *et al.*, 2007). Hence, monitoring of the Arctic sea ice cover is of paramount importance for our understanding of global warming. McIntire and Simpson (2002) used MLP NN to classify satellite image pixels into ice, clouds and water, thereby allowing monitoring of the Arctic sea ice extent.

Snow

The areal extent of snow in mountainous regions is an important piece of climate information. Furthermore, snow melt is a major source of water supply for many arid regions. Two approaches to identify clear land, cloud, and areal extent of snow in a satellite image of mid-latitude regions have been developed by Simpson and McIntire (2001). A feed-forward MLP NN is used to classify individual images, and a recurrent NN is used to classify sequences of images. Validation with independent *in situ* data found a classification accuracy of 94% for the feed-forward NN and 97% for the recurrent NN. Thus when there is rapid temporal sampling, e.g. images from the Geostationary Operational Environmental Satellites (GOES), the recurrent NN approach is preferred over the feed-forward NN. For other satellite images, e.g. AVHRR images from polar-orbiting satellites, which do not cover the same spatial area in rapid temporal sequences, the feed-forward NN classifier is to be used.

12.1.3 Passive microwave sensing

In satellite remote sensing, the shorter is the wavelength of the radiation, the finer the spatial resolution of the image. Thus microwaves, with their relatively long wavelengths are at a disadvantage compared to visible light and infrared sensing in terms of spatial resolution. However, microwaves from the Earth's surface generally can reach the satellite even with moderately cloudy skies, whereas visible light and infrared radiation cannot. Although microwaves can pass through a cloudy atmosphere, they interact with the water vapour and the liquid water in the atmosphere, thus making the inversion problem more complicated.

The Special Sensor Microwave Imager (SSM/I) is flown aboard the Defense Meteorological Satellite Program (DMSP) satellites. Stogryn *et al.* (1994) applied an MLP NN to retrieve surface *wind speed* from several channels of the SSM/I, where the NN has one hidden layer and a single output. Two NN models were developed, one for clear sky and one for cloudy sky conditions. Under cloudy skies, there was a factor of two improvement in the RMSE over standard retrieval methods. Krasnopolsky *et al.* (1995) showed that a single NN with the same architecture can generate the same accuracy as the two NNs for clear and cloudy skies.

Since the microwaves coming from the ocean surface were influenced by the surface wind speed, the columnar water vapour, the columnar liquid water and the SST, Krasnopolsky *et al.* (1999) retrieved these four parameters altogether, i.e. the NN had four outputs. The inputs were five SSM/I channels at the 19, 22 and 37 GHz frequency bands, where the 22 GHz band used only vertically polarized radiation, while the other two bands each separated into vertically and horizontally polarized channels. Since the four-output model retrieves wind speed better than the single output model, it is important to incorporate the influence of the columnar water vapour and liquid water, and the SST. Although the retrieved SST is not very accurate, having SST as an output improves the accuracy of the wind speed retrieval. This NN retrieval algorithm has been used as the operational algorithm since 1998 by NCEP/NOAA (National Center for Environmental Prediction/National Oceanic and Atmospheric Administration) in its global data assimilation system. Neural network models have also been developed to monitor *snow* characteristics on a global scale using SSM/I data (Cordisco *et al.*, 2006).

The SSM/I is a passive microwave sensor; we next discuss active microwave sensors, which are basically spaceborne radars. While SSM/I can measure wind speed, wind direction cannot be retrieved from the SSM/I data.

12.1.4 Active microwave sensing

Altimeter

The simplest active microwave sensor is the satellite *altimeter*. It emits a pulsed radar beam vertically downward from the satellite. The beam is strongly reflected by the sea surface, and the reflected pulse is measured by a receiver in the altimeter.

The travel time of the pulse (i.e. the difference between the time when the reflected pulse arrived and the time when the original pulse was emitted) gives the distance between the satellite and the sea level, since the pulse travelled at the speed of light. If the satellite orbital position is known, then this distance measured by the altimeter gives the sea level displacements. If the sea surface is smooth as a mirror, the relected pulse will be sharp, whereas if the surface is wavy, the reflected pulse will be stretched out in time, since the part of the pulse hitting the wave crest will reflect before the part hitting the wave trough. Hence, from the sharpness of the reflected pulse, the surface wave condition can be estimated. Furthermore, the surface wave condition is strongly related to the local wind condition, hence retrieval of surface *wind speed* is possible from the altimetry data. Gourrion *et al.* (2002) used MLP NN to retrieve surface wind speed from altimeter data, with the RMS wind error 10–15% lower than those in previous methods.

Scatterometer

Compared to surface wind speed retrieval, it is vastly more difficult to retrieve surface *wind direction* from satellite data. The *scatterometer* is designed to retrieve both wind speed and direction through the use of multiple radar beams. For instance in the European Remote-Sensing Satellites (ERS-1 and ERS-2), the scatterometer has three antennae, sending off three radar beams. The first beam (fore beam) is aimed at 45° from the forward direction, the second (mid beam) at 90°, and the third (aft beam) at 135° (i.e. 45° from the backward direction). As the satellite flies, the ocean surface to one side of the satellite path is illuminated by first the fore beam, then the mid beam, and finally the aft beam. From the power of the backscattered signal relative to that of the incident signal, the normalized radar cross section σ^0 is obtained. As the three antennae each pick up a backscattered signal, there are three cross sections (σ_1^0, σ_2^0, σ_3^0) for each small illuminated area (cell) on the Earth's surface. To first order, σ^0 depends on the sea surface roughness, which is related to the wind speed, the incidence angle θ (i.e. the angle between the incident radar beam and the vertical at the illuminated cell, Fig. 12.3), and the azimuth angle χ (measured on the horizontal plane between the wind vector and the incident radar beam).

The inversion problem of retrieving the wind speed and wind azimuth from σ^0 has been extensively studied (Thiria *et al.*, 1993; Mejia *et al.*, 1998; Chen *et al.*, 1999; Cornford *et al.*, 1999; Richaume *et al.*, 2000). In Richaume *et al.* (2000), two MLP NN models were used, the first to retrieve the wind speed, and the second to retrieve the wind azimuth. For inputs, a cell and 12 nearby cells are considered, each cell having three σ^0 values measured by the three antennae, hence a total of $13 \times 3 = 39$ input variables for the first NN model. This NN has two hidden layers, each with 26 neurons, and a single output (the wind speed). The

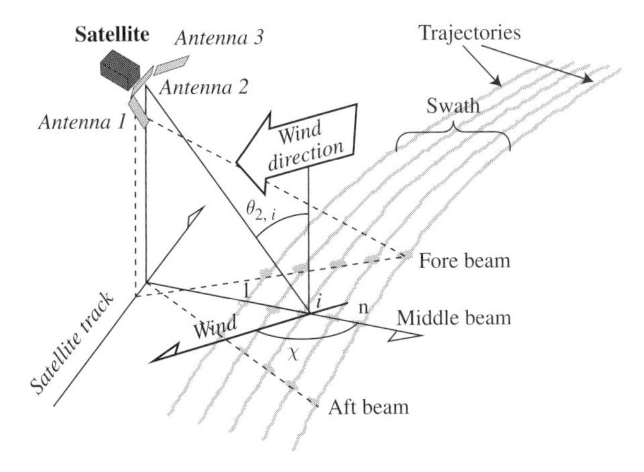

Fig. 12.3 A satellite scatterometer emitting three radar beams to retrieve surface wind, with θ the beam incidence angle, and χ the azimuth angle of the wind. (Reproduced from Mejia *et al.* (1998) with permission of the American Geophysical Union.)

second NN for the wind azimuth has the same 39 inputs plus the retrieved wind speed from the first NN for a total of 40 input variables. This NN has two hidden layers, each with 25 neurons. The output layer has 36 neurons, with the jth neuron representing the posterior probability of the wind azimuth being in the interval $[10(j-1)°, 10j°)$, $(j = 1, \ldots, 36)$, i.e. this second NN is a classifier, with the solution selected by the highest posterior probability among the 36 azimuth intervals. Unfortunately, the azimuth inversion problem is a very difficult problem with multiple solutions. In practice, additional information, usually output from a numerical weather prediction model, is used to resolve ambiguities in the wind azimuth. Richaume *et al.* (2000) found that the NN wind retrieval method for the ERS scatterometer performed well relative to existing methods.

Synthetic aperture radar

For active remote sensing, one is limited to microwaves since other radiation (e.g. visible, infrared, etc.) cannot generate a strongly reflected signal from the Earth's surface to travel back to the satellite. However, as the wavelengths of microwaves are much longer than the wavelengths of visible or infrared radiation, the spatial resolution of microwave images is far coarser than that of images generated using shorter wavelengths. To obtain high-resolution microwave images (with resolution down to about 10m), the *synthetic aperture radar* (SAR) has been invented.

For a radar operating at a given frequency, the larger the size of the receiver antenna, the finer is the observed spatial resolution. Since the satellite is flying, if an object on the ground is illuminated by the radar beam at time t_1 and continues to be illuminated till time t_2, then all the reflected data collected by the satellite between t_1 and t_2 can be combined as though the data have been collected by a very large antenna. A high resolution image can be constructed accurately provided that the object is stationary. If the object is moving, it is misplaced in the image, hence one finds moving trains displaced off their tracks in SAR images.

Horstmann *et al.* (2003) used MLP NN models with three hidden layers to retrieve wind speeds globally at about 30 m resolution from SAR data. An MLP NN classifier has been developed to detect oil spill on the ocean surface from SAR images (Del Frate *et al.*, 2000).

To retrieve forest biomass from SAR data, Del Frate and Solimini (2004) used an MLP NN with six inputs (the backscattering cross-sections for three different polarizations ($\sigma^0_{\rm hh}$, $\sigma^0_{\rm vv}$, $\sigma^0_{\rm hv}$) at two frequency bands), two hidden layers with 24 neurons each, and one output. A pruning procedure removed hidden neurons with connecting weights of small magnitude, resulting in an NN with only seven and four hidden neurons in the first and second hidden layers, respectively. The pruned network has a training RMSE almost 10% lower than the original. In agricultural applications of SAR data, Del Frate *et al.* (2004) used MLP NN to monitor the soil moisture and the growth cycle of wheat fields.

12.2 Oceanography

12.2.1 Sea level

Machine learning methods have been applied to forecast coastal sea level fluctuations. Arising from astronomical forcing, the tidal component of the sea level fluctuations is accurately forecast by harmonic analysis (Emery and Thomson, 1997). However, the non-tidal component, mainly forced by local meteorological conditions, is not easy to forecast, and without information on the non-tidal component, ships can be accidentally grounded in coastal waters. Cox *et al.* (2002) used MLP NN to forecast the non-tidal component at Galveston Bay, Texas, using the two wind components, barometric pressure, and the previously observed water level anomaly as predictors.

Coastal sea level is measured by tide gauges; however, long-term maintenance of tide gauges is costly. For instance, there are only about ten permanent tide gauge stations along the Canadian Atlantic coastline stretching 52000 km. Han and Shi (2008) proposed the training of MLP NN models to forecast hourly sea level fluctuations $h(t)$ at sites with only temporary tide gauges, using as predictors, the hourly sea level at neighbouring permanent tide gauge stations from time $t - 12$ h to t. Once the NN models have been trained, the temporary tide gauges can be removed, and their sea levels forecast using data from the neighbouring permanent stations. Requiring only one month of training data from a temporary gauge (plus data from the permanent stations), their model was capable of accurate and robust long-term predictions of both the tidal and non-tidal components of the sea level fluctuations. An offshore drilling platform can monitor the sea level fluctuations for a month and then use this method to predict the sea level at the platform from the permanent tide gauge stations in the future.

Nonlinear principal component analysis (NLPCA) (with a circular bottleneck node) has been used to analyze the tidal cycle off the German North Sea coast (Herman, 2007).

12.2.2 Equation of state of sea water

For sea water, the equation of state describes the density ρ as a function of the temperature T, salinity S and pressure p. The equation is nonlinear and can only be described empirically. The standard used by oceanographers has been the UNESCO International Equation of State for Seawater (UES) (UNESCO, 1981), where

$$\rho(T, S, p) = \rho(T, S, 0)/[1 - p/K(T, S, p)], \tag{12.3}$$

with ρ in kg m^{-3}, T in °C, S in practical salinity units (psu), p in bar, and $K(T, S, p)$ is a bulk modulus. The formula is defined over the range $-2 < T < 40\,°\mathrm{C}$, $0 < S < 40$ (psu), and $0 < p < 1000$ bar, encompassing the whole range of conditions encountered in the global oceans. Unfortunately, both $\rho(T, S, 0)$ and $K(T, S, p)$ are given by complicated empirical expressions (involving polynomials) (see e.g. Gill, 1982, Appendix 3), so the density in (12.3) is actually given by a cumbersome formula with over 40 parameters.

There are two disadvantages with the UES formula. First, in ocean general circulation models, the density at every grid point needs to be updated regularly as the model is integrated forward in time. Evaluating the UES formula at every grid point in the model could consume up to 40% of the total computing time (Krasnopolsky *et al.*, 2002; Krasnopolsky and Chevallier, 2003). Second, in data assimilation applications, observed data are assimilated into the ocean model to guide its evolution. As temperature data are far more plentiful than salinity data, one generally has temperature data, but not salinity data, available for assimilation into the model. If temperature data are assimilated into an ocean model using the full equation of state, and if the corresponding salinity values are not adjusted, then in some circumstances the vertical density structure in the water column can become gravitationally unstable. To adjust the salinity, one needs to calculate S from the UES formula as a function of T, ρ and p (or depth z). This involves an iterative procedure, which can be several orders of magnitude more costly computationally than solving the UES itself.

In numerical models using the depth z as the vertical coordinate, one can easily convert from p to z by assuming hydrostatic balance, i.e.

$$\frac{\partial p}{\partial z} = -\rho g, \tag{12.4}$$

with g the gravitational constant. Hence the UES can be regarded as defining the relation

$$\rho = \rho(T, S, z), \tag{12.5}$$

and through its inversion, the salinity equation

$$S = S(T, \rho, z). \tag{12.6}$$

The MLP NN models were developed (Krasnopolsky *et al.*, 2002; Krasnopolsky and Chevallier, 2003) as low-cost clones of these ρ and S equations. To develop the ρ clone, 4000 points of T, S and z were selected over a 3-dimensional grid, encompassing the full range of oceanographic conditions encountered in the global oceans. The NN has three inputs (T, S, z), a single hidden layer with three hidden neurons (using the tanh activation function), and a single output ρ (using the linear activation function). After training using the 4000 observations, the NN gives the relation

$$\rho = \rho_{\rm NN}(T, S, z). \tag{12.7}$$

Similarly, one can obtain an NN relation

$$S = S_{\rm NN}(T, \rho, z). \tag{12.8}$$

These NN equations were able to simulate the original UES relations to high accuracy at low costs – the density calculation by NN is about two times faster than the UES formula, while the salinity calculation by NN is several hundred times faster than an iterative numerical inversion of the UES.

For further computational cost reduction, the small density changes from one time step to the next can be approximated by the simple differential relation

$$\Delta\rho = \frac{\partial \rho}{\partial T}\Delta T + \frac{\partial \rho}{\partial S}\Delta S, \tag{12.9}$$

where ΔT and ΔS are increments of T and S. Differentiating the analytic expressions for $\rho_{\rm NN}$ and $S_{\rm NN}$ handily provides analytic expressions for $\partial\rho/\partial T$ and $\partial\rho/\partial S$, which allow inexpensive computation of (12.9). Thus (12.9) is used to estimate the small changes in ρ over a number of time steps (usually several tens of steps), before the UES or NN approximation of the UES are used for a more accurate update.

Overall, use of NN (and its derivatives) as a low-cost substitute for the UES in the high resolution Multiscale Ocean Forecast system has accelerated density calculations by a factor of 10, with the errors in the density calculations not exceeding the natural uncertainty of 0.1 kg m^{-3}. Hence the computational cost of the density calculations has dropped from 40% to 4–5% of the total cost (Krasnopolsky and Chevallier, 2003).

12.2.3 Wind wave modelling

Incoming water waves with periods of several seconds observed by sunbathers on a beach are called *wind waves*, as they are surface gravity waves (LeBlond and Mysak, 1978; Gill, 1982) generated by the wind. As the ocean surface allows efficient propagation of wind waves, they are usually generated by distant storms, often thousands of kilometres away from the calm sunny beach.

The wave energy spectrum F on the ocean surface is a function of the 2-dimensional horizontal wave vector $\mathbf{k}$. We recall that the wavelength is $2\pi\,\|\mathbf{k}\|^{-1}$, while the wave's phase propagation is along the direction of $\mathbf{k}$. The evolution of $F(\mathbf{k})$ is described by

$$\frac{\mathrm{d}F}{\mathrm{d}t} = S_{\mathrm{in}} + S_{\mathrm{nl}} + S_{\mathrm{ds}} + S_{\mathrm{sw}}, \tag{12.10}$$

where S_{in} is the input source term, S_{nl} the nonlinear wave–wave interaction term, S_{ds} the dissipation term, and S_{sw} is the term incorporating shallow-water effects, with S_{nl} being the most complicated one. For surface gravity waves, resonant nonlinear wave–wave interactions involve four waves satisfying the resonance conditions

$$\mathbf{k}_1 + \mathbf{k}_2 = \mathbf{k}_3 + \mathbf{k}_4, \quad \text{and} \quad \omega_1 + \omega_2 = \omega_3 + \omega_4, \tag{12.11}$$

where $\mathbf{k}_i$ and ω_i $(i = 1, \ldots, 4)$ are, respectively, the wave vector and angular frequency of the ith wave. Letting $\mathbf{k} = \mathbf{k}_4$ and $\omega = \omega_4$, S_{nl} in its full form can be written as (Hasselmann and Hasselmann, 1985)

$$S_{\mathrm{nl}}(\mathbf{k}) = \omega \int \sigma(\mathbf{k}_1, \mathbf{k}_2, \mathbf{k}_3, \mathbf{k})\, \delta(\mathbf{k}_1 + \mathbf{k}_2 - \mathbf{k}_3 - \mathbf{k})\, \delta(\omega_1 + \omega_2 - \omega_3 - \omega) \times [n_1 n_2 (n_3 + n) + n_3 n (n_1 + n_2)]\, \mathrm{d}\mathbf{k}_1 \mathrm{d}\mathbf{k}_2 \mathrm{d}\mathbf{k}_3, \tag{12.12}$$

with $n_i = F(\mathbf{k}_i)/\omega_i$, σ a complicated net scattering coefficient and δ the Dirac delta function which ensures that the resonance conditions (12.11) are satisfied during the 6-dimensional integration in (12.12). As this numerical integration requires 10^3–10^4 more computation than all other terms in the model, an approximation, e.g. the Discrete Interaction Approximation (DIA) (Hasselmann *et al.*, 1985), has to be made to reduce the amount of computation for operational wind wave forecast models.

Since (12.12) is in effect a map from $F(\mathbf{k})$ to $S_{nl}(\mathbf{k})$, Krasnopolsky *et al.* (2002) and Tolman *et al.* (2005) proposed the use of MLP NN to map from $F(\mathbf{k})$ to $S_{nl}(\mathbf{k})$ using training data generated from (12.12). Since $\mathbf{k}$ is 2-dimensional, both F and S_{nl} are 2-dimensional fields, containing of the order of 10^3 grid points in the $\mathbf{k}$-space. The computational burden is reduced by using PCA on $F(\mathbf{k})$ and $S_{nl}(\mathbf{k})$ and retaining only the leading PCs (about 20-50 for F and 100-150 for S_{nl}) before training the NN model. Once the NN model has been trained, computation of S_{nl} from F can be obtained from the NN model instead of from the original (12.12).

The NN model is nearly ten times more accurate than the DIA. It is about 10^5 times faster than the original approach using (12.12) and only seven times slower than DIA (Krasnopolsky, 2007).

12.2.4 Ocean temperature and heat content

Due to the ocean's vast heat storage capacity, upper ocean temperature and heat content anomalies have a major influence on global climate variability. The best known large-scale interannual variability in sea surface temperature (SST) is the El Niño-Southern Oscillation (ENSO), a coupled ocean–atmosphere interaction involving the oceanic phenomenon El Niño in the tropical Pacific, and the associated atmospheric phenomenon, the Southern Oscillation (Philander, 1990; Diaz and Markgraf, 2000). The coupled interaction results in anomalously warm SST in the eastern equatorial Pacific during El Niño episodes, and cool SST in the central equatorial Pacific during La Niña episodes. The ENSO is an irregular oscillation, but spectral analysis does reveal a broad spectral peak around the 4–5 year period. Because El Niño is associated with the collapse in the Peruvian anchovy fishery, and ENSO has a significant impact on extra-tropical climate variability, forecasting tropical Pacific SST anomalies at the seasonal to interannual time scale (Goddard *et al.*, 2001) is of great scientific and economic interest.

Tangang *et al.* (1997) used MLP NN to forecast the SST anomalies at the Niño3.4 region (see Fig. 2.3 for location). For predictors, the NN model used the seven leading principal components (PCs) of the tropical Pacific wind stress anomalies (up to four seasons prior) and the Niño3.4 SST anomaly itself, hence a total of $7 \times 4 + 1$ predictors. Tangang *et al.* (1998a) found that NN using the leading PCs of the tropical sea level pressure (SLP) anomalies forecast better than that using the PCs of the wind stress anomalies. Tangang *et al.* (1998b) found that using PCs from extended empirical orthogonal function analysis (also called space–time PCA or singular spectrum analysis) led to far fewer predictor variables, hence a much smaller NN model. Tang *et al.* (2000) compared the SST forecasts by MLP

NN (actually the ensemble average forecast of 20 NN models trained with different initial weights), linear regression (LR) and canonical correlation analysis (CCA), but using a cross-validated t-test, they were unable to find a statistically significant difference at the 5% level between the NN and LR models and between the NN and CCA models.

Yuval (2000) introduced generalized cross-validation (GCV) (Golub *et al.*, 1979; Haber and Oldenburg, 2000) to control overfitting/underfitting automatically in MLP NN and applied the method to forecasting tropical Pacific SST anomalies. Yuval (2001) used bootstrap resampling of the data to generate an ensemble of MLP NN models and used the ensemble spread to estimate the forecast uncertainty.

Thus far, the SST forecasts have been restricted to specific regions in the tropical Pacific, e.g. the Niño3.4 region. Wu *et al.* (2006b) developed an MLP NN model to forecast the SST anomalies of the whole tropical Pacific. This was achieved by forecasting the five leading PCs of the SST anomalies separately, then combining the contributions from the individual modes. For each PC, an ensemble of 30 NN models combined to give an ensemble-averaged forecast. For the first SST PC, NN failed to beat LR at lead times of 6 months or less in cross-validated correlation scores, and NN was only slightly ahead of LR at longer lead times. For PCs 2, 3 and 4, NN was ahead of LR at all lead times from 3–15 months. However, since PC 1 contains far more variance than the higher modes, the inability of NN to do well for PC 1, especially at shorter lead times, explains why overall NN is not vastly better than LR. Aguilar-Martinez (2008) tested support vector regression (SVR) and MLP NN in forecasting the tropical Pacific SST anomalies. The western warm water volume anomaly calculated by Meinen and McPhaden (2000) was also found to be helpful as an extra predictor.

Garcia-Gorriz and Garcia-Sanchez (2007) used MLP NN models to predict SST in the western Mediterranean Sea. The NNs were trained with monthly meteorological variables (SLP, 10 m zonal and meridional wind components, 2 m air temperature, 2 m dewpoint temperature, and total cloud cover) as input and satellite-derived SST as target output. The trained NNs predicted both the seasonal and the inter-annual variability of SST well in that region. The NNs were also used to reconstruct SST fields from satellite SST images with missing pixels.

Tropical Pacific SST anomalies have also been analyzed by nonlinear principal component analysis (NLPCA) (Monahan, 2001; Hsieh, 2001b, 2007) (see Section 10.1.2), with the first NLPCA mode extracting the ENSO signal in the SST anomalies. The NLPCA(cir) method, i.e. NLPCA with a circular bottleneck node (see Section 10.1.4), has also been applied to the tropical Pacific SST anomalies (Hsieh, 2001b), and to thermocline depth anomalies (An *et al.*, 2005). The closed curve solution of NLPCA(cir) for the thermocline depth anomalies is consistent with the recharge–discharge oscillator theory (Jin, 1997a,b) which has the recharge and discharge of equatorial heat content in the upper ocean causing the coupled ocean–atmosphere system to oscillate as ENSO. In An *et al.* (2006), combined PCA was first performed on the anomalies of sea level height, SST and the upper 50 m zonal current (i.e. the eastward component of the current), then the six leading PCs were input to an NLPCA(cir) model to extract the first combined nonlinear mode. The NLPCA and NLPCA(cir) models have also been used to analyze the observed tropical Pacific upper ocean heat content anomalies for nonlinear modes of decadal and inter-annual variability (Tang and Hsieh, 2003b).

Singular spectrum analysis (SSA) is an extension of PCA to include time lags (see Sections 3.4 and 3.5), hence it is also called space–time PCA or extended empirical orthogonal function analysis. Nonlinear SSA (NLSSA) has been applied to analyze the tropical Pacific SST anomalies and the SLP anomalies (Wu and Hsieh, 2002) (see Section 10.6).

Nonlinear canonical correlation analysis (NLCCA) has been applied to the tropical Pacific (see Section 11.1.1) to extract the correlated modes between sea level pressure (SLP) anomalies and SST anomalies (Hsieh, 2001a), and between wind stress and SST anomalies (Wu and Hsieh, 2002, 2003), with more nonlinear relations found during 1981–1999 than during 1961–1975. Collins *et al.* (2004) studied the predictability of the Indian Ocean SST anomalies by performing CCA and NLCCA on SST and SLP anomalies.

12.3 Atmospheric science

12.3.1 Hybrid coupled modelling of the tropical Pacific

In Section 12.2.4, we have reviewed use of machine learning methods for forecasting tropical Pacific SST anomalies, as the El Niño-Southern Oscillation (ENSO) generates warm El Niño states and cool La Niña states aperiodically. Forecasting of the ENSO has been done using three main types of model: dynamical models, statistical or empirical models, and hybrid models which use a combined dynamical–statistical approach (Barnston *et al.*, 1994).

In the dynamical approach, as there are fast waves in the atmosphere, small time steps (hence large computational efforts) are needed to keep the dynamical model stable. The hybrid approach replaces the dynamical atmosphere in the coupled ocean–atmosphere system with a statistical/empirical model, i.e. a dynamical ocean model is coupled to an inexpensive statistical/empirical atmospheric model. Originally only linear statistical models were used to predict wind stress from upper ocean variables.

Tang *et al.* (2001) used MLP NN to build a nonlinear regression model for the tropical wind stress anomalies. A six-layer dynamical ocean model of the tropical

Pacific was driven by observed wind stress. Leading principal components (PC) of the model upper ocean heat content anomalies and PCs of the observed wind stress anomalies were computed, and NN models were built using the heat content PCs as inputs and the wind stress PCs as outputs, so that given a heat content anomaly field in the ocean model, a simultaneous wind stress anomaly field can be predicted.

Next, the empirical atmospheric model using NN was coupled to the dynamical ocean model, with the upper ocean heat content giving an estimate of the wind stress via the NN model, and the wind stress in turn driving the dynamical ocean model (Tang, 2002). When used for ENSO forecasting, the ocean model was first driven by the observed wind stress to time t_1, then to start forecasting at t_1, the NN atmospheric model was coupled to the ocean model, and the coupled model was run for 15 months (Tang and Hsieh, 2002). Adding data assimilation further improved the forecasting capability of this hybrid coupled model (Tang and Hsieh, 2003a).

In the hybrid coupled model of Li *et al.* (2005), the dynamical ocean model from Zebiak and Cane (1987) was coupled to an NN model of the wind stress. The original ocean model had a rather simple parameterization of $T_{\rm sub}$ (the ocean temperature anomaly below the mixed layer) in terms of the thermocline depth anomaly h. In Li *et al.* (2005), the parameterization was improved by using an NN model to estimate $T_{\rm sub}$ from h. A similar hybrid coupled model was used to study how ENSO properties changed when the background climate state changed (Ye and Hsieh, 2006).

12.3.2 Climate variability and climate change

There are many known modes of climate variability, the most famous of which is the El Niño-Southern Oscillation (ENSO), which has been presented in Sections 12.2.4 and 12.3.1. Here we will review the application of machine learning methods to several other modes of climate variability – the Arctic Oscillation, the Pacific–North American teleconnection, the stratospheric Quasi-Biennial Oscillation, the Madden–Julian Oscillation and the Indian summer monsoon – as well as to anthropogenic climate change.

Arctic Oscillation (AO)

In the atmosphere, the most prominent mode in the Northern Hemisphere is the *Arctic Oscillation* (AO), also commonly referred to as the *North Atlantic Oscillation* (NAO), due to the stronger impact of the AO in the North Atlantic sector. The AO was defined by Thompson and Wallace (1998) as the first empirical orthogonal function (EOF) mode (i.e. the spatial pattern of the first mode from principal component analysis) of wintertime sea level pressure (SLP) anomalies over the extra-tropical Northern Hemisphere. The AO is a seesaw pattern in atmospheric pressure between the Arctic basin and middle latitudes, exhibiting lower than normal pressure over the polar region and higher than normal pressure at mid latitudes (about 37°N– 45°N) in its positive phase, and the reverse in its negative phase. Due to its strong zonal symmetry, the AO is also often referred to as the *northern annular mode* (NAM) (Thompson and Wallace, 2000). During the positive phase of AO, below normal Arctic SLP, enhanced surface westerlies in the North Atlantic, and warmer and wetter than normal conditions in northern Europe tend to prevail, together with a warmer winter in much of the USA east of the Rocky Mountains and central Canada, but colder in Greenland and Newfoundland.

The standardized leading principal component (PC) of the winter SLP anomalies is commonly taken to be an AO *index*, which is a time series indicating when the AO mode is excited. The usual assumption is that the atmospheric climate anomalies associated with positive and negative phases of the AO are opposite to each other, i.e. the impact of the AO on the climate is linear. There is, however, some evidence indicating a nonlinear relationship between the AO and northern winter climate (Pozo-Vazquez *et al.*, 2001).

The nonlinear association between the AO and North American winter climate variation has been examined by *nonlinear projection* (NLP) (Hsieh *et al.*, 2006). The NLP is easily performed by an MLP NN with a single input x and multiple outputs $\mathbf{y}$, i.e. $\mathbf{y} = \mathbf{f}(x)$. When $\mathbf{f}$ is linear, one simply has linear projection (LP), which is commonly used (Deser and Blackmon, 1995). The NLP can be separated into a linear component, which is simply LP, and a nonlinear component, which is NLP – LP. With x being the AO index, NLP were performed separately with $\mathbf{y}$ being the leading PCs of SLP, surface air temperature (SAT), precipitation and Z500 (500 hPa geopotential height) anomalies (Hsieh *et al.*, 2006) over the winter extra-tropical Northern Hemisphere. While with LP, the AO effects were much weaker in the Pacific sector than the Atlantic sector, such is not the case for the nonlinear component of NLP, e.g. the nonlinear component for SLP is stronger in the Pacific sector than the Atlantic sector.

The robustness of the nonlinear association between AO and North America winter climate found in the observed data was further investigated using output from a coupled general circulation model (Wu *et al.*, 2006a). As the nonlinear patterns of North American Z500 and SAT anomalies associated with the AO generated in the model basically agreed with the observational results, this provides support that the nonlinear behaviour of the North American winter climate with respect to the AO is real as this phenomenon was found in the observed data and in a climate simulation which was completely independent of the observations.

The application of NLPCA to the Northern Hemisphere winter climate data by Monahan *et al.* (2000), Monahan *et al.* (2001), Monahan *et al.* (2003) and

Teng *et al.* (2004) has led to some controversy regarding the robustness of the results (Christiansen, 2005; Monahan and Fyfe, 2007; Christiansen, 2007). Recall in Section 10.1.3, it was explained why NLPCA is fundamentally more prone to overfitting than nonlinear regression (which has NLP as a special case), and that using an additional information criterion helps.

Pacific–North American (PNA) teleconnection

The term 'teleconnection' refers to a recurring and persistent, large-scale pattern of atmospheric anomalies spanning vast geographical areas. The Pacific–North American teleconnection (PNA) is one of the most prominent modes of low-frequency variability in the extra-tropical Northern Hemisphere (Wallace and Gutzler, 1981). It stretches from the Pacific to North America in a wavy pattern of four alternating positive and negative anomaly cells as seen in the geopotential height anomaly fields. The positive phase of the PNA pattern shows positive geopotential height anomalies around Hawaii, negative anomalies south of the Aleutian Islands, positive anomalies over the inter-mountain region of North America, and negative anomalies over southeastern USA. The PNA is also associated with precipitation and surface temperature anomalies – e.g. the positive phase of the PNA pattern is associated with positive temperature anomalies over western Canada and the extreme western USA, and negative temperature anomalies across the south-central and southeastern USA. The PNA effects are mainly manifested in the winter months.

Although PNA is a natural internal mode of climate variability in the mid-latitude atmosphere, it is also strongly influenced by the El Niño-Southern Oscillation (ENSO), the coupled ocean–atmosphere oscillation centred in the equatorial Pacific. During ENSO warm episodes (El Niño), the positive phase of the PNA tends to be excited, while during ENSO cold episodes (La Niña), the negative phase of the PNA tends to be excited. Hence the extra-tropical effects of ENSO tend to be manifested by the PNA teleconnection.

The ENSO sea surface temperature (SST) index was defined as the standardized first principal component (PC) of the winter (November–March) SST anomalies over the tropical Pacific. A nonlinear projection (NLP) of this SST index onto the Northern Hemisphere winter sea level pressure (SLP) anomalies by MLP NN was performed by Wu and Hsieh (2004a) to investigate the nonlinear association between ENSO and the Euro–Atlantic winter climate. While the linear impact of ENSO on the Euro–Atlantic winter SLP is weak, the NLP revealed statistically significant SLP anomalies over the Euro–Atlantic sector during both extreme cold and warm ENSO episodes, suggesting that the Euro–Atlantic climate mainly responds to ENSO nonlinearly. The nonlinear response, mainly a quadratic response to the SST index, reveals that regardless of the sign of the SST index, positive SLP anomalies are found over the North Atlantic, stretching from eastern Canada to Europe (with the anomaly centre located just northwestward of Portugal), and negative anomalies centred over Scandinavia and the Norwegian Sea, consistent with the excitation of the positive North Atlantic Oscillation (NAO) pattern. A similar effect was found in the 500 hPa geopotential height anomalies (Wu and Hsieh, 2004b).

The NLP approach was also used to study ENSO effects on the winter surface air temperature and precipitation over North America (Wu *et al.*, 2005) and over the whole extra-tropical Northern Hemisphere (Hsieh *et al.*, 2006).

Relations between the tropical Pacific SST anomalies and the Northern Hemisphere mid-latitude winter atmospheric variability simulated in an atmospheric general circulation model (GCM) have also been explored using nonlinear canonical correlation analysis (NLCCA) by Wu *et al.* (2003) (see Section 11.1.2).

Quasi-Biennial Oscillation (QBO)

In the equatorial stratosphere, the zonal wind (i.e. the east-west component of the wind) manifests a quasi-biennial oscillation (QBO) (Naujokat, 1986; Hamilton, 1998; Baldwin *et al.*, 2001). The QBO dominates over the annual cycle or other variations in the equatorial stratosphere, with the period of oscillation varying roughly between 22 and 32 months, the mean period being about 28 months. In Section 10.1.4, we have seen the application of the NLPCA(cir) model (i.e. NLPCA with one circular bottleneck node) to the equatorial stratospheric zonal wind to identify the structure of the QBO. After the 45 year means were removed, the zonal wind u at seven vertical levels in the stratosphere became the seven inputs to the NLPCA(cir) network (Hamilton and Hsieh, 2002; Hsieh, 2007), and the NLPCA(cir) mode 1 solution was a closed curve in a 7-dimensional space (Fig. 10.8). The system goes around the closed curve once during one cycle of the QBO. The observed strong asymmetries between the easterly and westerly phases of the QBO (Hamilton, 1998; Baldwin *et al.*, 2001) are well captured by the nonlinear mode (Fig. 10.9).

The actual time series of the wind measured at a particular height level is somewhat noisy and it is often desirable to have a smoother representation of the QBO time series which captures the essential features at all vertical levels. Also, reversal of the wind from westerly to easterly and vice versa occurs at different times for different height levels, rendering it difficult to define the phase of the QBO. Hamilton and Hsieh (2002) found that the phase of the QBO as defined by the NLPC θ is more accurate than previous attempts to characterize the phase, leading to a stronger link between the QBO and the Northern Hemisphere polar stratospheric temperatures in winter (the Holton-Tan effect) (Holton and Tan, 1980) than previously found.

Nonlinear singular spectrum analysis (NLSSA) has also been applied to study the QBO (Hsieh and Hamilton, 2003).

Madden–Julian Oscillation (MJO)

The *Madden–Julian Oscillation* (MJO) is the dominant component of the intra-seasonal (30–90 day time scale) variability in the tropical atmosphere. It consists of large-scale coupled patterns in atmospheric circulation and deep convection, all propagating slowly eastward over the Indian and Pacific Oceans where the sea surface is warm (Zhang, 2005). Association between this tropical oscillation and the mid-latitude winter atmospheric conditions has been found (Vecchi and Bond, 2004). Using MLP NN, nonlinear projection (NLP) of an MJO index on to the precipitation and 200 hPa wind anomalies in the northeast Pacific during January–March by Jamet and Hsieh (2005) shows asymmetric atmospheric patterns associated with different phases of the MJO, with strong nonlinearity found for precipitation anomalies and moderate nonlinearity for wind anomalies.

Indian summer monsoon

The Great Indian Desert and adjoining areas of the northern and central Indian subcontinent heat up during summer, producing a low pressure area over the northern and central Indian subcontinent. As a result, moisture-laden winds from the Indian Ocean rush in to the subcontinent. As the air flows towards the Himalayas, it is forced to rise and precipitation occurs. The southwest monsoon is generally expected to begin around the start of June and dies down by the end of September. Failure of the Indian summer monsoon to deliver the normal rainfall would bring drought and hardship for the Indian economy. Since the late 1800s, several studies have attempted long-range prediction of the Indian summer monsoon rainfall. Cannon and McKendry (1999, 2002) applied MLP NN to forecast the rainfall using regional circulation PCs as predictors.

Climate change

There is even longer time scale variability or change in the climate system – e.g. the Pacific Decadal Oscillation (PDO) has main time scales around 15–25 years and 50–70 years (Mantua and Hare, 2002). The increase of anthropogenic greenhouse gases in the atmosphere may also cause long-term climate change. There have not been many applications of machine learning methods to climate change since the observed records are relatively short for nonlinear signal analysis. Nevertheless, NN methods have been applied to anthropogenic climate change problems (Walter *et al.*, 1998; Walter and Schonwiese, 2002; Pasini *et al.*, 2006).

In Walter *et al.* (1998), NN methods were used to simulate the observed global (and hemispheric) annual mean surface air temperature variations during 1874–1993 using anthropogenic and natural forcing mechanisms as predictors. The two anthropogenic forcings were equivalent CO_2 concentrations and tropospheric sulfate aerosol concentrations, while the three natural forcings were volcanism, solar activity and ENSO. The NN explained up to 83% of the observed temperature variance, significantly more than by multiple regression analysis. On a global average, the greenhouse gas signal was assessed to be 0.9–1.3 K (warming), the sulfate signal 0.2–0.4 K (cooling), which were similar to the findings from GCM experiments. The related signals of the three natural forcing mechanisms each covered an amplitude of around 0.1–0.3 K.

12.3.3 Radiation in atmospheric models

The earth radiates primarily in the infrared frequency band. This outgoing radiation is called the *longwave radiation* (LWR), as the wavelengths are long relative to those of the incoming solar radiation. Greenhouse gases, (e.g. carbon dioxide, water vapour, methane and nitrous oxide) absorb certain wavelengths of the LWR, adding heat to the atmosphere and in turn causing the atmosphere to emit more radiation. Some of this radiation is directed back towards the Earth, hence warming the Earth's surface. The Earth's radiation balance is very closely achieved since the outgoing LWR very nearly equals the absorbed incoming *shortwave radiation* (SWR) from the sun (primarily as visible, near-ultraviolet and near-infrared radiation). Atmospheric general circulation models (GCM) typically spend a major part of their computational resources on calculating the LWR and SWR fluxes through the atmosphere.

A GCM computes the net LWR heat flux $F(p)$, where the pressure p serves as a vertical coordinate. The cooling rate $C_r(p)$ is simply proportional to $\partial F/\partial p$. Besides being a function of p, F is also a function of **S**, variables at the Earth's surface, **T**, the vertical temperature profile, **V**, vertical profiles of chemical concentrations (e.g. CO_2 concentration), and **C**, cloud variables.

Chevallier *et al.* (1998, 2000) developed MLP NN models to replace LWR fluxes in GCMs. For the flux F at the discretized pressure level p_j, the original GCM computed

$$F = \sum_i a_i(\mathbf{C}) F_i(\mathbf{S}, \mathbf{T}, \mathbf{V}), \tag{12.13}$$

where the summation over i is from the Earth's surface to the level p_j, F_i is the flux at level p_i without the cloud correction factor a_i. Neural network models $N_i(\mathbf{S}, \mathbf{T}, \mathbf{V})$ were developed to replace $F_i(\mathbf{S}, \mathbf{T}, \mathbf{V})$. This 'NeuroFlux' model was highly accurate, and has been implemented in the European Centre for Medium-Range Weather Forecasts (ECMWF) global atmospheric model, as it ran eight times faster than the original LWR code (Chevallier *et al.*, 2000).

In an alternative approach by Krasnopolsky *et al.* (2005b), MLP NN models of the form $N(\mathbf{S}, \mathbf{T}, \mathbf{V}, \mathbf{C})$ were developed to replace the cooling rates C_r at levels p_j and several radiation fluxes in the GCM. Discussions on the relative merits of the two approaches were given in Chevallier (2005) and Krasnopolsky *et al.* (2005a). This alternative NN approach is also highly accurate and has been implemented in the National Center for Atmospheric Research (NCAR) Community Atmospheric Model (CAM), and in the National Aeronautics and Space Administration (NASA) Natural Seasonal-to-Interannual Predictability Program (NSIPP) GCM (Krasnopolsky *et al.*, 2005b; Krasnopolsky and Fox-Rabinovitz, 2006; Krasnopolsky, 2007). For the NCAR CAM model, the NN model has 220 inputs and 33 outputs, and a single hidden layer with 50 neurons was found to be enough, giving an NN LWR code capable of running 150 times faster than the original LWR code (Krasnopolsky, 2007).

Similarly, this NN approach has been applied to replace the SWR codes in the NCAR CAM GCM (Krasnopolsky and Fox-Rabinovitz, 2006; Krasnopolsky, 2007). For the CAM3 model, the NN SWR model has 451 inputs and 33 outputs, and a single hidden layer with 55 neurons was found to be enough, yielding an NN SWR code capable of running about 20 times faster than the original SWR code (Krasnopolsky, 2007). Hence MLP NN models have been able to emulate the LWR and SWR codes in GCMs accurately, leading to substantial savings in computational resources.

12.3.4 Post-processing and downscaling of numerical model output

Sophisticated numerical models known as *general circulation models* (GCMs) are used both in weather forecasting and in climate research. When used in climate research, GCM also stands for '*global climate models*'. Post-processing of GCM output by multiple linear regression has been discussed in Section 1.4.5. The reason post-processing is needed is because GCMs do not have fine enough resolution to model smaller scale processes (e.g. individual clouds, or hills and valleys). Also, some local variables (e.g. atmospheric pollutant concentrations) may not be variables computed by the GCM. In *model output statistics* (MOS) (see Section 1.4.5), GCM output variables are the predictors for a multiple regression model, trained using predictands from observed data. Hence MOS allows for correction of GCM model output bias and extension of GCM forecasts to variables not computed by the GCM. Machine learning methods such as NN allow the development of nonlinear MOS. One disadvantage of MOS is that long records of GCM output need to be saved as training data. In most weather prediction centres, the GCM is upgraded frequently. After each upgrade, the GCM needs to be rerun for years to generate the long records needed for training the new MOS regression relations. As this is very costly, *updateable* MOS (UMOS) systems have been developed to alleviate the need for complete retraining after a model upgrade (Wilson and Vallée, 2002). Yuval and Hsieh (2003) developed an MLP NN model to perform nonlinear updateable MOS.

Global climate models are the main tools for estimating future climate conditions under increasing concentrations of atmospheric greenhouse gases. Unfortunately, the spatial resolution of GCMs used in climate research is coarse, typically around two or more degrees of latitude or longitude – far too coarse to resolve climate changes at the local scale (e.g. in a valley). *Downscaling* the GCM model output is therefore crucial for application of GCM to local scale and local variables.

There are two approaches to downscaling GCMs: (i) process-based techniques; and (ii) empirical techniques (Hewitson and Crane, 1996). In the process-based approach, commonly called *dynamical downscaling*, *regional climate models* (RCMs), i.e. higher resolution numerical models of a local region, are run using boundary conditions provided by the GCMs. In the empirical approach, commonly called *statistical downscaling*, statistical or machine learning methods are used to relate the GCM model output to the local variables. The GCMs are capable of resolving, e.g. mid-latitude weather systems (typically about 1000 km in diameter), a scale known as the *synoptic scale*. In empirical downscaling, it is implictly assumed that the GCM is accurate in forecasting the synoptic-scale atmospheric circulation, and that the local variables are strongly influenced by the synoptic-scale circulation.

Hewitson and Crane (1996) used MLP NN for precipitation forecasts with predictors from the GCM atmospheric data over southern Africa and the surrounding ocean. The six leading PCs (principal components) of the sea level pressure field and the seven leading PCs of the 500 hPa geopotential height field from the GCM were used as inputs to the NN, and the predictands were $1^\circ \times 1^\circ$ gridded daily precipitation data over southern Africa. The GCM was also run at double the present amount of CO_2 in the atmosphere (called the $2{\times}CO_2$ experiment), and the model output fed into the previously derived NN to predict precipitation in the $2{\times}CO_2$ atmosphere. The assumption is that the empirical relation derived between the local precipitation and the synoptic-scale atmospheric circulation for the $1{\times}CO_2$ atmosphere remains valid for the $2{\times}CO_2$ atmosphere. The danger of such an assumption will be discussed later in this subsection. Cavazos (1997) also used MLP NN to downscale GCM synoptic-scale atmospheric circulation to local $1^\circ \times 1^\circ$ gridded winter daily precipitation over northeastern Mexico. Instead of PCs, rotated PCs of the GCM data were used as inputs to the NN.

For precipitation downscaling, Olsson *et al.* (2004) found it advantageous to use two MLP NN models in sequence – the first performs classification into rain or no-rain, and if there is rain, the second NN predicts the intensity. McGinnis (1997) used MLP NN to downscale 5-day averaged snowfall, and Schoof and Pryor (2001) used them similarly to downscale daily minimum temperature (T_{min}) and maximum temperature (T_{max}), daily precipitation and monthly precipitation. Also, the radial basis function (RBF) NN was compared against a linear statistical model by Weichert and Bürger (1998) in downscaling daily temperature, precipitation and vapour pressure. Comparing several ways to fill in missing precipitation data, Coulibaly and Evora (2007) found that MLP did well. When downscaling daily T_{min} and T_{max} in Portugal with MLP NN, Trigo and Palutikof (1999) found that using no hidden layer did better than using one hidden layer. When downscaling monthly precipitation over Portugal, Trigo and Palutikof (2001) found the NN to be essentially linear, which is not surprising since averaging daily precipitation to monthly precipitation linearizes the predictor-predictand relation (Yuval and Hsieh, 2002), as discussed in Section 6.10. Cannon (2007) introduced a hybrid MLP–analogue method for downscaling, i.e. the outputs of an MLP are inputs to an analogue model which issues the prediction. For probabilistic multi-site precipitation downscaling, Cannon (2008) developed a conditional density network (CDN) model, where the MLP NN model outputs are the parameters of the Bernoulli-gamma distribution, and the objective function has constraint terms forcing the predicted between-site covariances to match the observed covariances.

For the seasonal climate in a particular year, the mean is the most commonly used statistic. However, there are statistics for severe weather in the season. For instance, for precipitation, possible statistics for the season are: (i) the average precipitation on days with >1 mm precipitation; (ii) the 90th percentile of daily precipitation; (iii) the maximum precipitation recorded in any 5 consecutive days; (iv) the fraction of total precipitation from events greater than the climatological 90th percentile; and (v) the number of events greater than the climatological 90th percentile.

Haylock *et al.* (2006) compared six statistical and two dynamical downscaling models with regard to their ability to downscale several seasonal indices of heavy precipitation for two station networks in northwest and southeast England. The skill among the eight downscaling models was high for those indices and seasons that had greater spatial coherence, i.e. winter generally showed the highest downscaling skill and summer the lowest. The six statistical models used included a canonical correlation analysis (CCA) model, three MLP NN models in different setups, a radial basis function (RBF) NN model, and SDSM (a statistical downscaling method using a 2-step conditional resampling) (Wilby *et al.*,

2002). The ranking of the models based on their correlation skills (with Spearman rank correlation used) revealed the NN models performing well. However, as the NN models were designed to reproduce the conditional mean precipitation for each day, there was a tendency to underestimate extremes. The rainfall indices indicative of rainfall occurrence were better modelled than those indicative of intensity.

Six of the models were then applied to the Hadley Centre global circulation model HadAM3P forced by emissions according to two different scenarios for the projected period of 2071–2100. The inter-model differences between the future changes in the downscaled precipitation indices were at least as large as the differences between the emission scenarios for a single model. Haylock *et al.* (2006) cautioned against interpreting the output from a single model or a single type of model (e.g. regional climate models) and emphasized the advantage of including as many different types of downscaling model, global model and emission scenario as possible when developing climate-change projections at the local scale.

In Dibike and Coulibaly (2006), MLP NN was compared with the popular multiple regression-based model SDSM (Wilby *et al.*, 2002) in downscaling daily precipitation and temperatures ($T_{\min}$ and $T_{\max}$) over a region in northern Quebec, Canada. For predictors, 25 large-scale meteorological variables, plus the predictors itself at earlier times, were used. In the NN downscaling of precipitation, a time lag of 6 days for the predictors and 20 neurons in the single hidden layer gave the best performing network, whereas for temperature downscaling, a time lag of 3 days and 12 hidden neurons did best, indicating that the predictand–predictors relationship is more complex for precipitation than for temperature. For temperature downscaling, NN and SDSM were comparable in performance, while NN was generally better than SDSM for precipitation. For the 90th percentile of daily precipitation calculated over a season, NN did better than SDSM for all seasons except summer (Fig. 12.4).

The SVR approach has been applied to precipitation downscaling over India by Tripathi *et al.* (2006), although the error function used was the MSE instead of the more robust ϵ-insensitive error function. The SVR outperformed the MLP NN in this study.

Variance

A common problem with a statistically downscaled variable is that its variance is generally smaller than the observed variance (Zorita and von Storch, 1999). This is because the influence of the large-scale variables can only account for a portion of the variance of the local variable. Various methods have been used to boost the variance of the downscaled variable to match the observed variance. The simplest

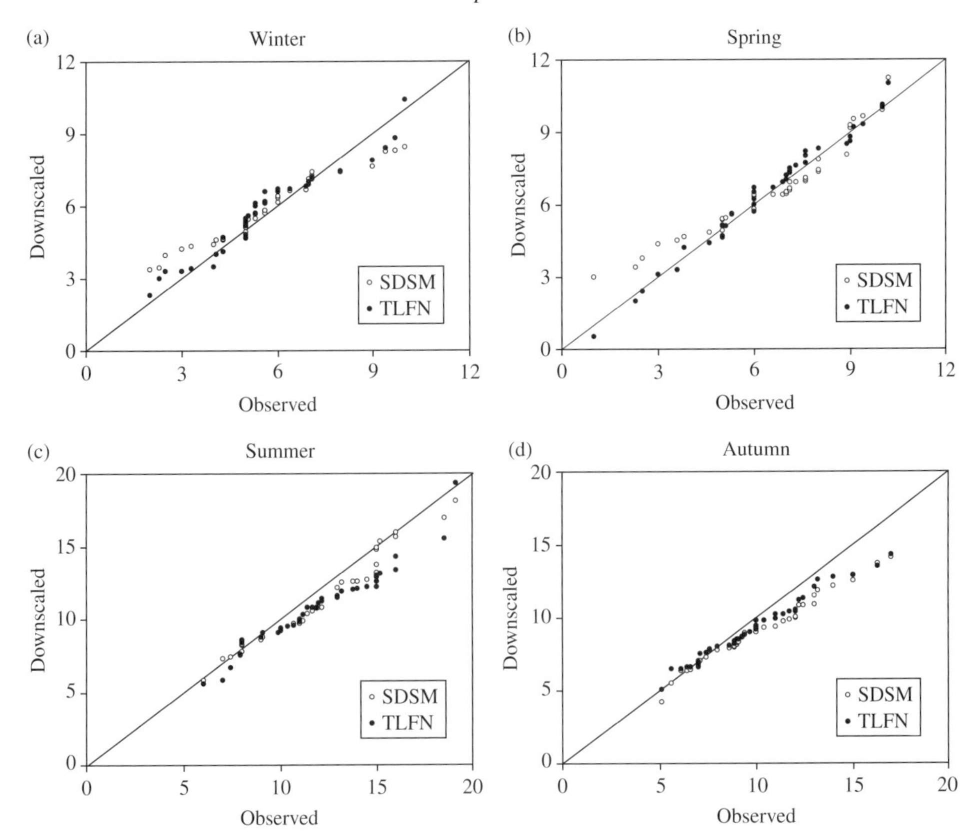

Fig. 12.4 Observed versus downscaled values of the 90th percentile precipitation (mm day^{-1}) for (a) winter, (b) spring, (c) summer and (d) autumn. Solid circles correspond to the downscaled output from the NN model (TLFN), while the open circles correspond to the downscaled output from SDSM. The diagonal line indicates perfect agreement between downscaled and observed values. (Reproduced from Dibike and Coulibaly (2006) with permission from Elsevier.)

method is *inflation* which linearly rescales the downscaled variable to match the observed variance (Karl *et al.*, 1990; von Storch, 1999). The second method is *randomization*, which adds some random noise to the downscaled variable (von Storch, 1999). The third method is *expansion*, which adds a constraint term to the objective function, forcing the predicted variance towards the observed variance during optimization (Bürger, 1996, 2002). If downscaling is for variables at multiple sites, then the constraint term is for the covariance matrix of the predicted variables to match that of the observations.

Extrapolation

Properly trained NN models (i.e. neither overfitting nor underfitting) perform nonlinear interpolation well. However when presented with new data where the predictor lies beyond the range of (predictor) values used in model training, the NN model is then *extrapolating* instead of interpolating. Let us illustrate the extrapolation behaviour with a simple test problem.

Let the signal be

$$y = x + \frac{1}{5}x^2. \tag{12.14}$$

We choose 300 observations, with x having unit standard deviation and Gaussian distribution, and y given by (12.14) plus Gaussian noise (with the noise standard deviation the same as the signal standard deviation). With six hidden neurons, the Bayesian NN model (BNN) from the *Matlab* Neural Network Toolbox was used to solve this nonlinear regression problem. In Fig. 12.5(a), upon comparing with the true signal (dashed curve), it is clear that the BNN model interpolated better than the linear regression (LR) model (solid line), but for large x values, BNN extrapolated worse than LR. Figure 12.5(b) shows the same data fitted by a fourth order polynomial, where for large positive and negative x values, the polynomial extrapolated worse than LR. Hence nonlinear models which interpolate better than LR provide no guarantee that they extrapolate better than LR.

How the nonlinear model extrapolates is dependent on the type of nonlinear model used. With a polynomial fit, as $|x| \to \infty$, $|y| \to \infty$. However, for NN models (with one hidden layer $\mathbf{h}$), where the kth hidden neuron

$$h_k = \tanh((\mathbf{W}\mathbf{x} + \mathbf{b})_k), \quad \text{and} \quad y = \widetilde{\mathbf{w}} \cdot \mathbf{h} + \widetilde{b}, \tag{12.15}$$

once the model has been trained, then as $\|\mathbf{x}\| \to \infty$, the tanh function remains bounded within ± 1, hence y remains bounded – in sharp contrast to the unbounded behaviour with polynomial extrapolation (Fig. 12.6). Extrapolations using SVR and Gaussian process (GP) models are also shown in Fig. 12.5 and 12.6, with the Gaussian or RBF kernel used in both models. With Gaussian kernels, it is straightforward to show that as $\|\mathbf{x}\| \to \infty$, $y \to$ constant for SVR, and $y \to 0$ for GP (see Exercises 12.1 and 12.2), but polynomial kernels would lead to unbounded asymptotic behaviour.

In statistical downscaling for climate change studies, the distribution of the predictors in the increased greenhouse gas climate system is usually shifted from the present day distribution. Applying the statistical downscaling model developed under the present climate to the changed climate system usually requires extrapolation of the nonlinear empirical relations. This is dangerous since we now know that nonlinear empirical models can extrapolate very differently from one another and often extrapolate worse than LR.

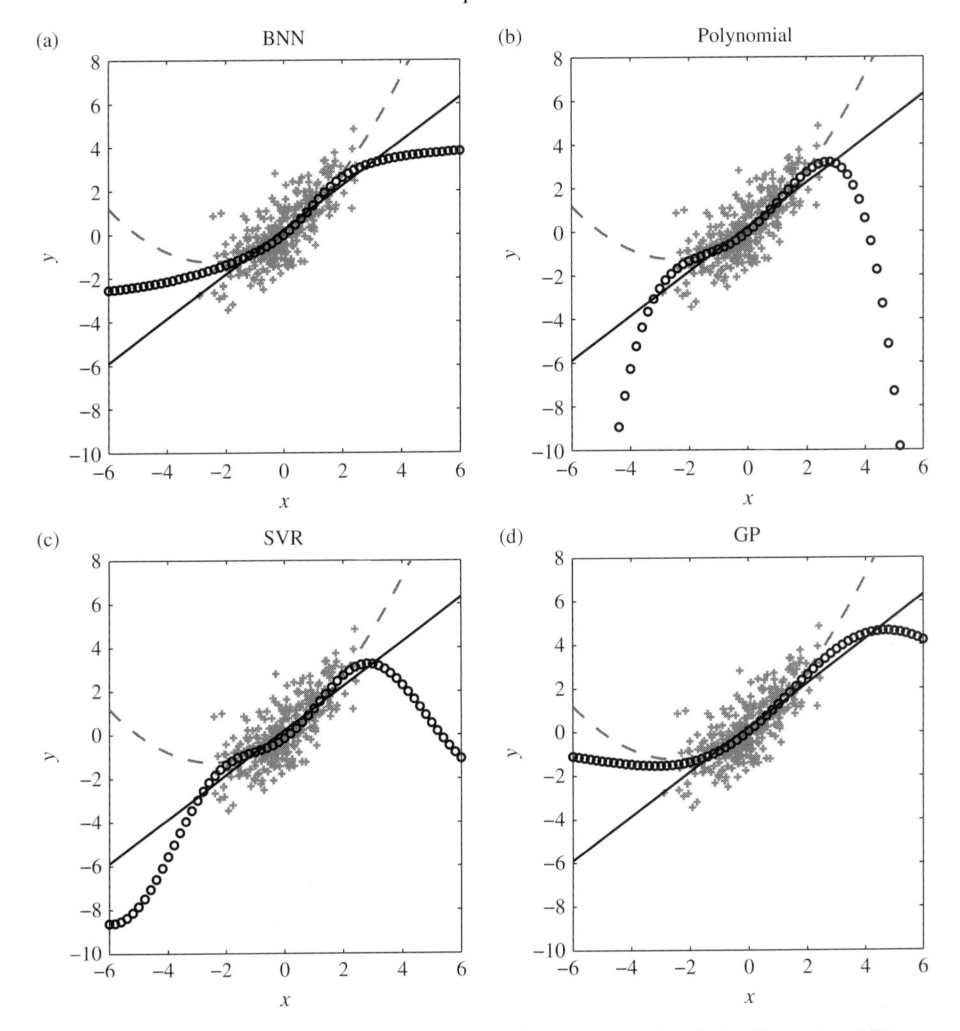

Fig. 12.5 Extrapolation behaviour. (a) Nonlinear regression fit by Bayesian NN. The data are indicated by crosses and the Bayesian NN solution by circles. Dashed curve indicates the theoretical signal and solid line the LR solution. Fit to the same dataset but by: (b) A fourth order polynomial. (c) Support vector regression (with Gaussian kernel). (d) Gaussian process (with Gaussian kernel).

12.3.5 Severe weather forecasting

Tornado

A tornado is a rapidly rotating column of air, with the bottom touching the surface of the earth. The air flow is around an intense low pressure centre, with the diameter of the tornado typically between 100–600 m, much smaller than its height. Tornadoes occur in many parts of the world, but no country experiences more tornadoes than the USA. Almost 20 000 deaths have been reported from more than 3600

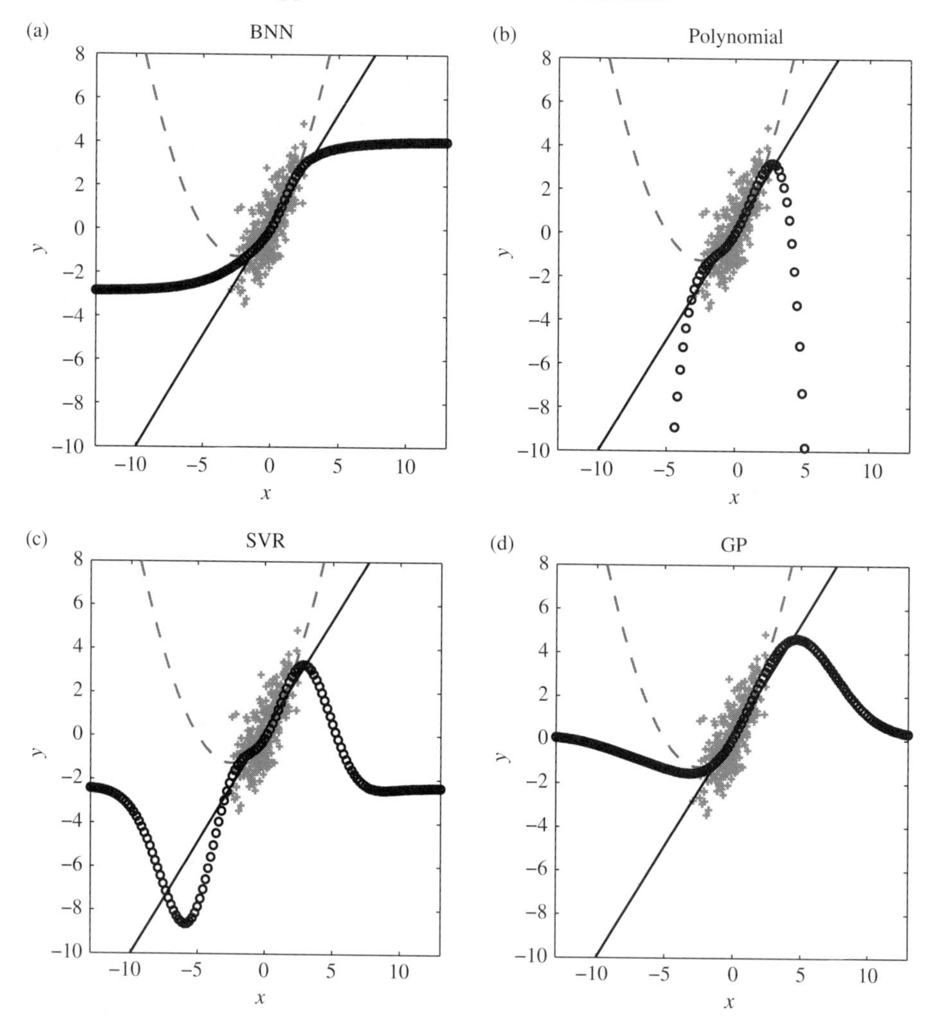

Fig. 12.6 Same as Fig. 12.5, but with the extrapolation proceeding further in both the $+x$ and $-x$ directions to show the asymptotic behaviour.

tornadoes in the USA since 1680 (Brooks and Doswell, 2002), hence the need to develop forecast capability for tornadoes in the USA. With 23 input variables characterizing the atmospheric circulation extracted from Doppler weather radar data, Marzban and Stumpf (1996) used an MLP NN classifier to forecast 'tornado' or 'no tornado' at a lead time of 20 minutes. Their NN method was found to outperform the rule-based algorithm used by the US National Severe Storms Laboratory, as well as other statistical techniques (discriminant analysis and logistic regression). Their work has been extended to prediction of damaging wind (defined as the existence of either or both of the following conditions: tornado, wind gust ≥ 25 m s^{-1}) (Marzban and Stumpf, 1998).

Tropical cyclone

Much larger on the horizontal scale and far more damaging than tornadoes are tropical cyclones, commonly called hurricanes in the Atlantic and typhoons in the Pacific. Seen from space, tropical cyclones have typical diameters of about 500 km, with spiralling bands of cloud swirling into the centre of the storm, where the air pressure is very low. Tropical cyclones are generated over warm tropical oceans where the sea surface temperature is above 26.5 °C, the winds are weak and the humidity is high. Tropical depressions are weaker low pressure systems than tropical cyclones, but some of them can strengthen into tropical cyclones. Hennon *et al.* (2005) evaluated a binary NN classifier against linear discriminant analysis in forecasting tropical cyclogenesis. With eight large-scale predictors from the National Center for Environmental Prediction–National Center for Atmospheric Research (NCEP–NCAR) reanalysis dataset, a dataset of cloud clusters during the 1998–2001 Atlantic hurricane seasons was classified into 'developing' and 'non-developing' cloud clusters, where 'developing' means a tropical depression forms within 48 hours. The model yielded 6–48 hr probability forecasts for genesis at 6 hr intervals, with the NN classifier performing comparably to or better than linear discriminant analysis on all performance measures examined, including probability of detection, Heidke skill score, etc.

Hail

Hail is precipitation in the form of ice particles (called hailstones) ranging in size from that of small peas to that of baseballs or bigger. Large hailstones can cause not only extensive property damage (breaking windows, denting cars and wrecking roofs), but also destruction of crops and livestocks, as well as human injuries and deaths. Marzban and Witt (2001) used an NN classifier to predict the size of severe hail, given that severe hail has occurred or is expected to occur. For predictors, they used four variables from Doppler weather radar and five other environmental variables. A 1-of-c coding is used for the three classes of severe hail, i.e. hailstones of coin size (class 1), golf-ball size (class 2) and baseball size (class 3). The NN model outperformed the existing method for predicting severe-hail size. The relative operating characteristic (ROC) diagrams (see Section 8.5) for the three classes (Fig. 12.7) indicate class 3 forecasts to be the best, and class 2 forecasts to be the worst, i.e. the model is least accurate in predicting mid-sized severe hailstones.

12.3.6 Air quality

Many substances in the air impair the health of humans, plants and animals, or reduce visibility. These arise both from natural processes (e.g. volcanic eruptions) and human activities (e.g. automobile and powerplant emissions). Atmospheric

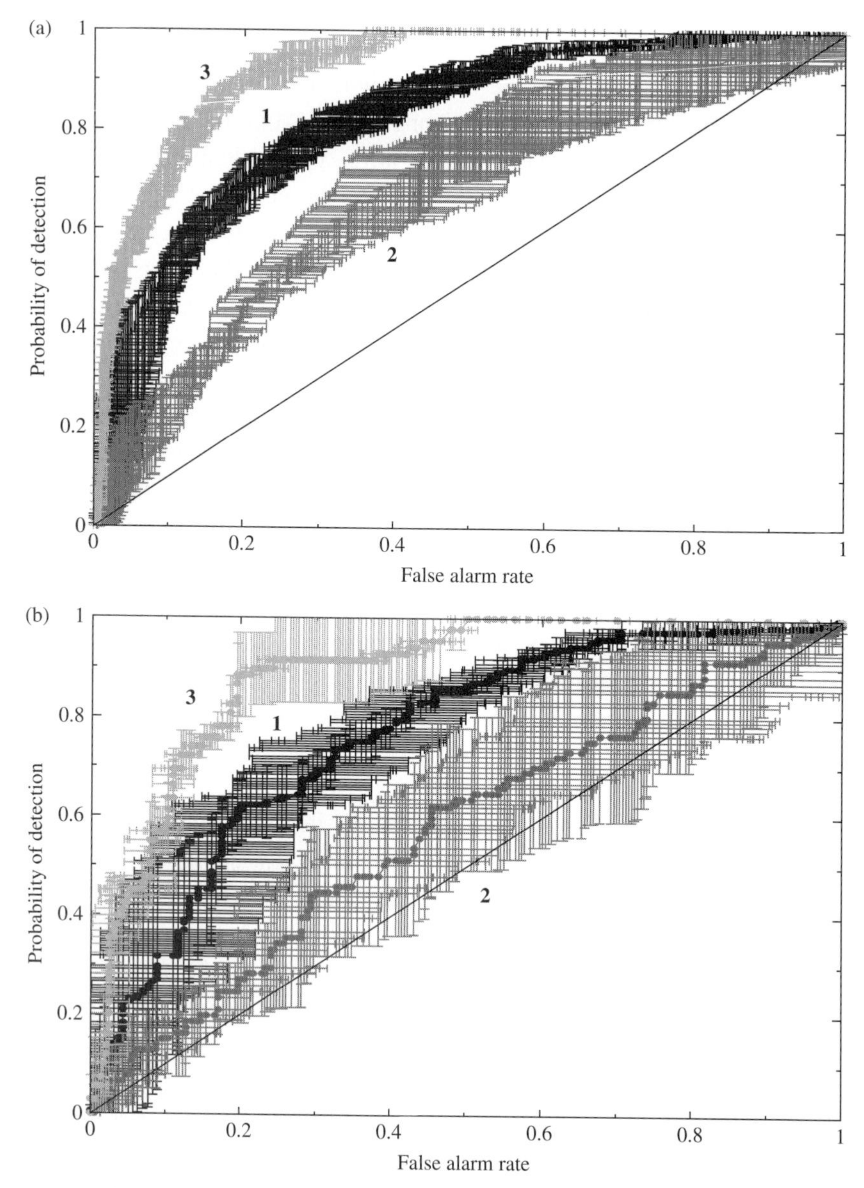

Fig. 12.7 ROC diagrams for hailstone class 1, 2, and 3 forecasts, for (a) the training data, and (b) the validation data. The error bars in the horizontal and vertical directions are the one standard deviation intervals based on bootstrapping. The diagonal line (i.e. probability of detection = false alarm rate) indicates a model with zero skill. (Reproduced from Marzban and Witt (2001) with permission from the American Meteorological Society.)

pollutants can be classified as either primary or secondary. Primary pollutants are substances directly emitted from a process, such as ash from a volcanic eruption or the carbon monoxide gas from an automobile exhaust. Secondary pollutants are not emitted directly, but are formed in the atmosphere when primary pollutants react or interact. A notorious secondary pollutant is ground-level ozone (O_3), which irritates the mucous membranes of the respiratory system. Some pollutants may be both primary and secondary, i.e. they are both emitted directly and produced by other primary pollutants.

The pollutants most commonly investigated in air quality studies include: ozone, nitrogen oxides (denoted by the symbol NO_x to include nitric oxide (NO) and nitrogen dioxide (NO_2)), sulfur oxides (SO_x, especially sulfur dioxide SO_2), and particulate matter (PM), measured as smoke and dust. The quantity PM10 is the fraction of suspended particles with diameter 10 micrometers (μm) or less, while PM2.5 relates to a maximum diameter of 2.5 μm.

Ground-level ozone, the main component of photochemical smog, is formed from the interaction of NO_x and volatile organic compounds (VOC) (e.g. hydrocarbon fuel vapours and solvents) in the presence of sunlight (hence the term 'photochemical'). It is ironic that ground-level ozone is a dreaded pollutant, while ozone high up in the stratosphere forms a protective shield against harmful ultraviolet radiation from the sun.

Yi and Prybutok (1996) applied MLP NN to forecast the daily maximum ozone concentration in an industrialized urban area in Texas. The nine predictors used were: an indicator for holiday versus working day (as traffic emissions are lighter on holidays), ozone level at 9:00 a.m., maximum daily temperature, CO_2 concentration, NO concentration, NO_2 concentration, oxides of nitrogen concentration, surface wind speed and wind direction (all measured on the same day as the predictand). The NN had four hidden neurons and one output. The NN model was found to outperform linear regression (LR) and the Box–Jenkins ARIMA model (Box and Jenkins, 1976).

Comrie (1997) also applied MLP NN to forecast the daily maximum 1 hour ozone concentrations from May–September for eight cities in the USA. The four predictors used were: daily maximum temperature, average daily dewpoint temperature, average daily wind speed and daily total sunshine. The NN had six hidden neurons and one output. A second NN model had a fifth predictor (the maximum 1 hour ozone concentration from the previous day) and six hidden neurons. Generally, the NN models slightly outperformed LR.

Using MLP NN to forecast the hourly ozone concentration at five cities in the UK, Gardner and Dorling (2000) found that NN outperformed both CART (classification and regression tree) (see Section 9.2) and LR. The eight predictors used included the amount of low cloud, base of lowest cloud, visibility, dry bulb temperature, vapour pressure, wind speed and direction. Their MLP model 1 architecture was 8:20:20:1, i.e. eight inputs, two hidden layers each with 20 neurons and one output. To account for seasonal effects, model 2 had two extra predictors, $\sin(2\pi d/365)$ and $\cos(2\pi d/365)$, with d the Julian day of the year, thereby informing the model where in the annual cycle the forecast was to be made. In model 3, two more predictors, $\sin(2\pi h/24)$ and $\cos(2\pi h/24)$, with h the hour of the day, were added to indicate the time of day. Thus the NN architecture was 10:20:20:1 for model 2 and 12:20:20:1 for model 3.

A useful measure of model performance used by Gardner and Dorling (2000) is the 'index of agreement' (Willmott, 1982), defined as

$$d_\alpha = 1 - \frac{[\sum_i |f_i - o_i|^\alpha]}{[\sum_i (|f_i - \bar{o}| + |o_i - \bar{o}|)^\alpha]}, \tag{12.16}$$

where f_i and o_i are the forecast and observed values, respectively, $\bar{o}$ the observed mean, and α can be 1 or 2. A larger d_α indicates better forecast performance. Evaluating forecasts over independent test data, Fig. 12.8 shows d_2 attained by MLP NN models, CART and LR at five cities (Gardner and Dorling, 2000). The CART and LR approaches were comparable in performance since LR has the disadvantage of being linear, while CART models nonlinear relations crudely by steps, and MLP outperformed both. Similar results were found using other measures of forecast performance (e.g. MAE, RMSE, correlation squared, etc.).

Neural network ozone forecast models have been built in many parts of the world, e.g. over western Canada (Cannon and Lord, 2000), Kuwait (Abdul-Wahab and Al-Alawi, 2002), Hong Kong (Wang *et al.*, 2003), Spain (Agirre-Basurko *et al.*, 2006) and France (Dutot *et al.*, 2007). Dorling *et al.* (2003) evaluated different objective functions for MLP ozone forecast models. Schlink *et al.* (2003) tested 15 different statistical methods for ozone forecasting over ten cities in Europe, and found that MLP NN models and generalized additive models performed best. Among MLP models, Bayesian NN models (using the Laplace approximation) underformed MLP models using early stopping to control overfitting.

When forecasting high ozone events, it is desirable to have a reliable predictive probability distribution for these extreme events. Cai *et al.* (2008) compared five probabilistic models forecasting predictive distributions of daily maximum ozone concentrations at two sites in British Columbia, Canada, with local meteorological variables used as predictors. The models were of two types, *conditional density network* (CDN) models (see Section 4.7) and Bayesian models. The Bayesian models, especially the Gaussian process (GP) model (Section 9.3), gave better forecasts for extreme events, namely poor air quality events defined as having ozone concentration $\geq$82 ppb (parts per billion). The main difference between the two types is

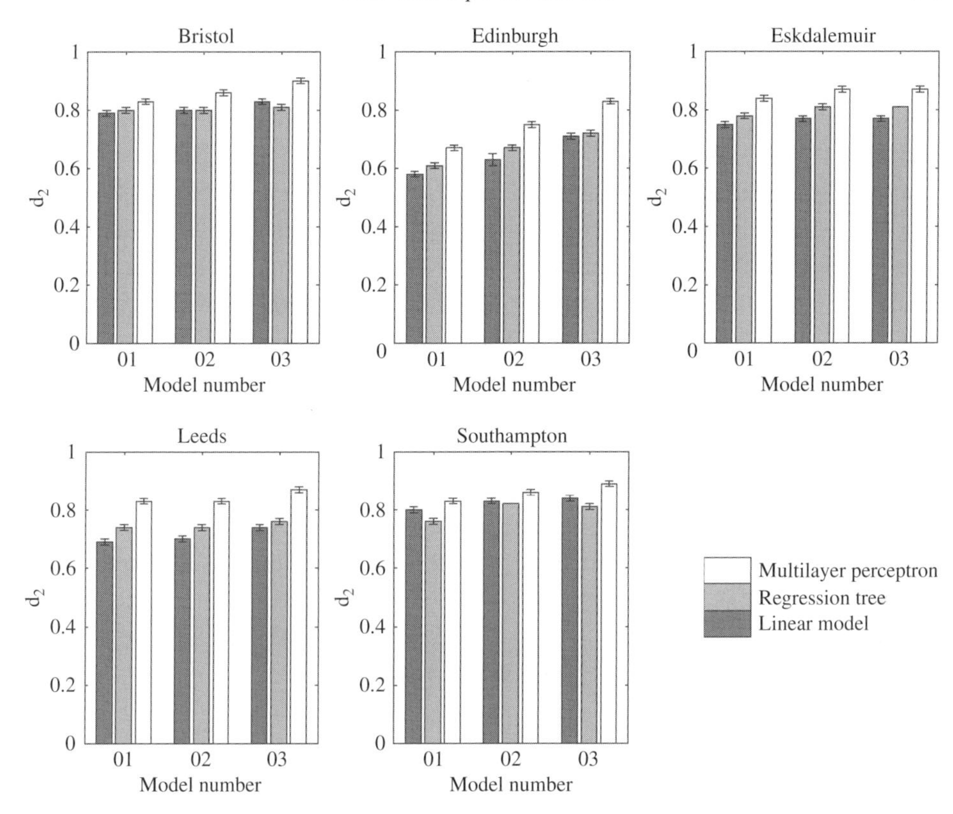

Fig. 12.8 Index of agreement (d_2) for ground-level ozone concentration forecasts by MLP NN model (white bar), regression tree (CART) (grey) and linear regression (black) at five cities in the UK. Models 1, 2 and 3 differ in the predictors used (with extra predictors in model 2 indicating the Julian day of the year, and in model 3 indicating both the day of the year and the hour of the day). The 95% confidence intervals from bootstrapping are plotted. (Reproduced from Gardner and Dorling (2000) with permission from Elsevier.)

that a CDN model uses a single optimal function (based on maximum likelihood) to forecast while a Bayesian model gives all probable functions a non-zero probability and integrates over all of them to obtain the forecast (see Section 6.9).

Cai *et al.* (2008) argued that by including low probability functions, the Bayesian approach forecasts extreme events better. In theory, Bayesian models can give an accurate measure of the predictive uncertainty arising from (a) the uncertainty of the noise process, and (b) the uncertainty of the model weights (i.e. parameters) due to finite sample size. In contrast, the CDN models only estimate the predictive uncertainty arising from the noise process, without taking into account the uncertainty of the model weights. In the observed data, most of the test data points have good similarities with the training data, so for these points, the uncertainty of the model weights (or uncertainty of the underlying function) is low. Using the weights found by maximizing likelihood, CDN models tend to find a function which is quite close to the true underlying function, thereby giving good prediction for these test points. For the relatively few points (usually the extreme events) which have little similarity with the training data, the uncertainty of the underlying function (hence the model weights) is high. The CDN models simply decide on one function and rule out other functions, while Bayesian models give all possible functions a non-zero probability, and integrate over all of them to obtain the forecast. Thus, in general, the Bayesian models have better performance over highly uncertain events. This explains why Bayesian models may have similar overall scores compared to CDN models, but outperform them over the extreme events.

For other pollutants, SO_2 forecasts by MLP models were developed by Boznar *et al.* (1993) and Nunnari *et al.* (2004), NO_x by Gardner and Dorling (1999), NO_2 by Kolehmainen *et al.* (2001) and Agirre-Basurko *et al.* (2006), NO_2 and PM10 by Kukkonen *et al.* (2003). Comparing SVR and RBF NN in forecasting various atmospheric pollutants, Lu and Wang (2005) concluded that SVR outperformed RBF NN.

12.4 Hydrology

When precipitation falls on the land surface, a portion of the water soaks into the ground. This portion, known as *infiltration*, is determined by the permeability of the soils/rocks and the surface vegetation. If the land is unable to absorb all the water, water flows on the surface as *runoff*, eventually reaching streams and lakes. The land area which contributes surface runoff to a stream or lake is called the *watershed*, which can be a few hectares in size to thousands of square kilometres. Streamflow is fed by surface runoff, interflow (the flow below the surface but above groundwater) and groundwater.

To visualize the streamflow at a location changing with time, a *hydrograph* plots the *discharge* (in volume per second) as a function of time. Hydrographs can be plotted for an individual storm event or for a longer period, e.g. one year (annual hydrograph).

The relation between streamflow and precipitation is very complicated, since the water from the precipitation is affected by the type of soil and vegetation in the watershed, before it eventually feeds into the streamflow. Because of this complexity, 'conceptual' or physical models, which try to model the physical mechanisms of the hydrological processes, are not very skillful in forecasting streamflow from precipitation data. In the last decade, machine learning methods such as MLP NN models have become popular in hydrology, as they are much simpler to develop than the physical models, yet offer better skills in modelling the precipitation-streamflow relation. Maier and Dandy (2000) reviewed 43 papers applying NN methods to hydrological problems.

Hsu *et al.* (1995) modelled the daily rainfall–runoff relation in the Leaf River Basin in Mississippi. With $x(t)$ the rainfall over the basin at day t and $y(t)$ the runoff, the MLP NN model with the single output $y(t)$ had inputs $y(t-1)$, $\ldots, y(t-n_a)$, and $x(t-1), \ldots, x(t-n_b)$, where the integers n_a and n_b are the number of lags used, i.e. the runoff on day t depends on the runoff on previous days up to $t-n_a$ and on the rainfall up to $t-n_b$. Comparisons with linear auto-regressive moving average models with exogenous inputs (ARMAX) (Box and Jenkins, 1976) and with a physical-based model showed that in 1 day leadtime runoff forecasting the NN models generally performed the best. Other studies of the rainfall–runoff relation using MLP NN include Minns and Hall (1996), Sajikumar and Thandaveswara (1999) and Dibike and Solomatine (2001). A review of over 50 publications on using NN models for hydrological modelling was provided by Dawson and Wilby (2001).

Cannon and Whitfield (2002) applied MLP NN to downscale GCM output to 5 day (pentad) averaged river discharge in 21 watersheds in British Columbia, Canada. For the 21 watersheds, there are three types of annual hydrograph: (a) *glacial* systems exhibiting a broad snow and ice melt peak in the hydrograph which persists through summer; (b) *nival* systems displaying a narrower melt peak in summer from the snowmelt; and (c) *pluvial* systems along the west coast showing strong streamflow in winter from heavy rainfall and low flow during the dry summer. The GCM data were from the NCEP/NCAR reanalysis project (Kalnay *et al.*, 1996) – reanalysis data are outputs from a high resolution atmospheric model run with extensive assimilation of observed data, hence results from the reanalysis data can be regarded as coming from an ideal GCM. The reanalysis data were smoothed to 10×10 degree grids, and four meteorological fields (sea level pressure, 500 hPa geopotential height, 850 hPa specific humidity and 1000–500 hPa thickness fields) over 7×7 grids were the predictor fields used in the downscaling. Time-lagged versions of the predictors and the predictand (the discharge) and sine and cosine of the day-of-year (to indicate the time of the year) were also used as predictors. The NN models generally outperformed stepwise multiple linear regression.

Direct measurements of river discharge are difficult, so hydrologists often infer the discharge from the measured water level (called the *stage*) by using a function (called the *rating curve*) relating the stage to the discharge. A plot of the stage as a function of time is called a stage hydrograph. By using the rating curve, one can then obtain the discharge hydrograph.

The most common form used for the stage–discharge relation is

$$Q = a(h - h_0)^b, \tag{12.17}$$

where Q is the discharge, h is the stage, h_0 is the minimum stage below which there is no discharge, and a and b are constant parameters. Bhattacharya and Solomatine

(2005) modelled the stage–discharge relation by the MLP NN model and by the M5 tree (Quinlan, 1993) model, which uses piecewise linear functions to represent the nonlinear relation. The model inputs were h_t, h_{t-1} and h_{t-2}, (i.e. h at time t, $t-1$ and $t-2$ hours, respectively) and Q_{t-1}, and the output was Q_t. The performances of MLP and M5 over independent test data were similar, though the M5 model was easier to interpret than MLP. Both MLP and M5 easily outperformed the traditional rating curve model (12.17), with the RMSE over test data being 69.7, 70.5 and 111.2 for the M5 model, the MLP and the conventional rating curve, respectively.

The MLP and M5 models were also compared in discharge prediction 1 to 6 hours ahead using effective rainfall and discharge as predictors in Solomatine and Dulal (2003), where MLP performed slightly better than M5, but M5 gave easily interpreted relations. Solomatine and Xue (2004) built models for predicting the flood discharge of Huai River in China 1 day ahead using discharge, rainfall, and evaporation as predictors. An interesting hybrid M5-MLP approach was proposed, i.e. since the M5 performed a classification of the inputs and then assigned a linear function for each class, the hybrid approach instead assigned an MLP model for each class. This hybrid approach was found to outperform both the M5 approach and the MLP approach. Application of M5 and MLP to model sediment transport in rivers has also been made (Bhattacharya *et al.*, 2007). Adding data assimilation capability to NN models in hydrology has been proposed by van den Boogaard and Mynett (2004).

Kernel methods have been used in hydrological research in recent years. Dibike *et al.* (2001) compared SVR with MLP in rainfall–runoff modelling for three rivers (in China, Vietnam and Nepal), and found SVR to outperform MLP, similar to the conclusion reached by Khan and Coulibaly (2006) in long-term prediction of lake water level in Lake Erie. Yu *et al.* (2006) used SVR for flood stage forecasting, while Yu and Liong (2007) applied kernel ridge regression to runoff forecasting. Bürger *et al.* (2007) modelled the rainfall–runoff relation in Spain using SVR and the relevance vector machine (RVM), and found RVM to slightly outperform SVR, similarly to the conclusion reached by Ghosh and Mujumdar (2008) in their statistical downscaling of GCM output to streamflow in India.

12.5 Ecology

Ecology studies the interrelationship between organisms and their environments. As such relations are highly complex, machine learning methods offer an effective way to model them.

A central problem in ecology is how species richness (i.e. the number of species in an area) is affected by enviromental variables. Guégan *et al.* (1998) built an MLP NN model for predicting species richness in 183 rivers over the world. The three predictors were: the total surface area of the drainage area, the mean annual flow regime (in m^3s^{-1}) at the river mouth, and the net terrestrial primary productivity (i.e. the rate of energy flow through the plants of the region where the river is located). Their sensitivity studies found that variations in species richness were most strongly influenced by the net terrestrial primary production, then by the flow regime and least by the area of the drainage area. They concluded that: (i) the greater the available energy, the more fish species the aquatic environment can support; (ii) as rivers with high flow regime generally contain a greater array of habitat configurations, for regions with the same energy input, the habitat with the greater heterogeneity may support more fish species.

A surprisingly large portion of the world's photosynthesis is carried out by algae. Algal blooms can cause a variety of problems: (i) water discolouration, with the water colour turning to red, brown or green; (ii) release of toxins harmful to humans by some algal species; (iii) killing of fish and invertebrates by oxygen depletion or clogging of fish gills, etc.; and (iv) impairment of water treatment by taste, odour or biomass. Recknagel *et al.* (1997) studied use of MLP NN for modelling and predicting algal blooms in three lakes (two in Japan and one in Finland) and one river (in Australia). The predictors numbered about ten, being various nutrient and dissolved oxygen concentrations, water properties (e.g. pH, water temperature) and meteorological variables. The predictands were ten or so biological species (algal and non-algal). Since the predictions over independent data seemed to be considerably worse than the performance over training data, it seemed that overfitting might not have been fully addressed in this early application of NN to algal bloom forecasting.

Olden and Jackson (2002) examined four methods for predicting the presence or absence of 27 fish species in 286 temperate lakes located in south-central Ontario, Canada. The 13 predictors were environmental variables (lake area, lake volume, total shoreline perimeter, maximum depth, altitude, pH, total dissolved solids, growing degree days, occurrence of summer stratification, etc.), with 27 classification models built for the 27 fish species. The four methods used to perform the classification were logistic regression analysis, linear discriminant analysis, classification trees (CART) and MLP NN, with the results from leave-one-out cross-validation shown in Fig. 12.9. On average, MLP outperformed the other methods in predicting the presence/absence of fish species from environmental variables, though all methods had moderate to excellent results.

De'ath and Fabricius (2000) advocated the application of CART to ecological problems. They illustrated the CART method by relating physical and spatial environmental variables to the abundances of soft coral in the Australian Great Barrier Reef.

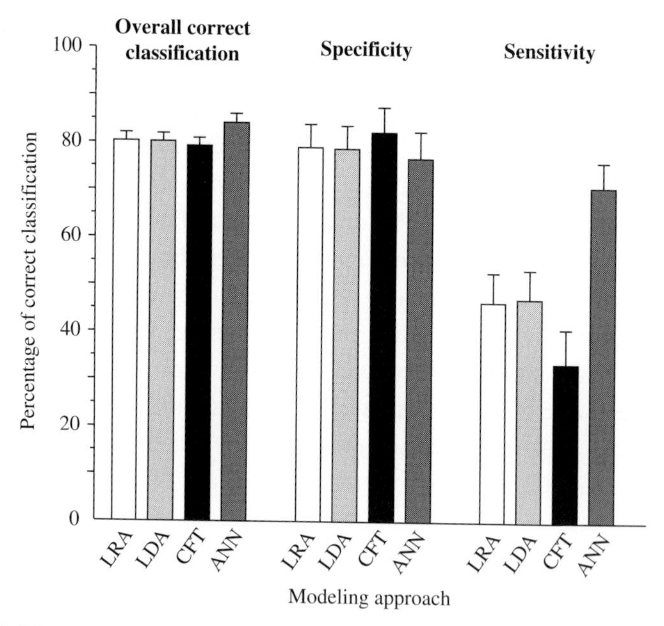

Fig. 12.9 Mean (and one standard deviation) for overall correct classification, specificity and sensitivity based on 27 fish-habitat models using logistic regression analysis (LRA), linear discriminant analysis (LDA), classification trees (CFT) and MLP NN (ANN). Overall classification performance of a model is the percentage of lakes where the model correctly predicted the presence or absence of the particular fish species. Model sensitivity is the ability correctly to predict species' presence, while model specificity is the ability correctly to predict species' absence. (Reproduced from Olden and Jackson (2002) with permission from Blackwell.)

Park *et al.* (2003) applied self-organizing map (SOM) and MLP for patterning and predicting aquatic insect species richness in a stream system in France. With 155 samples taken from the stream system, a 2-dimensional SOM was fitted to the species richness of four aquatic insect orders. For prediction of the species richness, four environmental variables (elevation, stream order, distance from the source, and maximum water temperature) were used as predictors in an MLP NN.

The otolith is a bone-like structure found in the inner ear of many species of fish. Fish otoliths accrete layers of calcium carbonate and gelatinous matrix throughout their lives. The accretion rate varies with the growth rate of the fish, often less in winter and more in summer, resulting in the appearance of rings similar to tree rings. By counting the rings in a 2-D cross-sectional image of an otolith, it is possible to determine the age of the fish in years. As this is a tedious task for humans, automation by machine learning methods is needed. Fablet and Le Josse (2005) compared two classifiers (MLP NN and SVM) for estimating the age of the fish from otolith images and other fish information (mainly the length, sex and catch date), and found that MLP significantly underperformed SVM.

Exercises

(12.1) With the Gaussian or RBF kernel (7.49) used, show that for support vector regression (9.18), as the predictor $\|\mathbf{x}\| \to \infty$, the predictand $y \to$ constant.

(12.2) With the Gaussian kernel (9.35) used, show that for the Gaussian process model (9.43), as the predictor $\|\mathbf{x}\| \to \infty$, the mean $\mu(\mathbf{x}) \to 0$.

(12.3) Suppose the predictand y is modelled by the linear regression relation $y = ax$, where x is the predictor, the variables have been scaled to have zero mean and $\mathrm{var}(x) = \mathrm{var}(y) = 1$, and $0 < a < 1$. Suppose the observed predictand y_{d} is given by $y_{\mathrm{d}} = ax + \epsilon$, where ϵ is Gaussian noise with standard deviation σ. To inflate the variance of y to match the observed variance, we introduce an inflated predictand

$$\tilde{y} = \sqrt{\frac{\mathrm{var}(y_{\mathrm{d}})}{\mathrm{var}(y)}}\, y.$$

Show that $\mathrm{var}(\tilde{y} - y_{\mathrm{d}}) > \mathrm{var}(y - y_{\mathrm{d}})$ (i.e. the MSE has increased from inflating the predictand).

Appendices

A Sources for data and codes

The book website is `www.cambridge.org/hsieh`.

Data:

Machine Learning Repository, Center for Machine Learning and Intelligent Systems, University of California, Irvine
`http://archive.ics.uci.edu/ml/`

Oceanic and atmospheric climate data. Climate Prediction Center, National Weather Service, NOAA.
`http://www.cpc.ncep.noaa.gov/products/monitoring_and_data/oadata.shtml`
In particular, monthly atmospheric and sea surface temperature indices (for the tropics),
`http://www.cpc.ncep.noaa.gov/data/indices/`

Gridded climate datasets. Earth System Research Laboratory, NOAA.
`http://www.cdc.noaa.gov/PublicData/`
In particular, the NCEP/NCAR Reanalysis data,
`http://www.cdc.noaa.gov/cdc/data.ncep.reanalysis.html`

Sea surface temperatures. National Climate Data Center, NOAA.
`http://lwf.ncdc.noaa.gov/oa/climate/research/sst/sst.php`

Climate data. Climate Research Unit, University of East Anglia.
`http://www.cru.uea.ac.uk/cru/data/`

Intergovernmental Panel on Climate Change (IPCC) Data Distribution Centre.
`http://www.ipcc-data.org/`

Computer codes:

Matlab neural network toolbox
`http://www.mathworks.com/products/neuralnet/`

Also *Matlab* statistical toolbox (contains classification and regression trees, etc.)
`http://www.mathworks.com/products/statistics/`

Netlab neural network software (by I. T. Nabney, written in *Matlab*)
`http://www.ncrg.aston.ac.uk/netlab/index.php`

Library for Support Vector Machines (LIBSVM) (by C.-C. Chang and C.-J. Lin)
`http://www.csie.ntu.edu.tw/~cjlin/libsvm/`

Gaussian Processes for Machine Learning (GPML) (by C. E. Rasmussen and C. K. I. Williams, written in *Matlab*)
`http://www.gaussianprocess.org/gpml/code/matlab/doc/`

Nonlinear principal component analysis and nonlinear canonical correlation analysis (by W. W. Hsieh, written in *Matlab*)
`http://www.ocgy.ubc.ca/projects/clim.pred/download.html`

B Lagrange multipliers

Let us consider the problem of finding the maximum of a function $f(\mathbf{x})$ subject to constraints on $\mathbf{x}$. (If we need to find the minimum of $f(\mathbf{x})$, the problem is equivalent to finding the maximum of $-f(\mathbf{x})$, so we need only to consider the case of finding the maximum.) We will consider two types of constraint: (a) equality constraints like $g(\mathbf{x}) = 0$; and (b) inequality constraints like $g(\mathbf{x}) \geq 0$.

First consider type (a), with the constraint $g(\mathbf{x}) = 0$ describing a surface in the $\mathbf{x}$-space (Fig. B.1). The gradient vector $\nabla g(\mathbf{x})$ is normal to this surface. Suppose at the point $\mathbf{x}_0$ lying on this surface, the maximum value of f occurs. The gradient vector $\nabla f(\mathbf{x})$ must also be normal to the surface at $\mathbf{x}_0$; otherwise f can be increased if we move on the surface to a point $\mathbf{x}_1$ slightly to the side of $\mathbf{x}_0$, which contradicts

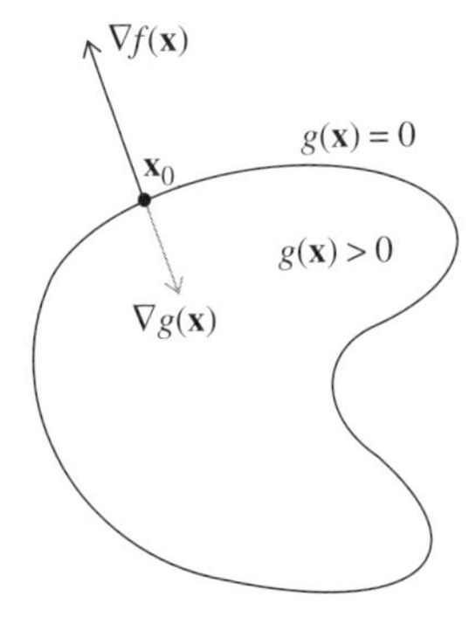

Fig. B.1 Illustrating the situation where the maximum of $f(\mathbf{x})$ occurs at a point $\mathbf{x}_0$ on the surface described by $g(\mathbf{x}) = 0$. Both gradient vectors $\nabla f(\mathbf{x})$ and $\nabla g(\mathbf{x})$ are normal to the surface at $\mathbf{x}_0$. Here the interior is assumed to have $g(\mathbf{x}) > 0$, so the gradient vector ∇g points to the interior.

the assumption that f at $\mathbf{x}_0$ is the maximum on the surface. This means ∇f and ∇g are parallel to each other at $\mathbf{x}_0$ (but may point in opposite directions), hence

$$\nabla f + \lambda \nabla g = 0, \tag{B.1}$$

for some $\lambda \neq 0$. This λ parameter is called a *Lagrange multiplier* and can have either positive or negative sign.

The *Lagrangian function* is defined by

$$L(\mathbf{x}, \lambda) = f(\mathbf{x}) + \lambda\, g(\mathbf{x}). \tag{B.2}$$

From $\nabla_{\mathbf{x}} L = 0$, we obtain (B.1), while $\partial L/\partial \lambda = 0$ gives the original constraint $g(\mathbf{x}) = 0$. If $\mathbf{x}$ is of dimension m, the original constrained maximization problem is solved by finding the stationary point of L with respect to $\mathbf{x}$ and to λ, i.e. use $\nabla_{\mathbf{x}} L = 0$ and $\partial L/\partial \lambda = 0$ to provide $m + 1$ equations for solving the values of the stationary point $\mathbf{x}_0$ and λ.

Next consider a type (b) constraint, i.e. $g(\mathbf{x}) \geq 0$. There are actually two situations to consider. The first situation is when the constrained stationary point lies on the boundary $g(\mathbf{x}) = 0$, while in the second situation, the point lies in the region $g(\mathbf{x}) > 0$. In the first situation, we are back to the previous case of a type (a) constraint. However, this time λ is not free to take on either sign. If $f(\mathbf{x})$ is a maximum at a point $\mathbf{x}_0$ on the surface $g(\mathbf{x}) = 0$, its gradient ∇f must point opposite to ∇g (Fig. B.1) (otherwise f increases in the region $g(\mathbf{x}) > 0$ contradicting that f is maximum on the boundary surface). Hence $\nabla f(\mathbf{x}) = -\lambda \nabla g(\mathbf{x})$ for some $\lambda > 0$. In the second situation, $g(\mathbf{x})$ does not affect the maximization of $f(\mathbf{x})$, so $\lambda = 0$ and $\nabla_{\mathbf{x}} L = 0$ gives back $\nabla f(\mathbf{x}) = 0$.

In either situation, $\lambda\, g(\mathbf{x}) = 0$. Hence the problem of maximizing $f(\mathbf{x}) = 0$ subject to $g(\mathbf{x}) \geq 0$ is solved by finding the stationary point of the Lagrangian (B.2) with respect to $\mathbf{x}$ and λ, subject to the *Karush–Kuhn–Tucker* (KKT) *conditions* (Karush, 1939; Kuhn and Tucker, 1951)

$$\lambda \geq 0, \tag{B.3}$$

$$g(\mathbf{x}) \geq 0, \tag{B.4}$$

$$\lambda\, g(\mathbf{x}) = 0. \tag{B.5}$$

Next, instead of maximization, *minimization* of $f(\mathbf{x})$ subject to $g(\mathbf{x}) \geq 0$ is sought. In the situation where the stationary point is on the boundary surface $g(\mathbf{x}) = 0$, ∇f must point in the same direction as ∇g, i.e. $\nabla f = \lambda \nabla g$, with λ positive. Hence the Lagrangian function to be used for the minimization problem with inequality constraint is

$$L(\mathbf{x}, \lambda) = f(\mathbf{x}) - \lambda\, g(\mathbf{x}), \tag{B.6}$$

with $\lambda \geq 0$.

Finally, if $f(\mathbf{x})$ is to be maximized subject to multiple constraints, $g_i(\mathbf{x}) = 0$ $(i = 1, \ldots, I)$ and $h_j(\mathbf{x}) \geq 0$ $(j = 1, \ldots, J)$, then the Lagrangian function becomes

$$L(\mathbf{x}, \boldsymbol{\lambda}, \boldsymbol{\mu}) = f(\mathbf{x}) + \sum_i \lambda_i g_i(\mathbf{x}) + \sum_j \mu_j h_j(\mathbf{x}). \tag{B.7}$$

The KKT conditions also give $\mu_j \geq 0$ and $\mu_j h_j(\mathbf{x}) = 0$ for $j = 1, \ldots, J$.

References

Abdul-Wahab, S. A. and Al-Alawi, S. M. (2002). Assessment and prediction of tropospheric ozone concentration levels using artificial neural networks. *Environmental Modelling & Software*, **17**(3):219–28.

Agirre-Basurko, E., Ibarra-Berastegi, G. and Madariaga, I. (2006). Regression and multilayer perceptron-based models to forecast hourly O_3 and NO_2 levels in the Bilbao area. *Environmental Modelling & Software*, **21**(4):430–46.

Aguilar-Martinez, S. (2008). *Forecasts of Tropical Pacific Sea Surface Temperatures by Neural Networks and Support Vector Regression*. M.Sc. thesis, University of British Columbia.

Aires, F. (2004). Neural network uncertainty assessment using Bayesian statistics with application to remote sensing: 1. Network weights. *Journal of Geophysical Research*, **109**. D10303, doi:10.1029/2003JD004173.

Aires, F., Chedin, A. and Nadal, J. P. (2000). Independent component analysis of multivariate time series: Application to the tropical SST variability. *Journal of Geophysical Research*, **105**(D13):17437–55.

Aires, F., Prigent, C. and Rossow, W. B. (2004a). Neural network uncertainty assessment using Bayesian statistics with application to remote sensing: 2. Output errors. *Journal of Geophysical Research*, **109**. D10304, doi:10.1029/2003JD004174.

Aires, F., Prigent, C. and Rossow, W. B. (2004b). Neural network uncertainty assessment using Bayesian statistics with application to remote sensing: 3. Network Jacobians. *Journal of Geophysical Research*, **109**. D10305, doi:10.1029/2003JD004175.

Amari, S., Murata, N., Müller, K.-R., Finke, M. and Yang, H. (1996). Statistical theory of overtraining – is cross validation asymptotically effective? *Advances in Neural Information Processing Systems*, **8**:176–182.

An, S. I., Hsieh, W. W. and Jin, F. F. (2005). A nonlinear analysis of the ENSO cycle and its interdecadal changes. *Journal of Climate*, **18**(16):3229–39.

An, S. I., Ye, Z. Q. and Hsieh, W. W. (2006). Changes in the leading ENSO modes associated with the late 1970s climate shift: Role of surface zonal current. *Geophysical Research Letters*, **33**(14). L14609, doi:10.1029/2006GL026604.

Bakir, G. H., Weston, J. and Schölkopf, B. (2004). Learning to find pre-images. *Advances in Neural Information Processing Systems*, **16**:449–56.

Balakrishnan, P. V., Cooper, M. C., Jacob, V. S. and Lewis, P. A. (1994). A study of the classification capabilities of neural networks using unsupervised learning – a comparison with k-means clustering. *Psychometrika*, **59**(4):509–25.

Baldwin, M., Gray, L., Dunkerton, T. *et al.* (2001). The Quasi-Biennial Oscillation. *Reviews of Geophysics*, **39**:179–229.

Bankert, R. L. (1994). Cloud classification of AVHRR imagery in maritime regions using a probabilistic neural network. *Journal of Applied Meteorology*, **33**(8):909–18.

Barnett, T. P. (1983). Interaction of the monsoon and Pacific trade wind system at interannual time scales Part I: The equatorial zone. *Monthly Weather Review*, **111**(4):756–73.

Barnett, T. P. and Preisendorfer, R. (1987). Origins and levels of monthly and seasonal forecast skill for United States surface air temperatures determined by canonical correlation analysis. *Monthly Weather Review*, **115**(9):1825–50.

Barnston, A. G. and Ropelewski, C. F. (1992). Prediction of ENSO episodes using canonical correlation analysis. *Journal of Climate*, **5**:1316–45.

Barnston, A. G., van den Dool, H. M., Zebiak, S. E. *et al.* (1994). Long-lead seasonal forecasts – where do we stand? *Bulletin of the American Meteorological Society*, **75**:2097–114.

Barron, A. R. (1993). Universal approximation bounds for superposition of a sigmoidal function. *IEEE Transactions on Information Theory*, **39**(3):930–45.

Benediktsson, J. A., Swain, P. H. and Ersoy, O. K. (1990). Neural network approaches versus statistical-methods in classification of multisource remote-sensing data. *IEEE Transactions on Geoscience and Remote Sensing*, **28**(4):540–52.

Bhattacharya, B., Price, R. K. and Solomatine, D. P. (2007). Machine learning approach to modeling sediment transport. *Journal of Hydraulic Engineering*, **133**(4):440–50.

Bhattacharya, B. and Solomatine, D. P. (2005). Neural networks and M5 model trees in modelling water level – discharge relationship. *Neurocomputing*, **63**:381–96.

Bickel, P. J. and Doksum, K. A. (1977). *Mathematical Statistics: Basic Ideas and Selected Topics*. Oakland, CA: Holden-Day.

Bishop, C. M. (1995). *Neural Networks for Pattern Recognition*. Oxford: Clarendon Press.

Bishop, C. M. (2006). *Pattern Recognition and Machine Learning*. New York: Springer.

Boser, B. E., Guyon, I. M. and Vapnik, V. N. (1992). A training algorithm for optimal margin classifiers. In Haussler, D., ed., *Proceedings of the 5th Annual ACM Workshop on Computational Learning Theory*, pp. 144–52. New York: ACM Press.

Box, G. P. E. and Jenkins, G. M. (1976). *Time Series Analysis: Forecasting and Control*. Oakland, CA: Holden-Day.

Boyle, P. and Frean, M. (2005). Dependent Gaussian processes. In Saul, L., Weiss, Y. and Bottou, L., eds., *Advances in Neural Information Processing Systems*, volume 17, pp. 217–24. Cambridge, MA: MIT Press.

Boznar, M., Lesjak, M. and Mlakar, P. (1993). A neural-network-based method for short-term predictions of ambient SO_2 concentrations in highly polluted industrial-areas of complex terrain. *Atmospheric Environment Part B-Urban Atmosphere*, **27**(2):221–30.

Brajard, J., Jamet, C., Moulin, C. and Thiria, S. (2006). Use of a neuro-variational inversion for retrieving oceanic and atmospheric constituents from satellite ocean colour sensor: Application to absorbing aerosols. *Neural Networks*, **19**:178–85.

Breiman, L. (1996). Bagging predictions. *Machine Learning*, **24**:123–40.

Breiman, L. (2001). Random forests. *Machine Learning*, **45**:5–32.

Breiman, L. and Friedman, J. H. (1985). Estimating optimal transformations for multiple regression and correlation. *Journal of the American Statistical Association*, **80**:580–98.

Breiman, L., Friedman, J., Olshen, R. A. and Stone, C. (1984). *Classification and Regression Trees*. New York: Chapman and Hall.

Brent, R. P. (1973). *Algorithms for Minimization without Derivatives*. Englewood Cliffs, New Jersey: Prentice-Hall.

Bretherton, C. S., Smith, C. and Wallace, J. M. (1992). An intercomparison of methods for finding coupled patterns in climate data. *Journal of Climate*, **5**:541–60.

Brier, W. G. (1950). Verification of forecasts expressed in terms of probabilities. *Monthly Weather Review*, **78**:1–3.

Brooks, H. E. and Doswell, C. A. (2002). Deaths in the 3 May 1999 Oklahoma City tornado from a historical perspective. *Weather and Forecasting*, **17**(3):354–61.

Broyden, C. G. (1970). The convergence of a class of double-rank minimization algorithms. *Journal of the Institute of Mathematics and Its Applications*, **6**:76–90.

Bürger, C. M., Kolditz, O., Fowler, H. J. and Blenkinsop, S. (2007). Future climate scenarios and rainfall-runoff modelling in the Upper Gallego catchment (Spain). *Environmental Pollution*, **148**(3):842–54.

Bürger, G. (1996). Expanded downscaling for generating local weather scenarios. *Climate Research*, **7**(2):111–28.

Bürger, G. (2002). Selected precipitation scenarios across Europe. *Journal of Hydrology*, **262**(1-4):99–110.

Burrows, W. R. (1991). Objective guidance for 0–24 hour and 24–48 hour mesoscale forecasts of lake-effect snow using CART. *Weather and Forecasting*, **6**:357–78.

Burrows, W. R. (1997). CART regression models for predicting UV radiation at the ground in the presence of cloud and other environmental factors. *Journal of Applied Meteorology*, **36**:531–44.

Burrows, W. R. (1999). Combining classification and regression trees and the neuro-fuzzy inference system for environmental data modeling. In *18th International Conference of the North American Fuzzy Information Processing Society - NAFIPS*, pp. 695–99. New York, NY: NAFIPS.

Burrows, W. R., Benjamin, M., Beauchamp, S. *et al.* CART decision-tree statistical analysis and prediction of summer season maximum surface ozone for the Vancouver, Montreal, and Atlantic regions of Canada. *Journal of Applied Meteorology*, **34**:1848–62.

Burrows, W. R., Price, C. and Wilson, L. J. (2005). Warm season lightning probability prediction for Canada and the northern United States. *Weather and Forecasting*, **20**:971–88.

Cai, S., Hsieh, W. W. and Cannon, A. J. (2008). A comparison of Bayesian and conditional density models in probabilistic ozone forecasting. In *Proceedings of the 2008 IEEE World Congress in Computational Intelligence*, Hong Kong. (See: `http://ieeexplore.ieee.org/xpls/abs_all.jsp?arnumber=4634117`)

Camps-Valls, G., Bruzzone, L., Rojo-Alvarez, J. L. and Melgani, F. (2006). Robust support vector regression for biophysical variable estimation from remotely sensed images. *IEEE Geoscience and Remote Sensing Letters*, **3**(3):339–43.

Cannon, A. J. (2006). Nonlinear principal predictor analysis: Application to the Lorenz system. *Journal of Climate*, **19**:579–89.

Cannon, A. J. (2007). Nonlinear analog predictor analysis: A coupled neural network/analog model for climate downscaling. *Neural Networks*, **20**:444–53. doi:10.1016/j.neunet.2007.04.002.

Cannon, A. J. (2008). Probabilistic multi-site precipitation downscaling by an expanded Bernoulli-gamma density network. *Journal of Hydrometeorology*, **9**:1284–300.

Cannon, A. J. and Hsieh, W. W. (2008). Robust nonlinear canonical correlation analysis: application to seasonal climate forecasting. *Nonlinear Processes in Geophysics*, **12**:221–32.

Cannon, A. J. and Lord, E. R. (2000). Forecasting summertime surface-level ozone concentrations in the Lower Fraser Valley of British Columbia: An ensemble neural network approach. *Journal of the Air and Waste Management Association*, **50**:322–39.

Cannon, A. J. and McKendry, I. G. (1999). Forecasting all-India summer monsoon rainfall using regional circulation principal components: A comparison between neural network and multiple regression models. *International Journal of Climatology*, **19**(14):1561–78.

Cannon, A. J. and McKendry, I. G. (2002). A graphical sensitivity analysis for statistical climate models: application to Indian monsoon rainfall prediction by artificial neural networks and multiple linear regression models. *International Journal of Climatology*, **22**:1687–708.

Cannon, A. J. and Whitfield, P. H. (2002). Downscaling recent streamflow conditions in British Columbia, Canada using ensemble neural network models. *Journal of Hydrology*, **259**(1-4):136–51.

Cavazos, T. (1997). Downscaling large-scale circulation to local winter rainfall in northeastern Mexico. *International Journal of Climatology*, **17**(10):1069–82.

Cavazos, T. (1999). Large-scale circulation anomalies conducive to extreme precipitation events and derivation of daily rainfall in northeastern Mexico and southeastern Texas. *Journal of Climate*, **12**:1506–23.

Cawley, G. C., Janacek, G. J., Haylock, M. R. and Dorling, S. R. (2007). Predictive uncertainty in environmental modelling. *Neural Networks*, **20**:537–49.

Chang, C.-C. and Lin, C.-J. (2001). LIBSVM: A library for support vector machines. Software available at `http://www.csie.ntu.edu.tw/~cjlin/libsvm`.

Chen, K. S., Tzeng, Y. C. and Chen, P. C. (1999). Retrieval of ocean winds from satellite scatterometer by a neural network. *IEEE Transactions on Geoscience and Remote Sensing*, **37**(1):247–56.

Chen, X. L., Li, Y. S., Liu, Z. G. *et al.* (2004). Integration of multi-source data for water quality classification in the Pearl River estuary and its adjacent coastal waters of Hong Kong. *Continental Shelf Research*, **24**(16):1827–43.

Cherkassky, V. and Mulier, F. (1998). *Learning from Data.* New York: Wiley.

Chevallier, F. (2005). Comments on 'New approach to calculation of atmospheric model physics: Accurate and fast neural network emulation of longwave radiation in a climate model'. *Monthly Weather Review*, **133**(12):3721–3.

Chevallier, F., Cheruy, F., Scott, N. A. and Chedin, A. (1998). A neural network approach for a fast and accurate computation of a longwave radiative budget. *Journal of Applied Meteorology*, **37**(11):1385–97.

Chevallier, F., Morcrette, J. J., Cheruy, F. and Scott, N. A. (2000). Use of a neural-network-based long-wave radiative-transfer scheme in the ECMWF atmospheric model. *Quarterly Journal of the Royal Meteorological Society*, **126**(563 Part B):761–76.

Christiansen, B. (2005). The shortcomings of nonlinear principal component analysis in identifying circulation regimes. *Journal of Climate*, **18**(22):4814–23.

Christiansen, B. (2007). Reply. *Journal of Climate*, **20**:378–9.

Chu, W., Keerthi, S. S. and Ong, C. J. (2004). Bayesian support vector regression using a unified loss function. *IEEE Transactions on Neural Networks*, **15**(1):29–44.

Clarke, T. (1990). Generalization of neural network to the complex plane. *Proceedings of International Joint Conference on Neural Networks*, **2**:435–40.

Collins, D. C., Reason, C. J. C. and Tangang, F. (2004). Predictability of Indian Ocean sea surface temperature using canonical correlation analysis. *Climate Dynamics*, **22**(5):481–97.

Comon, P. (1994). Independent component analysis – a new concept? *Signal Processing*, **36**:287–314.

Comrie, A. C. (1997). Comparing neural networks and regression models for ozone forecasting. *Journal of the Air and Waste Management Association*, **47**(6): 653–63.

Cordisco, E., Prigent, C. and Aires, F. (2006). Snow characterization at a global scale with passive microwave satellite observations. *Journal of Geophysical Research*, **111**(D19). D19102, doi:10.1029/2005JD006773.

Cornford, D., Nabney, I. T. and Bishop, C. M. (1999). Neural network-based wind vector retrieval from satellite scatterometer data. *Neural Computing and Applications*, **8**:206–17. doi:10.1007/s005210050023.

Cortes, C. and Vapnik, V. (1995). Support vector networks. *Machine Learning*, **20**:273–97.

Coulibaly, P. and Evora, N. D. (2007). Comparison of neural network methods for infilling missing daily weather records. *Journal of Hydrology*, **341**(1-2):27–41.

Cox, D. T., Tissot, P. and Michaud, P. (2002). Water level observations and short-term predictions including meteorological events for entrance of Galveston Bay, Texas. *Journal of Waterway, Port, Coastal and Ocean Engineering*, **128**(1):21–9.

Cressie, N. (1993). *Statistics for Spatial Data*. New York: Wiley.

Cristianini, N. and Shawe-Taylor, J. (2000). *An Introduction to Support Vector Machines and Other Kernel-based Methods*. Cambridge, UK: Cambridge University Press.

Cybenko, G. (1989). Approximation by superpositions of a sigmoidal function. *Mathematics of Control, Signals, and Systems*, **2**:303–14.

Dash, J., Mathur, A., Foody, G. M. *et al.* (2007). Land cover classification using multi-temporal MERIS vegetation indices. *International Journal of Remote Sensing*, **28**(6):1137–59.

Davidon, W. C. (1959). Variable metric methods for minimization. A.E.C.Res. and Develop. Report ANL-5990, Argonne National Lab.

Dawson, C. W. and Wilby, R. L. (2001). Hydrological modelling using artificial neural networks. *Progress in Physical Geography*, **25**(1):80–108.

De'ath, G. and Fabricius, K. E. (2000). Classification and regression trees: A powerful yet simple technique for ecological data analysis. *Ecology*, **81**(11):3178–92.

Del Frate, F., Ferrazzoli, P., Guerriero, L. *et al.* (2004). Wheat cycle monitoring using radar data and a neural network trained by a model. *IEEE Transactions on Geoscience and Remote Sensing*, **42**(1):35–44.

Del Frate, F., Pacifici, F., Schiavon, G. and Solimini, C. (2007). Use of neural networks for automatic classification from high-resolution images. *IEEE Transactions on Geoscience and Remote Sensing*, **45**(4):800–9.

Del Frate, F., Petrocchi, A., Lichtenegger, J. and Calabresi, G. (2000). Neural networks for oil spill detection using ERS-SAR data. *IEEE Transactions on Geoscience and Remote Sensing*, **38**(5):2282–7.

Del Frate, F. and Schiavon, G. (1999). Nonlinear principal component analysis for the radiometric inversion of atmospheric profiles by using neural networks. *IEEE Transactions on Geoscience and Remote Sensing*, **37**(5):2335–42.

Del Frate, F. and Solimini, D. (2004). On neural network algorithms for retrieving forest biomass from SAR data. *IEEE Transactions on Geoscience and Remote Sensing*, **42**(1):24–34.

Deser, C. and Blackmon, M. L. (1995). On the relationship between tropical and North Pacific sea surface temperature variations. *Journal of Climate*, **8**(6):1677–80.

Diaz, H. F. and Markgraf, V., eds. (2000). *El Niño and the Southern Oscillation: Multiscale Variability and Global and Regional Impacts*. Cambridge, UK: Cambridge University Press.

Dibike, Y. B. and Coulibaly, P. (2006). Temporal neural networks for downscaling climate variability and extremes. *Neural Networks*, **19**(2):135–44.

Dibike, Y. B. and Solomatine, D. P. (2001). River flow forecasting using artificial neural networks. *Physics and Chemistry of the Earth, Part B - Hydrology, Oceans and Atmosphere*, **26**(1):1–7.

Dibike, Y. B., Velickov, S., Solomatine, D. and Abbott, M. B. (2001). Model induction with support vector machines: Introduction and applications. *Journal of Computing In Civil Engineering*, **15**(3):208–16.

Dong, D. and McAvoy, T. J. (1996). Nonlinear principal component analysis based on principal curves and neural networks. *Computers and Chemical Engineering*, **20**:65–78.

Dorling, S. R., Foxall, R. J., Mandic, D. P. and Cawley, G. C. (2003). Maximum likelihood cost functions for neural network models of air quality data. *Atmospheric Environment*, **37**:3435–43. doi:10.1016/S1352-2310(03)00323-6.

Draper, N. R. and Smith, H. (1981). *Applied Regression Analysis*, 2nd edn. New York: Wiley.

Duda, R. O., Hart, P. E. and Stork, D. G. (2001). *Pattern Classification*, 2nd edn. New York: Wiley.

Dutot, A. L., Rynkiewicz, J., Steiner, F. E. and Rude, J. (2007). A 24 hr forecast of ozone peaks and exceedance levels using neural classifiers and weather predictions. *Environmental Modelling & Software*, **22**(9):1261–9.

Efron, B. (1979). Bootstrap methods: another look at the jackknife. *Annals of Statistics*, **7**:1–26.

Efron, B. and Tibshirani, R. J. (1993). *An Introduction to the Bootstrap*. Boca Raton, Florida: CRC Press.

Elachi, C. and van Zyl, J. (2006). *Introduction To The Physics and Techniques of Remote Sensing*, 2nd edn. Hoboken, NJ: Wiley-Interscience.

Elsner, J. B. and Tsonis, A. A. (1996). *Singular Spectrum Analysis*. New York: Plenum.

Emery, W. J. and Thomson, R. E. (1997). *Data Analysis Methods in Physical Oceanography*. Oxford: Pergamon.

Essenreiter, R., Karrenbach, M. and Treitel, S. (2001). Identification and classification of multiple reflections with self-organizing maps. *Geophysical Prospecting*, **49**(3):341–52.

Fablet, R. and Le Josse, N. (2005). Automated fish age estimation from otolith images using statistical learning. *Fisheries Research*, **72**(2-3):279–90.

Fang, W. and Hsieh, W. W. (1993). Summer sea surface temperature variability off Vancouver Island from satellite data. *Journal of Geophysical Research*, **98**(C8):14391–400.

Faucher, M., Burrows, W. R. and Pandolfo, L. (1999). Empirical-statistical reconstruction of surface marine winds along the western coast of Canada. *Climate Research*, **11**(3):173–90.

Fletcher, R. (1970). A new approach to variable metric algorithms. *Computer Journal*, **13**:317–22.

Fletcher, R. and Powell, M. J. D. (1963). A rapidly convergent descent method for minization. *Computer Journal*, **6**:163–8.

Fletcher, R. and Reeves, C. M. (1964). Function minimization by conjugate gradients. *Computer Journal*, **7**:149–54.

Fogel, D. (2005). *Evolutionary Computation: Toward a New Philosophy of Machine Intelligence*, 3rd edn. Hoboken, NJ: Wiley-IEEE.

Foody, G. M. and Mathur, A. (2004). A relative evaluation of multiclass image classification by support vector machines. *IEEE Transactions on Geoscience and Remote Sensing*, **42**(6):1335–43.

Foresee, F. D. and Hagan, M. T. (1997). Gauss–Newton approximation to Bayesian regularization. In *Proceedings of the 1997 International Joint Conference on Neural Networks*. (See: `http://ieeexplore.ieee.org/xpls/abs_all.jsp?arnumber=614194`)

Fraser, R. H. and Li, Z. (2002). Estimating fire-related parameters in boreal forest using spot vegetation. *Remote Sensing of Environment*, **82**(1):95–110.

Freund, Y. and Schapire, R. E. (1997). A decision-theoretical generalization of on-line learning and an application to boosting. *Journal of Computer System Sciences*, **55**:119–39.

Galton, F. J. (1885). Regression towards mediocrity in hereditary stature. *Journal of the Anthropological Institute*, **15**:246–63.

Garcia-Gorriz, E. and Garcia-Sanchez, J. (2007). Prediction of sea surface temperatures in the western Mediterranean Sea by neural networks using satellite observations. *Geophysical Research Letters*, **34**. L11603, doi:10.1029/2007GL029888.

Gardner, M. W. and Dorling, S. R. (1999). Neural network modelling and prediction of hourly NO_x and NO_2 concentrations in urban air in London. *Atmospheric Environment*, **33**:709–19.

Gardner, M. W. and Dorling, S. R. (2000). Statistical surface ozone models: an improved methodology to account for non-linear behaviour. *Atmospheric Environment*, **34**:21–34.

Georgiou, G. and Koutsougeras, C. (1992). Complex domain backpropagation. *IEEE Trans. Circuits and Systems II*, **39**:330–4.

Ghil, M., Allen, M. R., Dettinger, M. D. *et al.* (2002). Advanced spectral methods for climatic time series. *Reviews of Geophysics*, **40**. 1003, DOI: 10.1029/2000RG000092.

Ghosh, S. and Mujumdar, P. P. (2008). Statistical downscaling of GCM simulations to streamflow using relevance vector machine. *Advances in Water Resources*, **31**(1):132–46.

Gill, A. E. (1982). *Atmosphere-Ocean Dynamics*. Orlando Florida: Academic Press.

Gill, P. E., Murray, W. and Wright, M. H. (1981). *Practical Optimization*. London: Academic Press.

Gneiting, T., Raftery, A. E., Westveld, A. H. I. and Goldman, T. (2005). Calibrated probabilistic forecasting using ensemble model output statistics and minimum CRPS estimation. *Monthly Weather Review*, **133**:1098–118.

Goddard, L., Mason, S. J., Zebiak, S. E. *et al.* (2001). Current approaches to seasonal-to-interannual climate predictions. *International Journal of Climatology*, **21**(9):1111–52.

Goldfarb, F. (1970). A family of variable metric methods derived by variational means. *Mathematics of Computation*, **24**:23–6.

Golub, G. H., Heath, M. and Wahba, G. (1979). Generalized cross-validation as a method for choosing a good ridge parameter. *Technometrics*, **21**:215–23.

Gopal, S. and Woodcock, C. (1996). Remote sensing of forest change using artificial neural networks. *IEEE Transactions on Geoscience and Remote Sensing*, **34**:398–404.

Gourrion, J., Vandemark, D., Bailey, S. *et al.* (2002). A two-parameter wind speed algorithm for Ku-band altimeters. *Journal of Atmospheric and Oceanic Technology*, **19**(12):2030–48.

Grieger, B. and Latif, M. (1994). Reconstruction of the El Niño attractor with neural networks. *Climate Dynamics*, **10**(6):267–76.

Gross, L., Thiria, S., Frouin, R. and Greg, M. B. (2000). Artificial neural networks for modeling the transfer function between marine reflectance and phytoplankton pigment concentration. *Journal of Geophysical Research*, **105**(C2):3483–3496. doi: 10.1029/1999JC900278.

Guégan, J. F., Lek, S. and Oberdorff, T. (1998). Energy availability and habitat heterogeneity predict global riverine fish diversity. *Nature*, **391**:382–4.

Gull, S. F. (1989). Developments in maximum entropy data analysis. In Skilling, J., ed. *Maximum Entropy and Bayesian Methods*, pp. 53–71. Dordrecht: Kluwer.

Haber, E. and Oldenburg, D. (2000). A GCV based method for nonlinear ill-posed problems. *Computational Geoscience*, **4**:41–63.

Hamilton, K. (1988). A detailed examination of the extratropical response to tropical El Niño/Southern Oscillation events. *Journal of Climatology*, **8**:67–86.

Hamilton, K. (1998). Dynamics of the tropical middle atmosphere: A tutorial review. *Atmosphere-Ocean*, **36**(4):319–54.

Hamilton, K. and Hsieh, W. W. (2002). Representation of the QBO in the tropical stratospheric wind by nonlinear principal component analysis. *Journal of Geophysical Research*, **107**(D15). 4232, doi: 10.1029/2001JD001250.

Han, G. and Shi, Y. (2008). Development of an Atlantic Canadian coastal water level neural network model (ACCSLENNT). *Journal of Atmospheric and Oceanic Technology*, **25**:2117–32.

Hardman-Mountford, N. J., Richardson, A. J., Boyer, D. C., Kreiner, A. and Boyer, H. J. (2003). Relating sardine recruitment in the Northern Benguela to satellite-derived sea surface height using a neural network pattern recognition approach. *Progress in Oceanography*, **59**:241–55.

Hardoon, D. R., Szedmak, S. and Shawe-Taylor, J. (2004). Canonical correlation analysis: An overview with application to learning methods. *Neural Computation*, **16**:2639–64.

Hardy, D. M. (1977). Empirical eigenvector analysis of vector wind measurements. *Geophysical Research Letters*, **4**:319–20.

Hardy, D. M. and Walton, J. J. (1978). Principal component analysis of vector wind measurements. *Journal of Applied Meteorology*, **17**:1153–62.

Hasselmann, K. (1988). PIPs and POPs – a general formalism for the reduction of dynamical systems in terms of Principal Interaction Patterns and Principal Oscillation Patterns. *Journal of Geophysical Research*, **93**:11015–20.

Hasselmann, S. and Hasselmann, K. (1985). Computations and parameterizations of the nonlinear energy-transfer in a gravity-wave spectrum. Part I: A new method for efficient computations of the exact nonlinear transfer integral. *Journal of Physical Oceanography*, **15**(11):1369–77.

Hasselmann, S., Hasselmann, K., Allender, J. H. and Barnett, T. P. (1985). Computations and parameterizations of the nonlinear energy-transfer in a gravity-wave spectrum.

Part II: Parameterizations of the nonlinear energy-transfer for application in wave models. *Journal of Physical Oceanography*, **15**(11):1378–91.

Hastie, T. and Stuetzle, W. (1989). Principal curves. *Journal of the American Statistical Association*, **84**:502–16.

Hastie, T., Tibshirani, R. and Friedman, J. (2001). *Elements of Statistical Learning: Data Mining, Inference and Prediction*. New York: Springer-Verlag.

Haupt, R. L. and Haupt, S. E. (2004). *Practical Genetic Algorithms*. New York: Wiley.

Haupt, S. E., Pasini, A. and Marzban, C., eds. (2009). *Artificial Intelligence Methods in the Environmental Sciences*. Springer.

Haykin, S. (1999). *Neural Networks: A Comprehensive Foundation*. New York: Prentice Hall.

Haylock, M. R., Cawley, G. C., Harpham, C., Wilby, R. L. and Goodess, C. M. (2006). Downscaling heavy precipitation over the United Kingdom: A comparison of dynamical and statistical methods and their future scenarios. *International Journal of Climatology*, **26**(10):1397–415. doi: 10.1002/joc.1318.

Heidke, P. (1926). Berechnung des Erfolges und der Güte der Windstärkevorhersagen in Sturmwarnungsdienst. *Geografiska Annaler*, **8**:310–49.

Hennon, C. C., Marzban, C. and Hobgood, J. S. (2005). Improving tropical cyclogenesis statistical model forecasts through the application of a neural network classifier. *Weather and Forecasting*, **20**:1073–1083. doi: 10.1175/WAF890.1.

Herman, A. (2007). Nonlinear principal component analysis of the tidal dynamics in a shallow sea. *Geophysical Research Letters*, **34**(2).

Hestenes, M. R. and Stiefel, E. (1952). Methods of conjugate gradients for solving linear systems. *Journal of Research of the National Bureau of Standards*, **49**(6):409–36.

Hewitson, B. C. and Crane, R. G. (1996). Climate downscaling: Techniques and application. *Climate Research*, **7**(2):85–95.

Hewitson, B. C. and Crane, R. G. (2002). Self-organizing maps: applications to synoptic climatology. *Climate Research*, **22**(1):13–26.

Hirose, A. (1992). Continuous complex-valued backpropagation learning. *Electronic Letters*, **28**:1854–5.

Hoerling, M. P., Kumar, A. and Zhong, M. (1997). El Niño, La Niña and the nonlinearity of their teleconnections. *Journal of Climate*, **10**:1769–86.

Holton, J. R. and Tan, H.-C. (1980). The influence of the equatorial quasi-biennial oscillation on the global circulation at 50 mb. *Journal of the Atmospheric Sciences*, **37**:2200–8.

Hong, Y., Hsu, K. L., Sorooshian, S. and Gao, X. G. (2004). Precipitation estimation from remotely sensed imagery using an artificial neural network cloud classification system. *Journal of Applied Meteorology*, **43**(12):1834–52.

Horel, J. D. (1981). A rotated principal component analysis of the interannual variability of the Northern Hemisphere 500 mb height field. *Monthly Weather Review*, **109**:2080–92.

Horel, J. D. (1984). Complex principal component analysis: Theory and examples. *Journal of Climate and Applied Meteorology*, **23**:1660–73.

Horel, J. D. and Wallace, J. M. (1981). Planetary-scale atmospheric phenomena associated with the Southern Oscillation. *Monthly Weather Review*, **109**:813–29.

Hornik, K. (1991). Approximation capabilities of multilayer feedforward networks. *Neural Networks*, **4**:252–7.

Hornik, K., Stinchcombe, M. and White, H. (1989). Multilayer feedforward networks are universal approximators. *Neural Networks*, **2**:359–66.

Horstmann, J., Schiller, H., Schulz-Stellenfleth, J. and Lehner, S. (2003). Global wind speed retrieval from SAR. *IEEE Transactions on Geoscience and Remote Sensing*, **41**(10):2277–86.

Hotelling, H. (1933). Analysis of a complex of statistical variables into principal components. *Journal of Educational Psychology*, **24**:417–41.

Hotelling, H. (1936). Relations between two sets of variates. *Biometrika*, **28**:321–77.

Hou, Z. and Koh, T. S. (2004). Image denoising using robust regression. *IEEE Signal Processing Letters*, **11**:234–46.

Hsieh, W. W. (2000). Nonlinear canonical correlation analysis by neural networks. *Neural Networks*, **13**:1095–105.

Hsieh, W. W. (2001a). Nonlinear canonical correlation analysis of the tropical Pacific climate variability using a neural network approach. *Journal of Climate*, **14**:2528–39.

Hsieh, W. W. (2001b). Nonlinear principal component analysis by neural networks. *Tellus*, **53A**:599–615.

Hsieh, W. W. (2004). Nonlinear multivariate and time series analysis by neural network methods. *Reviews of Geophysics*, **42**. RG1003, doi:10.1029/2002RG000112.

Hsieh, W. W. (2007). Nonlinear principal component analysis of noisy data. *Neural Networks*, **20**:434–43.

Hsieh, W. W. and Cannon, A. J. (2008). Towards robust nonlinear multivariate analysis by neural network methods. In Donner, R. and Barbosa, S., eds., *Nonlinear Time Series Analysis in the Geosciences – Applications in Climatology, Geodynamics, and Solar-Terrestrial Physics*, pp. 97–124. Berlin: Springer.

Hsieh, W. W. and Hamilton, K. (2003). Nonlinear singular spectrum analysis of the tropical stratospheric wind. *Quarterly Journal of the Royal Meteorological Society*, **129**:2367–82.

Hsieh, W. W. and Tang, B. (1998). Applying neural network models to prediction and data analysis in meteorology and oceanography. *Bulletin of the American Meteorological Society*, **79**:1855–70.

Hsieh, W. W., Tang, B. and Garnett, E. R. (1999). Teleconnections between Pacific sea surface temperatures and Canadian prairie wheat yield. *Agricultural and Forest Meteorology*, **96**:209–17.

Hsieh, W. W. and Wu, A. (2002). Nonlinear multichannel singular spectrum analysis of the tropical Pacific climate variability using a neural network approach. *Journal of Geophysical Research*, **107**(C7). doi: 10.1029/2001JC000957.

Hsieh, W. W., Wu, A. and Shabbar, A. (2006). Nonlinear atmospheric teleconnections. *Geophysical Research Letters*, **33**. L07714, doi:10.1029/2005GL025471.

Hsu, K. L., Gao, X. G., Sorooshian, S. and Gupta, H. V. (1997). Precipitation estimation from remotely sensed information using artificial neural networks. *Journal of Applied Meteorology*, **36**(9):1176–90.

Hsu, K. L., Gupta, H. V. and Sorooshian, S. (1995). Artificial neural-network modeling of the rainfall-runoff process. *Water Resources Research*, **31**(10):2517–30.

Huang, C., Davis, L. S. and Townshend, J. R. G. (2002). An assessment of support vector machines for land cover classification. *International Journal of Remote Sensing*, **23**(4):725–49.

Huber, P. J. (1964). Robust estimation of a location parameter. *The Annals of Mathematical Statistics*, **35**:73–101.

Hyvärinen, A., Karhunen, J. and Oja, E. (2001). *Independent Component Analysis*. New York: Wiley.

IPCC (2007). *Climate Change 2007: The Physical Science Basis. Contribution of Working Group I to the Fourth Assessment Report of the Intergovernmental Panel on Climate Change.* Cambridge, UK: Cambridge University Press.

Jamet, C. and Hsieh, W. W. (2005). The nonlinear atmospheric variability in the winter northeast Pacific associated with the Madden–Julian Oscillation. *Geophysical Research Letters*, **32**(13). L13820, doi: 10.1029/2005GL023533.

Jamet, C., Thiria, S., Moulin, C. and Crepon, M. (2005). Use of a neurovariational inversion for retrieving oceanic and atmospheric constituents from ocean color imagery: A feasibility study. *Journal of Atmospheric and Oceanic Technology*, **22**(4):460–75.

Jaynes, E. T. (2003). *Probability theory: the logic of science.* Cambridge, UK: Cambridge University Press.

Jenkins, G. M. and Watts, D. G. (1968). *Spectral Analysis and Its Applications.* San Francisco: Holden-Day.

Jin, F. F. (1997a). An equatorial ocean recharge paradigm for ENSO. Part I: Conceptual model. *Journal of the Atmospheric Sciences*, **54**(7):811–29.

Jin, F. F. (1997b). An equatorial ocean recharge paradigm for ENSO. Part II: A stripped-down coupled model. *Journal of the Atmospheric Sciences*, **54**(7):830–47.

Jolliffe, I. T. (2002). *Principal Component Analysis.* New York: Springer.

Jolliffe, I. T. and Stephenson, D. B., eds., (2003). *Forecast Verification: A Practitioner's Guide in Atmospheric Science.* Chichester: Wiley.

Kaiser, H. F. (1958). The varimax criterion for analytic rotation in factor analysis. *Psychometrika*, **23**:187–200.

Kalnay, E., Kanamitsu, M., Kistler, R. *et al.* (1996). The NCEP/NCAR 40 year reanalysis project. *Bulletin of the American Meteorological Society*, **77**(3):437–71.

Kaplan, A., Kushnir, Y. and Cane, M. A. (2000). Reduced space optimal interpolation of historical marine sea level pressure: 1854-1992. *Journal of Climate*, **13**(16):2987–3002.

Karl, T. R., Wang, W. C., Schlesinger, M. E., Knight, R. W. and Portman, D. (1990). A method of relating general-circulation model simulated climate to the observed local climate. 1. Seasonal statistics. *Journal of Climate*, **3**(10):1053–79.

Karush, W. (1939). *Minima of functions of several variables with inequalities as side constraints.* M.Sc. thesis, University of Chicago.

Keiner, L. E. and Yan, X.-H. (1998). A neural network model for estimating sea surface chlorophyll and sediments from Thematic Mapper imagery. *Remote Sensing of Environment*, **66**:153–65.

Kelly, K. (1988). Comment on 'Empirical orthogonal function analysis of advanced very high resolution radiometer surface temperature patterns in Santa Barbara Channel' by G.S.E. Lagerloef and R.L. Bernstein. *Journal of Geophysical Research*, **93**(C12):15743–54.

Khan, M. S. and Coulibaly, P. (2006). Application of support vector machine in lake water level prediction. *Journal of Hydrologic Engineering*, **11**(3):199–205.

Kharin, V. V. and Zwiers, F. W. (2003). On the ROC score of probability forecasts. *Journal of Climate*, **16**(24):4145–50.

Kim, T. and Adali, T. (2002). Fully complex multi-layer perceptron network for nonlinear signal processing. *Journal of VLSI Signal Processing*, **32**:29–43.

Kirby, M. J. and Miranda, R. (1996). Circular nodes in neural networks. *Neural Computation*, **8**:390–402.

Kohonen, T. (1982). Self-organizing formation of topologically correct feature maps. *Biological Cybernetics*, **43**:59–69.

Kohonen, T. (2001). *Self-Organizing Maps*, 3rd edn. Berlin: Springer.

Kolehmainen, M., Martikainen, H. and Ruuskanen, J. (2001). Neural networks and periodic components used in air quality forecasting. *Atmospheric Environment*, **35**(5):815–25.

Kramer, M. A. (1991). Nonlinear principal component analysis using autoassociative neural networks. *AIChE Journal*, **37**:233–43.

Krasnopolsky, V. M. (2007). Neural network emulations for complex multidimensional geophysical mappings: Applications of neural network techniques to atmospheric and oceanic satellite retrievals and numerical modeling. *Reviews of Geophysics*, **45**(3). RG3009, doi:10.1029/2006RG000200.

Krasnopolsky, V. M., Breaker, L. C. and Gemmill, W. H. (1995). A neural network as a nonlinear transfer function model for retrieving surface wind speeds from the special sensor microwave imager. *Journal of Geophysical Research*, **100**(C6):11033–45.

Krasnopolsky, V. M., Chalikov, D. V. and Tolman, H. L. (2002). A neural network technique to improve computational efficiency of numerical oceanic models. *Ocean Modelling*, **4**:363–83.

Krasnopolsky, V. M. and Chevallier, F. (2003). Some neural network applications in environmental sciences. Part II: advancing computational efficiency in environmental numerical models. *Neural Networks*, **16**(3-4):335–48.

Krasnopolsky, V. M. and Fox-Rabinovitz, M. S. (2006). Complex hybrid models combining deterministic and machine learning components for numerical climate modeling and weather prediction. *Neural Networks*, **19**:122–34.

Krasnopolsky, V. M., Fox-Rabinovitz, M. S. and Chalikov, D. V. (2005a). Comments on 'New approach to calculation of atmospheric model physics: Accurate and fast neural network emulation of longwave radiation in a climate model' - Reply. *Monthly Weather Review*, **133**(12):3724–8.

Krasnopolsky, V. M., Fox-Rabinovitz, M. S. and Chalikov, D. V. (2005b). New approach to calculation of atmospheric model physics: Accurate and fast neural network emulation of longwave radiation in a climate model. *Monthly Weather Review*, **133**(5):1370–83.

Krasnopolsky, V. M., Gemmill, W. H. and Breaker, L. C. (1999). A multiparameter empirical ocean algorithm for SSM/I retrievals. *Canadian Journal of Remote Sensing*, **25**:486–503.

Kuhn, H. W. and Tucker, A. W. (1951). Nonlinear programming. In *Proceedings of the 2nd Berkeley Symposium on Mathematical Statistics and Probabilities*, pp. 481–92. University of California Press.

Kukkonen, J., Partanen, L., Karppinen, A. *et al.* (2003). Extensive evaluation of neural network models for the prediction of NO2 and PM10 concentrations, compared with a deterministic modelling system and measurements in central Helsinki. *Atmospheric Environment*, **37**(32):4539–50.

Kwiatkowska, E. J. and Fargion, G. S. (2003). Application of machine-learning techniques toward the creation of a consistent and calibrated global chlorophyll concentration baseline dataset using remotely sensed ocean color data. *IEEE Transactions on Geoscience and Remote Sensing*, **41**(12):2844–60.

Kwok, J. T.-Y. and Tsang, I. W.-H. (2004). The pre-image problem in kernel methods. *IEEE Transactions on Neural Networks*, **15**:1517–25.

Lai, P. L. and Fyfe, C. (1999). A neural implementation of canonical correlation analysis. *Neural Networks*, **12**:1391–7.

Lai, P. L. and Fyfe, F. (2000). Kernel and non-linear canonical correlation analysis. *International Journal of Neural Systems*, **10**:365–77.

Lambert, S. J. and Fyfe, J. C. (2006). Changes in winter cyclone frequencies and strengths simulated in enhanced greenhouse warming experiments: results from the models participating in the IPCC diagnostic exercise. *Climate Dynamics*, **26**(7-8):713–28.

Lax, D. A. (1985). Robust estimators of scale: Finite-sample performance in long-tailed symmetric distributions. *Journal of the American Statistical Association*, **80**:736–41.

Le, N. D. and Zidek, J. V. (2006). *Statistical Analysis of Environmental Space-Time Processes.* New York: Springer.

LeBlond, P. H. and Mysak, L. A. (1978). *Waves in the Ocean.* Amsterdam: Elsevier.

LeCun, Y., Kanter, I. and Solla, S. A. (1991). Second order properties of error surfaces: Learning time and generalization. In *Advances in Neural Information Processing Systems*, volume 3, pp. 918–24. Cambridge, MA: MIT Press.

Lee, J., Weger, R. C., Sengupta, S. K. and Welch, R. M. (1990). A neural network approach to cloud classification. *IEEE Transactions on Geoscience and Remote Sensing*, **28**(5):846–55.

Lee, Y., Wahba, G. and Ackerman, S. A. (2004). Cloud classification of satellite radiance data by multicategory support vector machines. *Journal of Atmospheric and Oceanic Technology*, **21**(2):159–69.

Legler, D. M. (1983). Empirical orthogonal function analysis of wind vectors over the tropical Pacific region. *Bulletin of the American Meteorological Society*, **64**(3):234–41.

Levenberg, K. (1944). A method for the solution of certain non-linear problems in least squares. *Quarterly Journal of Applied Mathematics*, **2**:164–8.

Li, S., Hsieh, W. W. and Wu, A. (2005). Hybrid coupled modeling of the tropical Pacific using neural networks. *Journal of Geophysical Research*, **110**(C09024). doi: 10.1029/2004JC002595.

Liu, Y. G. and Weisberg, R. H. (2005). Patterns of ocean current variability on the West Florida Shelf using the self-organizing map. *Journal Of Geophysical Research – Oceans*, **110**(C6). C06003, doi:10.1029/2004JC002786.

Liu, Y., Weisberg, R. H. and Mooers, C. N. K. (2006). Performance evaluation of the self-organizing map for feature extraction. *Journal of Geophysical Research*, **111** (C05018). doi:10.1029/2005JC003117.

Liu, Y. G., Weisberg, R. H. and Shay, L. K. (2007). Current patterns on the West Florida Shelf from joint self-organizing map analyses of HF radar and ADCP data. *Journal of Atmospheric and Oceanic Technology*, **24**(4):702–12.

Liu, Z. and Jin, L. (2008). LATTICESVM – A new method for multi-class support vector machines. In *Proceedings of the 2008 IEEE World Congress in Computational Intelligence*, Hong Kong. (See `http://ieeexplore.ieee.org/xpls/abs_all.jsp?arnumber=463387`)

Lorenz, E. N. (1956). Empirical orthogonal functions and statistical weather prediction. Sci. rep. no. 1, Statistical Forecasting Project, Dept. of Meteorology, MIT.

Lorenz, E. N. (1963). Deterministic nonperiodic flow. *Journal of the Atmospheric Sciences*, **20**:130–41.

Lu, W. Z. and Wang, W. J. (2005). Potential assessment of the 'support vector machine' method in forecasting ambient air pollutant trends. *Chemosphere*, **59**(5):693–701.

Luenberger, D. G. (1984). *Linear and Nonlinear Programming*, 2nd edn. Reading, MA: Addison-Wesley.

MacKay, D. J. C. (1992a). A practical Bayesian framework for backpropagation networks. *Neural Computation*, **4**(3):448–72.

MacKay, D. J. C. (1992b). Bayesian interpolation. *Neural Computation*, **4**(3):415–47.

MacKay, D. J. C. (1995). Probable networks and plausible predictions – a review of practical Bayesian methods for supervised neural networks. *Network: Computation in Neural Systems*, **6**:469–505.

MacKay, D. J. C. (2003). *Information Theory, Inference and Learning Algorithms*. Cambridge, UK: Cambridge University Press.

Maier, H. R. and Dandy, G. C. (2000). Neural networks for the prediction and forecasting of water resources variables: a review of modelling issues and applications. *Environmental Modelling and Software*, **15**:101–24.

Mantua, N. J. and Hare, S. R. (2002). The Pacific decadal oscillation. *Journal of Oceanography*, **58**(1):35–44.

Mardia, K. V., Kent, J. T. and Bibby, J. M. (1979). *Multivariate Analysis*. London: Academic Press.

Marquardt, D. (1963). An algorithm for least-squares estimation of nonlinear parameters. *SIAM Journal on Applied Mathematics*, **11**:431–41.

Marzban, C. (2004). The ROC curve and the area under it as performance measures. *Weather and Forecasting*, **19**(6):1106–14.

Marzban, C. and Stumpf, G. J. (1996). A neural network for tornado prediction based on doppler radar-derived attributes. *Journal of Applied Meteorology*, **35**(5):617–26.

Marzban, C. and Stumpf, G. J. (1998). A neural network for damaging wind prediction. *Weather and Forecasting*, **13**:151–63.

Marzban, C. and Witt, A. (2001). A Bayesian neural network for severe-hail size prediction. *Weather and Forecasting*, **16**(5):600–10.

Masters, T. (1995). *Advanced Algorithms for Neural Networks – A C++ Sourcebook*. New York: Wiley.

McCulloch, W. S. and Pitts, W. (1943). A logical calculus of the ideas immanent in neural nets. *Bulletin of Mathematical Biophysics*, **5**:115–37.

McGinnis, D. L. (1997). Estimating climate-change impacts on Colorado Plateau snowpack using downscaling methods. *Professional Geographer*, **49**(1):117–25.

McIntire, T. J. and Simpson, J. J. (2002). Arctic sea ice, cloud, water, and lead classification using neural networks and 1.61μm data. *IEEE Transactions on Geoscience and Remote Sensing*, **40**(9):1956–72.

Meinen, C. S. and McPhaden, M. J. (2000). Observations of warm water volume changes in the equatorial Pacific and their relationship to El Niño and La Niña. *Journal of Climate*, **13**:3551–9.

Mejia, C., Thiria, S., Tran, N., Crepon, M. and Badran, F. (1998). Determination of the geophysical model function of the ERS-1 scatterometer by the use of neural networks. *Journal of Geophysical Research*, **103**(C6):12853–68.

Miikkulainen, R., Bryant, B. D., Cornelius, R. *et al.* (2006). Computational intelligence in games. In Yen, G. and Fogel, D., eds., *Computational Intelligence: Principles and Practice*, pp. 155–191. IEEE Computational Intelligence Soc.

Mika, S., Schölkopf, B., Smola, A. J. *et al.* (1999). Kernel PCA and de-noising in feature spaces. In Kearns, M., Solla, S., and Cohn, D., eds., *Advances in Neural Information Processing Systems*, volume 11, pp. 536–42. Cambridge, MA: MIT Press.

Miller, S. W. and Emery, W. J. (1997). An automated neural network cloud classifier for use over land and ocean surfaces. *Journal of Applied Meteorology*, **36**(10): 1346–62.

Mingoti, S. A. and Lima, J. O. (2006). Comparing SOM neural network with Fuzzy c-means, K-means and traditional hierarchical clustering algorithms. *European Journal of Operational Research*, **174**(3):1742–59.

Minns, A. W. and Hall, M. J. (1996). Artificial neural networks as rainfall-runoff models. *Hydrological Sciences Journal – Journal Des Sciences Hydrologiques*, **41**(3):399–417.

Minsky, M. and Papert, S. (1969). *Perceptrons*. Cambridge, MA: MIT Press.

Monahan, A. H. (2000). Nonlinear principal component analysis by neural networks: theory and application to the Lorenz system. *Journal of Climate*, **13**:821–35.

Monahan, A. H. (2001). Nonlinear principal component analysis: Tropical Indo-Pacific sea surface temperature and sea level pressure. *Journal of Climate*, **14**:219–33.

Monahan, A. H. and Fyfe, J. C. (2007). Comment on 'The shortcomings of nonlinear principal component analysis in identifying circulation regimes'. *Journal of Climate*, **20**:375–77.

Monahan, A. H., Fyfe, J. C. and Flato, G. M. (2000). A regime view of northern hemisphere atmospheric variability and change under global warming. *Geophysical Research Letters*, **27**:1139–42.

Monahan, A. H., Fyfe, J. C. and Pandolfo, L. (2003). The vertical structure of wintertime climate regimes of the northern hemisphere extratropical atmosphere. *Journal of Climate*, **16**:2005–21.

Monahan, A. H., Pandolfo, L. and Fyfe, J. C. (2001). The preferred structure of variability of the northern hemisphere atmospheric circulation. *Geophysical Research Letters*, **28**:1019–22.

Monahan, A. H., Tangang, F. T. and Hsieh, W. W. (1999). A potential problem with extended EOF analysis of standing wave fields. *Atmosphere-Ocean*, **37**:241–54.

Moody, J. and Darken, C. J. (1989). Fast learning in networks of locally-tuned processing units. *Neural Computation*, **1**:281–94.

Morel, A. (1988). Optical modeling of the upper ocean in relation to its biogenous matter content (case 1 waters). *Journal of Geophysical Research*, **93**:10749–68.

Mosteller, F. and Tukey, J. W. (1977). *Data Analysis and Regression: A Second Course in Statistics*. Addison-Wesley.

Nabney, I. T. (2002). *Netlab: Algorithms for Pattern Recognition*. London: Springer.

Naujokat, B. (1986). An update of the observed quasi-biennial oscillation of the stratospheric winds over the tropics. *Journal of the Atmospheric Sciences*, **43**:1873–7.

Neal, R. M. (1996). *Bayesian Learning for Neural Networks*, volume 118 of *Lecture Notes in Statistics*. New York: Springer.

Newbigging, S. C., Mysak, L. A. and Hsieh, W. W. (2003). Improvements to the non-linear principal component analysis method, with applications to ENSO and QBO. *Atmosphere-Ocean*, **41**(4):290–98.

Nguyen, D. and Widrow, B. (1990). Improving the learning speed of 2-layer neural networks by choosing initial values of the adaptive weights. In *International Joint Conference on Neural Networks*, volume 3, pp. 21–6. (See `http://ieeexplore.ieee.org/xpls/abs_all.jsp?arnumber=137819`)

Niang, A., Badran, A., Moulin, C., Crepon, M., and Thiria, S. (2006). Retrieval of aerosol type and optical thickness over the Mediterranean from SeaWiFS images using an automatic neural classification method. *Remote Sensing of Environment*, **100**:82–94.

Niermann, S. (2006). Evolutionary estimation of parameters of Johnson distributions. *Journal of Statistical Computation and Simulation*, **76**:185–93.

Nitta, T. (1997). An extension of the back-propagation algorithm to complex numbers. *Neural Networks*, **10**:1391–415.

North, G. R., Bell, T. L., Cahalan, R. F. and Moeng, F. J. (1982). Sampling errors in the estimation of empirical orthogonal functions. *Monthly Weather Review*, **110**:699–706.

Nunnari, G., Dorling, S., Schlink, U. *et al.* (2004). Modelling SO_2 concentration at a point with statistical approaches. *Environmental Modelling & Software*, **19**(10):887–905.

Oja, E. (1982). A simplified neuron model as a principal component analyzer. *Journal of Mathematical Biology*, **15**:267–73.

Olden, J. D. and Jackson, D. A. (2002). A comparison of statistical approaches for modelling fish species distributions. *Freshwater Biology*, **47**(10):1976–95.

Olsson, J., Uvo, C. B., Jinno, K. *et al.* (2004). Neural networks for rainfall forecasting by atmospheric downscaling. *Journal of Hydrologic Engineering*, **9**(1):1–12.

Pacifici, F., Del Frate, F., Solimini, C. and Emery, W. J. (2007). An innovative neural-net method to detect temporal changes in high-resolution optical satellite imagery. *IEEE Transactions on Geoscience and Remote Sensing*, **45**(9):2940–52.

Park, Y. S., Cereghino, R., Compin, A. and Lek, S. (2003). Applications of artificial neural networks for patterning and predicting aquatic insect species richness in running waters. *Ecological Modelling*, **160**(3):265–80.

Pasini, A., Lore, M. and Ameli, F. (2006). Neural network modelling for the analysis of forcings/temperatures relationships at different scales in the climate system. *Ecological Modelling*, **191**(1):58–67.

Pearson, K. (1901). On lines and planes of closest fit to systems of points in space. *Philosophical Magazine, Ser. 6*, **2**:559–72.

Peirce, C. S. (1884). The numerical measure of the success of predictions. *Science*, **4**:453–4.

Penland, C. and Magorian, T. (1993). Prediction of Nino-3 sea surface temperatures using linear inverse modeling. *Journal of Climate*, **6**(6):1067–76.

Philander, S. G. (1990). *El Niño, La Niña, and the Southern Oscillation*. San Diego: Academic Press.

Polak, E. (1971). *Computational Methods in Optimization: A Unified Approach*. New York: Academic Press.

Polak, E. and Ribiere, G. (1969). Note sur la convergence de methods de directions conjures. *Revue Francaise d'Informat. et de Recherche Operationnelle*, **16**:35–43.

Powell, M. J. D. (1987). Radial basis functions for multivariate interpolation: a review. In Mason, J. and Cox, M., eds., *Algorithms for Approximation*, pp. 143–67. Oxford: Clarendon Press.

Pozo-Vazquez, D., Esteban-Parra, M. J., Rodrigo, F. S. and Castro-Diez, Y. (2001). A study of NAO variability and its possible nonlinear influences on European surface temperature. *Climate Dynamics*, **17**:701–15.

Preisendorfer, R. W. (1988). *Principal Component Analysis in Meteorology and Oceanography*. New York: Elsevier.

Press, W. H., Flannery, B. P., Teukolsky, S. A. and Vetterling, W. T. (1986). *Numerical Recipes*. Cambridge, UK: Cambridge University Press.

Price, K. V., Storn, R. M. and Lampinen, J. A. (2005). *Differential Evolution: A Practical Approach to Global Optimization*. Berlin: Springer.

Pyper, B. J. and Peterman, R. M. (1998). Comparison of methods to account for autocorrelation in correlation analyses of fish data. *Canadian Journal of Fisheries and Aquatic Sciences*, **55**:2127–40.

Quinlan, J. R. (1993). *C4.5: Programs for Machine Learning*. San Mateo: Morgan Kaufmann.

Rasmussen, C. E. and Williams, C. K. I. (2006). *Gaussian Processes for Machine Learning*. Cambridge, MA: MIT Press.

Rattan, S. S. P. and Hsieh, W. W. (2004). Nonlinear complex principal component analysis of the tropical Pacific interannual wind variability. *Geophysical Research Letters*, **31**(21). L21201, doi:10.1029/2004GL020446.

Rattan, S. S. P. and Hsieh, W. W. (2005). Complex-valued neural networks for nonlinear complex principal component analysis. *Neural Networks*, **18**(1):61–9.

Rattan, S. S. P., Ruessink, B. G. and Hsieh, W. W. (2005). Non-linear complex principal component analysis of nearshore bathymetry. *Nonlinear Processes in Geophysics*, **12**(5):661–70.

Recknagel, F., French, M., Harkonen, P. and Yabunaka, K. (1997). Artificial neural network approach for modelling and prediction of algal blooms. *Ecological Modelling*, **96**(1-3):11–28.

Reynolds, R. W. and Smith, T. M. (1994). Improved global sea surface temperature analyses using optimum interpolation. *Journal of Climate*, **7**(6):929–48.

Richardson, A. J., Risien, C. and Shillington, F. A. (2003). Using self-organizing maps to identify patterns in satellite imagery. *Progress in Oceanography*, **59**(2-3):223–39.

Richaume, P., Badran, F., Crepon, M. *et al.* (2000). Neural network wind retrieval from ERS-1 scatterometer data. *Journal of Geophysical Research*, **105**(C4):8737–51.

Richman, M. B. (1986). Rotation of principal components. *Journal of Climatology*, **6**:293–335.

Rojas, R. (1996). *Neural Networks– A Systematic Introduction*. New York: Springer.

Rojo-Alvarez, J. L., Martinez-Ramon, M., Figueiras-Vidal, A. R., Garcia-Armada, A. and Artes-Rodriguez, A. (2003). A robust support vector algorithm for nonparametric spectral analysis. *IEEE Signal Processing Letters*, **10**(11):320–23.

Ropelewski, C. F. and Jones, P. D. (1987). An extension of the Tahiti-Darwin Southern Oscillation Index. *Monthly Weather Review*, **115**:2161–5.

Rosenblatt, F. (1958). The perceptron: A probabilistic model for information storage and organization in the brain. *Psychological Review*, **65**:386–408.

Rosenblatt, F. (1962). *Principles of Neurodynamics*. New York: Spartan.

Roweis, S. T. and Saul, L. K. (2000). Nonlinear dimensionality reduction by locally linear embedding. *Science*, **290**:2323–6.

Ruessink, B. G., van Enckevort, I. M. J. and Kuriyama, Y. (2004). Non-linear principal component analysis of nearshore bathymetry. *Marine Geology*, **203**:185–97.

Rumelhart, D. E., Hinton, G. E. and Williams, R. J. (1986). Learning internal representations by error propagation. In Rumelhart, D., McClelland, J. and Group, P. R., eds., *Parallel Distributed Processing*, volume 1, pp. 318–62. Cambridge, MA: MIT Press.

Saff, E. B. and Snider, A. D. (2003). *Fundamentals of Complex Analysis with Applications to Engineering and Science*. Englewood Cliffs, NJ: Prentice-Hall.

Sajikumar, N. and Thandaveswara, B. S. (1999). A non-linear rainfall-runoff model using an artificial neural network. *Journal of Hydrology*, **216**(1-2):32–55.

Sanger, T. D. (1989). Optimal unsupervised learning in a single-layer linear feedforward neural network. *Neural Networks*, **2**:459–73.

Schiller, H. (2007). Model inversion by parameter fit using NN emulating the forward model - Evaluation of indirect measurements. *Neural Networks*, **20**(4): 479–83.

Schiller, H. and Doerffer, R. (1999). Neural network for emulation of an inverse model – operational derivation of Case II water properties from MERIS data. *International Journal of Remote Sensing*, **20**:1735–46.

Schiller, H. and Doerffer, R. (2005). Improved determination of coastal water constituent concentrations from MERIS data. *IEEE Transactions on Geoscience and Remote Sensing*, **43**(7):1585–91.

Schlink, U., Dorling, S., Pelikan, E. *et al.* (2003). A rigorous inter-comparison of ground-level ozone predictions. *Atmospheric Environment*, **37**(23):3237–53.

Schölkopf, B., Smola, A. and Muller, K.-R. (1998). Nonlinear component analysis as a kernel eigenvalue problem. *Neural Computation*, **10**:1299–319.

Schölkopf, B., Smola, A., Williamson, R. and Bartlett, P. L. (2000). New support vector algorithms. *Neural Computation*, **12**:1207–45.

Schölkopf, B. and Smola, A. J. (2002). *Learning with Kernels: Support Vector Machines, Regularization, Optimization, and Beyond (Adaptive Computation and Machine Learning)*. Cambridge, MA: MIT Press.

Schölkopf, B., Sung, K. K., Burges, C. J. C. *et al.* (1997). Comparing support vector machines with Gaussian kernels to radial basis function classifiers. *IEEE Transactions on Signal Processing*, **45**(11):2758–65.

Schoof, J. T. and Pryor, S. C. (2001). Downscaling temperature and precipitation: A comparison of regression-based methods and artificial neural networks. *International Journal of Climatology*, **21**(7):773–90.

Shabbar, A. and Barnston, A. G. (1996). Skill of seasonal climate forecasts in Canada using canonical correlation analysis. *Monthly Weather Review*, **124**:2370–85.

Shanno, D. F. (1970). Conditioning of quasi-Newton methods for function minimization. *Mathematics of Computation*, **24**:647–57.

Shanno, D. F. (1978). Conjugate-gradient methods with inexact searches. *Mathematics of Operations Research*, **3**:244–56.

Shawe-Taylor, J. and Cristianini, N. (2004). *Kernel Methods for Pattern Analysis*. Cambridge, UK: Cambridge University Press.

Simpson, J. J. and McIntire, T. J. (2001). A recurrent neural network classifier for improved retrievals of areal extent of snow cover. *IEEE Transactions on Geoscience and Remote Sensing*, **39**(10):2135–47.

Simpson, J. J., Tsou, Y. L., Schmidt, A. and Harris, A. (2005). Analysis of along track scanning radiometer-2 (ATSR-2) data for clouds, glint and sea surface temperature using neural networks. *Remote Sensing of Environment*, **98**:152–81.

Smith, T. M., Reynolds, R. W., Livezey, R. E. and Stokes, D. C. (1996). Reconstruction of historical sea surface temperatures using empirical orthogonal functions. *Journal of Climate*, **9**(6):1403–20.

Solomatine, D. P. and Dulal, K. N. (2003). Model trees as an alternative to neural networks in rainfall-runoff modelling. *Hydrological Sciences Journal*, **48**: 399–411.

Solomatine, D. P. and Xue, Y. P. (2004). M5 model trees and neural networks: Application to flood forecasting in the upper reach of the Huai River in China. *Journal of Hydrologic Engineering*, **9**(6):491–501.

Sorooshian, S., Hsu, K. L., Gao, X. *et al.* (2000). Evaluation of PERSIANN system satellite-based estimates of tropical rainfall. *Bulletin of the American Meteorological Society*, **81**(9):2035–46.

Srivastava, A. N., Oza, N. C. and Stroeve, J. (2005). Virtual sensors: Using data mining techniques to efficiently estimate remote sensing spectra. *IEEE Transactions on Geoscience and Remote Sensing*, **43**(3):590–600.

Stacey, M. W., Pond, S. and LeBlond, P. H. (1986). A wind-forced Ekman spiral as a good statistical fit to low-frequency currents in a coastal strait. *Science*, **233**:470–2.

Stogryn, A. P., Butler, C. T. and Bartolac, T. J. (1994). Ocean surface wind retrievals from Special Sensor Microwave Imager data with neural networks. *Journal of Geophysical Research*, **99**(C1):981–4.

Storn, R. and Price, K. (1997). Differential evolution – A simple and efficient heuristic for global optimization over continuous spaces. *Journal of Global Optimization*, **11**:341–59. doi 10.1023/A:1008202821328.

Strang, G. (2005). *Linear Algebra and Its Applications*. Pacific Grove, CA: Brooks Cole.

Stroeve, J., Holland, M. M., Meier, W., Scambos, T. and Serreze, M. (2007). Arctic sea ice decline: Faster than forecast. *Geophysical Research Letters*, **34**. L09501, doi:10.1029/2007GL029703.

Suykens, J. A. K., Van Gestel, T., De Brabanter, J., De Moor, B. and Vandewalle, J. (2002). *Least Squares Support Vector Machines*. New Jersey: World Scientific.

Tag, P. M., Bankert, R. L. and Brody, L. R. (2000). An AVHRR multiple cloud-type classification package. *Journal of Applied Meteorology*, **39**(2):125–34.

Taner, M. T., Berge, T., Walls, J. A. *et al.* (2001). Well log calibration of Kohonen-classified seismic attributes using Bayesian logic. *Journal of Petroleum Geology*, **24**(4):405–16.

Tang, B. (1995). Periods of linear development of the ENSO cycle and POP forecast experiments. *Journal of Climate*, **8**:682–91.

Tang, B., Flato, G. M. and Holloway, G. (1994). A study of Arctic sea ice and sea-level pressure using POP and neural network methods. *Atmosphere-Ocean*, **32**:507–29.

Tang, B. and Mazzoni, D. (2006). Multiclass reduced-set support vector machines. In *Proceedings of the 23rd International Conference on Machine Learning (ICML 2006)*, Pittsburgh, PA. New York: ACM.

Tang, B. Y., Hsieh, W. W., Monahan, A. H. and Tangang, F. T. (2000). Skill comparisons between neural networks and canonical correlation analysis in predicting the equatorial Pacific sea surface temperatures. *Journal of Climate*, **13**(1):287–93.

Tang, Y. (2002). Hybrid coupled models of the tropical Pacific: I. Interannual variability. *Climate Dynamics*, **19**:331–42.

Tang, Y. and Hsieh, W. W. (2002). Hybrid coupled models of the tropical Pacific: II. ENSO prediction. *Climate Dynamics*, **19**:343–53.

Tang, Y. and Hsieh, W. W. (2003a). ENSO simulation and prediction in a hybrid coupled model with data assimilation. *Journal of the Meteorological Society of Japan*, **81**:1–19.

Tang, Y. and Hsieh, W. W. (2003b). Nonlinear modes of decadal and interannual variability of the subsurface thermal structure in the Pacific Ocean. *Journal of Geophysical Research*, **108**(C3). 3084, doi: 10.1029/2001JC001236.

Tang, Y., Hsieh, W. W., Tang, B. and Haines, K. (2001). A neural network atmospheric model for hybrid coupled modelling. *Climate Dynamics*, **17**:445–55.

Tangang, F. T., Hsieh, W. W. and Tang, B. (1997). Forecasting the equatorial Pacific sea surface temperatures by neural network models. *Climate Dynamics*, **13**:135–47.

Tangang, F. T., Hsieh, W. W. and Tang, B. (1998a). Forecasting the regional sea surface temperatures of the tropical Pacific by neural network models, with wind stress and sea level pressure as predictors. *Journal of Geophysical Research*, **103**(C4): 7511–22.

Tangang, F. T., Tang, B., Monahan, A. H. and Hsieh, W. W. (1998b). Forecasting ENSO events – a neural network-extended EOF approach. *Journal of Climate*, **11**:29–41.

Tenenbaum, J. B., de Silva, V. and Langford, J. C. (2000). A global geometric framework for nonlinear dimensionality reduction. *Science*, **290**:2319–23.

Teng, Q. B., Monahan, A. H., and Fyfe, J. C. (2004). Effects of time averaging on climate regimes. *Geophysical Research Letters*, **31**(22). L22203, doi:10.1029/2004GL020840.

Teschl, R. and Randeu, W. L. (2006). A neural network model for short term river flow prediction. *Natural Hazards and Earth System Sciences*, **6**:629–35.

Teschl, R., Randeu, W. L. and Teschl, F. (2007). Improving weather radar estimates of rainfall using feed-forward neural networks. *Neural Networks*, **20**:519–27.

Thiria, S., Mejia, C., Badran, F. and Crepon, M. (1993). A neural network approach for modeling nonlinear transfer functions – application for wind retrieval from spaceborne scatterometer data. *Journal of Geophysical Research*, **98**(C12):22827–41.

Thompson, D. W. J. and Wallace, J. M. (1998). The Arctic Oscillation signature in the wintertime geopotential height and temperature fields. *Geophysical Research Letters*, **25**:1297–300.

Thompson, D. W. J. and Wallace, J. M. (2000). Annular modes in the extratropical circulation. Part I: Month-to-month variability. *Journal of Climate*, **13**(5):1000–16.

Tipping, M. E. (2001). Sparse Bayesian learning and the relevance vector machine. *Journal of Machine Learning Research*, **1**:211–44.

Tolman, H. L., Krasnopolsky, V. M. and Chalikov, D. V. (2005). Neural network approximations for nonlinear interactions in wind wave spectra: direct mapping for wind seas in deep water. *Ocean Modelling*, **8**:252–78.

Trigo, R. M. and Palutikof, J. P. (1999). Simulation of daily temperatures for climate change scenarios over Portugal: a neural network model approach. *Climate Research*, **13**(1):45–59.

Trigo, R. M. and Palutikof, J. P. (2001). Precipitation scenarios over Iberia: A comparison between direct GCM output and different downscaling techniques. *Journal of Climate*, **14**(23):4422–46.

Tripathi, S., Srinivas, V. V. and Nanjundiah, R. S. (2006). Downscaling of precipitation for climate change scenarios: A support vector machine approach. *Journal of Hydrology*, **330**(3-4):621–40.

Troup, A. J. (1965). The 'southern oscillation'. *Quarterly Journal of the Royal Meteorological Society*, **91**:490–506.

UNESCO (1981). *The Practical Salinity Scale 1978 and the International Equation of State for Seawater 1980*. Tenth Report of the Joint Panel on Oceanographic Tables and Standards. Technical Report 36, UNESCO.

van den Boogaard, H. and Mynett, A. (2004). Dynamic neural networks with data assimilation. *Hydrological Processes*, **18**:1959–66.

van Laarhoven, P. J. and Aarts, E. H. (1987). *Simulated Annealing: Theory and Applications*. Dordrecht: Reidel.

Vapnik, V. N. (1995). *The Nature of Statistical Learning Theory*. Berlin: Springer Verlag.

Vapnik, V. N. (1998). *Statistical Learning Theory*. New York: Wiley.

Vecchi, G. A. and Bond, N. A. (2004). The Madden–Julian Oscillation (MJO) and northern high latitude wintertime surface air temperatures. *Geophysical Research Letters*, **31**. L04104, doi: 10.1029/2003GL018645.

Villmann, T., Merenyi, E. and Hammer, B. (2003). Neural maps in remote sensing image analysis. *Neural Networks*, **16**(3-4):389–403.

von Storch, H. (1999). On the use of 'inflation' in statistical downscaling. *Journal of Climate*, **12**(12):3505–6.

von Storch, H., Bruns, T., Fischer-Bruns, I. and Hasselman, K. (1988). Principal Oscillation Pattern Analysis of the 30 to 60 day oscillation in general circulation model equatorial troposphere. *Journal of Geophysical Research*, **93**(D9): 11021–36.

von Storch, H., Burger, G., Schnur, R. and von Storch, J.-S. (1995). Principal oscillation patterns: A review. *Journal of Climate*, **8**(3):377–400.

von Storch, H. and Zwiers, F. W. (1999). *Statistical Analysis in Climate Research*. Cambridge, UK: Cambridge University Press.

Wallace, J. M. (1972). Empirical orthogonal representation of time series in the frequency domain. Part II: Application to the study of tropical wave disturbances. *Journal of Applied Meteorology*, **11**:893–990.

Wallace, J. M. and Dickinson, R. E. (1972). Empirical orthogonal representation of time series in the frequency domain. Part I: Theoretical considerations. *Journal of Applied Meteorology*, **11**(6):887–92.

Wallace, J. M. and Gutzler, D. S. (1981). Teleconnections in the geopotential height fields during the northern hemisphere winter. *Monthly Weather Review*, **109**:784–812.

Wallace, J. M., Smith, C. and Bretherton, C. S. (1992). Singular value decomposition of wintertime sea surface temperature and 500 mb height anomalies. *Journal of Climate*, **5**(6):561–76.

Walter, A., Denhard, M. and Schonwiese, C.-D. (1998). Simulation of global and hemispheric temperature variations and signal detection studies using neural networks. *Meteorologische Zeitschrift*, **N.F.7**:171–80.

Walter, A. and Schonwiese, C. D. (2002). Attribution and detection of anthropogenic climate change using a backpropagation neural network. *Meteorologische Zeitschrift*, **11**(5):335–43.

Wang, W. J., Lu, W. Z., Wang, X. K. and Leung, A. Y. T. (2003). Prediction of maximum daily ozone level using combined neural network and statistical characteristics. *Environment International*, **29**(5):555–62.

Webb, A. R. (1999). A loss function approach to model selection in nonlinear principal components. *Neural Networks*, **12**:339–45.

Weichert, A. and Bürger, G. (1998). Linear versus nonlinear techniques in downscaling. *Climate Research*, **10**(2):83–93.

Weigend, A. S. and Gershenfeld, N. A., eds., (1994). *Time Series Prediction: Forecasting the Future and Understanding the Past*. Santa Fe Institute Studies in the Sciences of Complexity, Proceedings vol. XV. Addison-Wesley.

Werbos, P. J. (1974). *Beyond regression: new tools for prediction and analysis in the behavioural sciences*. Ph.D. thesis, Harvard University.

Widrow, B. and Hoff, M. E. (1960). Adaptive switching circuits. In *IRE WESCON Convention Record*, volume 4, pp. 96–104, New York.

Wilby, R. L., Dawson, C. W. and Barrow, E. M. (2002). SDSM – a decision support tool for the assessment of regional climate change impacts. *Environmental Modelling & Software*, **17**(2):147–59.

Wilcox, R. R. (2004). *Robust Estimation and Hypothesis Testing*. Amsterdam: Elsevier.

Wilks, D. S. (1995). *Statistical Methods in the Atmospheric Sciences*. San Diego: Academic Press.

Willmott, C. J. (1982). Some comments on the evaluation of model performance. *Bulletin of the American Meteorological Society*, **63**(11):1309–13.

Wilson, L. J. and Vallée, M. (2002). The Canadian Updateable Model Output Statistics (UMOS) system: Design and development tests. *Weather and Forecasting*, **17**(2): 206–22.

Woodruff, S. D., Slutz, R. J., Jenne, R. L. and Steurer, P. M. (1987). A comprehensive ocean-atmosphere data set. *Bulletin of the American Meteorological Society*, **68**: 1239–50.

Wu, A. and Hsieh, W. W. (2002). Nonlinear canonical correlation analysis of the tropical Pacific wind stress and sea surface temperature. *Climate Dynamics*, **19**:713–22. doi 10.1007/s00382-002-0262-8.

Wu, A. and Hsieh, W. W. (2003). Nonlinear interdecadal changes of the El Nino-Southern Oscillation. *Climate Dynamics*, **21**:719–30.

Wu, A. and Hsieh, W. W. (2004a). The nonlinear association between ENSO and the Euro-Atlantic winter sea level pressure. *Climate Dynamics*, **23**:859–68. doi: 10.1007/s00382-004-0470-5.

Wu, A. and Hsieh, W. W. (2004b). The nonlinear Northern Hemisphere atmospheric response to ENSO. *Geophysical Research Letters*, **31**. L02203, doi:10.1029/2003GL018885.

Wu, A., Hsieh, W. W. and Shabbar, A. (2005). The nonlinear patterns of North American winter temperature and precipitation associated with ENSO. *Journal of Climate*, **18**:1736–52.

Wu, A., Hsieh, W. W., Shabbar, A., Boer, G. J. and Zwiers, F. W. (2006a). The nonlinear association between the Arctic Oscillation and North American winter climate. *Climate Dynamics*, **26**:865–79.

Wu, A., Hsieh, W. W. and Tang, B. (2006b). Neural network forecasts of the tropical Pacific sea surface temperatures. *Neural Networks*, **19**:145–54.

Wu, A., Hsieh, W. W. and Zwiers, F. W. (2003). Nonlinear modes of North American winter climate variability detected from a general circulation model. *Journal of Climate*, **16**:2325–39.

Xu, J. S. (1992). On the relationship between the stratospheric Quasi-Biennial Oscillation and the tropospheric Southern Oscillation. *Journal of the Atmospheric Sciences*, **49**(9):725–34.

Xu, J.-S. and von Storch, H. (1990). Predicting the state of the Southern Oscillation using principal oscillation pattern analysis. *Journal of Climate*, **3**:1316–29.

Yacoub, M., Badran, F. and Thiria, S. (2001). A topological hierarchical clustering: Application to ocean color classification. In *Artificial Neural Networks-ICANN 2001, Proceedings. Lecture Notes in Computer Science.*, volume 2130, pp. 492–499. Berlin: Springer.

Ye, Z. and Hsieh, W. W. (2006). The influence of climate regime shift on ENSO. *Climate Dynamics*, **26**:823–33.

Yhann, S. R. and Simpson, J. J. (1995). Application of neural networks to AVHRR cloud segmentation. *IEEE Transactions on Geoscience and Remote Sensing*, **33**(3):590–604.

Yi, J. S. and Prybutok, V. R. (1996). A neural network model forecasting for prediction of daily maximum ozone concentration in an industrialized urban area. *Environmental Pollution*, **92**(3):349–57.

Yu, P. S., Chen, S. T. and Chang, I. F. (2006). Support vector regression for real-time flood stage forecasting. *Journal of Hydrology*, **328**(3-4):704–16.

Yu, X. Y. and Liong, S. Y. (2007). Forecasting of hydrologic time series with ridge regression in feature space. *Journal of Hydrology*, **332**(3-4):290–302.

Yuval (2000). Neural network training for prediction of climatological time series; regularized by minimization of the Generalized Cross Validation function. *Monthly Weather Review*, **128**:1456–73.

Yuval (2001). Enhancement and error estimation of neural network prediction of Niño 3.4 SST anomalies. *Journal of Climate*, **14**:2150–63.

Yuval and Hsieh, W. W. (2002). The impact of time-averaging on the detectability of nonlinear empirical relations. *Quarterly Journal of the Royal Meteorological Society*, **128**:1609–22.

Yuval and Hsieh, W. W. (2003). An adaptive nonlinear MOS scheme for precipitation forecasts using neural networks. *Weather and Forecasting*, **18**(2):303–10.

Zebiak, S. E. and Cane, M. A. (1987). A model El Niño – Southern Oscillation. *Monthly Weather Review*, **115**(10):2262–78.

Zhang, C. (2005). Madden–Julian Oscillation. *Reviews of Geophysics*, **43**. RG2003, doi:10.1029/2004RG000158.

Zhang, X., Hogg, W. D. and Mekis, E. (2001). Spatial and temporal characteristics of heavy precipitation events over Canada. *Journal of Climate*, **14**:1923–36.

Zorita, E. and von Storch, H. (1999). The analog method as a simple statistical downscaling technique: Comparison with more complicated methods. *Journal of Climate*, **12**(8):2474–89.

Index